高等职业教育计算机类专业规划教材

Java & JSP 应用程序实例开发

潘国荣　主　编
殷存举　徐　栋　副主编

电子工业出版社
Publishing House of Electronics Industry
北京·BEIJING

内 容 简 介

本书主要内容分两大部分，第一部分包括项目一至项目三，详细介绍了利用 Java 开发桌面数据库系统的方法和步骤；第二部分包括项目四至项目六，详细介绍了利用纯 JSP 页面模式、JSP＋JavaBean 模式、JSP＋JavaBean＋Servlet 模式、Struts 模式开发 JSP 应用程序的方法和步骤。每个项目均按"项目描述"、"项目预览"、"项目分析"、"项目设计"、"环境需求"、"项目知识点"和具体任务进行组织，其中"项目实施"以任务的形式给出了完整的实施步骤和代码。

本书是作者多年从事教学和开发经验的总结，项目实施步骤详尽，注重实用，非常适合作为高等职业教育计算机类专业的教材，也可以作为开发人员和自学爱好者的参考书。

未经许可，不得以任何方式复制或抄袭本书之部分或全部内容。
版权所有，侵权必究。

图书在版编目（CIP）数据

Java & JSP 应用程序实例开发/潘国荣主编.--北京：电子工业出版社，2014.10
ISBN 978-7-121-24111-6

Ⅰ.①J… Ⅱ.①潘… Ⅲ.①JAVA 语言—程序设计—高等学校—教材 Ⅳ.①TP312

中国版本图书馆 CIP 数据核字（2014）第 188923 号

策划编辑：朱怀永
责任编辑：朱怀永　　　　　　特约编辑：王 纲
印　　刷：三河市鑫金马印装有限公司
装　　订：三河市鑫金马印装有限公司
出版发行：电子工业出版社
　　　　　北京市海淀区万寿路 173 信箱　邮编　100036
开　　本：787×1092　1/16　印张：20.25　字数：518 千字
版　　次：2014 年 10 月第 1 版
印　　次：2014 年 10 月第 1 次印刷
印　　数：3000 册
定　　价：45.00 元

凡所购买电子工业出版社图书有缺损问题，请向购买书店调换，若书店售缺，请与本社发行部联系，联系及邮购电话：(010)88254888。
质量投诉请发邮件至 zlts@phei.com.cn，盗版侵权举报请发邮件至 dbqq@phei.com.cn。
服务热线：(010)88258888。

前　　言

　　Java 是一种可以撰写跨平台应用软件的程序设计语言,它具有面向对象、与平台无关、安全机制、高可靠性和支持网络编程的特性,自面世后就非常流行,发展迅速,对其他语言形成了有力冲击。Java 是被广泛使用的一种语言,应用于各种软件的开发中,具备显著优势和广阔前景。同时,随着网络技术的不断发展,大量软件需要采用 B/S 结构来实现,基于 Java 技术的动态网页开发技术 JSP 由于其跨平台、强大的可伸缩性、支持服务器端组件等特性而被广泛应用于 Web 应用程序的开发中。

　　目前,学习 Java 和 JSP 开发的人员众多、层次多样,讲述 Java 和 JSP 程序设计的教材也很多,并且各有特色。本书主要有以下几个特点:

　　1. 本书是作者多年教学和开发工作的总结,书中的项目都经过教学实际的检验,有些项目已经过多年的实际应用,具有较高的学习和参考价值。

　　2. 项目和任务的编排符合认知规律。项目由浅入深,由简单到复杂,任务的分解安排完全按照项目开发的实际过程进行,易于上手和理解。

　　3. 项目实用性强。项目中的很多模块和代码段是 Java 和 JSP 程序开发中非常常用的,可以方便地在其他项目开发中加以使用。

　　本书按照"项目—任务"的形式编写。全书分为两大部分,第一部分包括用户登录程序、简单的列表框程序、学生成绩管理系统三个项目,详细介绍了利用 Java 开发桌面数据库系统的方法和步骤;第二部分包括网上用户注册程序、学生基本信息维护系统、教师测评系统(Struts 模式)三个项目,详细介绍了利用纯 JSP 页面模式、JSP+JavaBean 模式、JSP+JavaBean+Servlet 模式、Struts 模式开发 JSP 应用程序的方法和步骤。每个项目均按"项目描述"、"项目预览"、"项目分析"、"项目设计"、"环境需求"、"项目知识点"和具体任务进行组织,其中"项目实施"以任务的形式给出了完整的实施步骤和代码。

　　本书由潘国荣主编。其中潘国荣编写项目三和项目六,并对全书进行统稿,徐栋编写项目一和项目二,殷存举编写项目四和项目五,在此对相关参与教师表示诚挚的感谢。

　　由于编写时间仓促和水平有限,书中难免存在不足之处,恳请读者赐教斧正。

目 录

项目一 用户登录程序 ……………………………………………………………………… 1

    任务一 准备数据库,定义窗体类,编制界面 …………………………………………… 2

    任务二 编制事件处理代码 ………………………………………………………………… 5

项目二 简单的列表框程序 …………………………………………………………………… 7

    任务一 定义窗体类,编制界面 …………………………………………………………… 7

    任务二 编制事件处理代码 ………………………………………………………………… 9

项目三 学生成绩管理系统 ………………………………………………………………… 11

    任务一 准备数据库,编制主窗体 ……………………………………………………… 14

    任务二 编制教师基本情况维护模块 …………………………………………………… 16

    任务三 编制课程基本情况维护模块 …………………………………………………… 26

    任务四 编制班级基本情况维护模块 …………………………………………………… 28

    任务五 表格及表格模型 ………………………………………………………………… 33

    任务六 编制学生基本情况维护模块 …………………………………………………… 40

    任务七 编制班级选课模块 ……………………………………………………………… 47

    任务八 编制单科方式录入学生成绩模块 ……………………………………………… 56

    任务九 编制多科方式录入学生成绩模块 ……………………………………………… 62

    任务十 编制单个学生成绩查询模块 …………………………………………………… 63

    任务十一 编制班级成绩明细查询模块 ………………………………………………… 67

    任务十二 编制班级成绩汇总数据查询模块 …………………………………………… 71

    任务十三 编制同学科成绩比较数据查询模块 ………………………………………… 75

    任务十四 编制班级不及格人数统计模块 ……………………………………………… 79

    任务十五 编制用户登录模块并打包 …………………………………………………… 83

项目四 网上用户注册程序 ………………………………………………………………… 86

    任务一 用纯JSP页面模式实现 ………………………………………………………… 88

    任务二 用JSP+JavaBean模式实现 …………………………………………………… 98

    任务三 用JSP+JavaBean+Servlet模式实现 ………………………………………… 105

    任务四 用Struts模式实现 ……………………………………………………………… 110

项目五 学生基本信息维护系统 …………………………………………………………… 118

    任务一 用JSP+JavaBean模式实现学生信息列表 …………………………………… 120

    任务二 用JSP+JavaBean模式实现学生信息查询 …………………………………… 127

    任务三 用JSP+JavaBean模式实现学生数据维护 …………………………………… 130

    任务四 用JSP+JavaBean模式实现学生数据分页显示 ……………………………… 142

任务五　用JSP＋JavaBean＋Servlet模式实现学生信息列表 …………………… 144

任务六　用JSP＋JavaBean＋Servlet模式实现学生信息查询 …………………… 148

任务七　用JSP＋JavaBean＋Servlet模式实现学生数据维护 …………………… 150

任务八　用Struts模式实现学生信息列表 ……………………………………… 154

任务九　用Struts模式实现学生信息查询 ……………………………………… 159

任务十　用Struts模式实现学生数据维护 ……………………………………… 161

项目六　教师测评系统（Struts模式） ……………………………………… 170

任务一　创建工程、准备数据库、编制首页面 ………………………………… 181

任务二　编制系统管理子系统登录模块 ………………………………………… 184

任务三　编制教师基本情况维护模块 …………………………………………… 193

任务四　编制课程基本情况维护模块 …………………………………………… 210

任务五　编制班级基本情况维护模块 …………………………………………… 213

任务六　编制班级科目维护模块 ………………………………………………… 232

任务七　编制用户情况维护模块 ………………………………………………… 248

任务八　编制系统初始化模块 …………………………………………………… 259

任务九　编制开放或禁止测评系统模块 ………………………………………… 263

任务十　编制汇总班主任测评数据模块 ………………………………………… 266

任务十一　编制汇总任课教师测评数据模块 …………………………………… 270

任务十二　编制测评班主任模块的学生登录部分 ……………………………… 273

任务十三　编制测评班主任模块 ………………………………………………… 281

任务十四　编制测评任课教师模块的学生登录部分 …………………………… 289

任务十五　编制测评任课教师模块 ……………………………………………… 293

任务十六　编制数据查询子系统登录模块 ……………………………………… 298

任务十七　编制查询班主任汇总数据模块 ……………………………………… 303

任务十八　编制按系部查询班主任汇总数据模块 ……………………………… 308

任务十九　编制按班级查询任课教师汇总数据模块 …………………………… 310

任务二十　编制按学科查询任课教师汇总数据模块 …………………………… 313

任务二十一　编制按班级查询意见、建议和简要评价模块 …………………… 315

项目一　用户登录程序

项目描述

用户登录程序是一个简单的 Java 桌面应用程序,它能够检查登录用户的合法性。

项目预览

运行用户登录程序时,出现如图 1-1 所示窗体。

输入用户名及口令后,窗体如图 1-2 所示。

 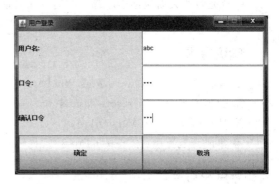

　　图 1-1　"用户登录"窗体　　　　　　　图 1-2　输入数据后的用户登录窗体

单击"确定"按钮后,在数据库表中查找是否有与该用户名、口令一致的记录,若存在,弹出"登录成功"的对话框,如图 1-3 所示。

若不存在,弹出"数据库中不存在该用户"的对话框,如图 1-4 所示。

　　图 1-3　登录成功对话框　　　　　　图 1-4　数据库中不存在该用户对话框

若未输入用户名,弹出一对话框,如图 1-5 所示。

若未输入口令,弹出一对话框,如图 1-6 所示。

　　图 1-5　提示用户名必须输入的对话框　　　图 1-6　提示密码必须输入的对话框

若两次输入的口令不一致,弹出一对话框,如图1-7所示。

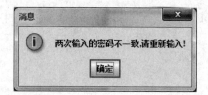

图1-7 两次输入的密码不一致时弹出的对话框

单击"取消"按钮后退出应用程序。

项目分析

一个完整的桌面应用程序都需要有登录功能。登录时,根据用户名及密码到数据库表中去核对是否存在该用户,主要涉及对数据库表进行查询操作。

环境需求

① 操作系统:Windows XP 或 Windows 7;
② 数据库平台:Microsoft Access;
③ JDK 版本:JDK Version1.6;
④ 集成开发环境:JCreator Pro 3.5。

项目知识点

① Swing 中常用容器 JFrame、JPanel 的使用方法;
② 布局管理器及常见布局方式;
③ 按钮组件 JButton 的使用方法;
④ 事件处理机制及动作事件的处理方法;
⑤ JDBC-ODBC 桥接方式连接数据库、查询数据库的一般方法。

任务一 准备数据库,定义窗体类,编制界面

【任务描述】

设计并创建一个 Access 数据库。从 JFrame 类继承出子类创建一个窗体,在其上生成界面。利用 JDBC-ODBC 桥接方式访问数据库。

【任务分析】

数据库中只需要建一张用户表。窗体上的组件可用网格布局方式组织在一个面板中。

【任务实施】

1. 准备数据库

① 在桌面上新建一个名称为 login 的文件夹;
② 在 Microsoft Access 中创建一个数据库,名称为 suserdb.mdb,保存在 login 文件夹中;
③ 在数据库中创建一个表,名称为 suser,表结构见表1-1。

项目一 用户登录程序

表 1-1 系统用户表 suser 的表结构

列名	数据类型	长度	允许为空	主键	含义
id	自动编号		否	是	id 号
username	文本	20	否		用户名
password	文本	20	否		密码

④ 在表中输入几条数据记录。

2. 编写程序框架代码

① 启动 JCreator Pro,新建一个名为 login.java 的源程序文件,保存在 login 文件夹中;
② 定义一个窗体类,类名为 login,该类从 JFrame 窗体类继承;
③ 定义构造方法,并设置窗体的大小及可见性;
④ 编写一个主方法(main()方法)。

实现代码如下:

```
//导入包
import java.awt.*;
import java.awt.event.*;
import java.sql.*;
import javax.swing.*;
//定义公共类
public class login extends JFrame
{
    //定义构造方法
    public login(String title)
    {
        super(title);

        setDefaultCloseOperation(JFrame.EXIT_ON_CLOSE);
        setSize(500,300);
        Dimension ss = Toolkit.getDefaultToolkit().getScreenSize();
        setLocation( ( ss.width - this.getWidth() )/2,(ss.height - this.getHeight() )/2);
        setResizable(false);
        setVisible(true);
    }
    //定义主方法
    public static void main(String [ ]args)
    {
        new login("用户登录");
    }
}
```

3. 编制界面

(1) 定义一个面板对象及相关组件对象

将它们定义为类的成员变量,组件包括三个标签组件(JLabel)、一个文本框组件(JTextField)、两个密码框组件(JPasswordField)、两个按钮组件(JButton)。

实现代码如下:

```
JPanel p;
JLabel l1,l2,l3;
JTextField tf1;
JPasswordField tf2,tf3;
JButton bt1,bt2;
```

(2) 在构造方法中创建组件并加入至窗体中

先创建面板,将面板的布局方式设定为网格方式(GridLayout)方式,有 4 行 2 列,再创建组件,将组件按顺序放置在面板,再将面板放置在窗体的内容窗格中。
实现代码如下:

```
p = new JPanel();
p.setLayout(new GridLayout(4,2));

l1 = new JLabel("用户名:");
l2 = new JLabel("口令:");
l3 = new JLabel("确认口令");

tf1 = new JTextField();
tf2 = new JPasswordField();
tf3 = new JPasswordField();

bt1 = new JButton("确定");
bt2 = new JButton("取消");

p.add(l1);
p.add(tf1);
p.add(l2);
p.add(tf2);
p.add(l3);
p.add(tf3);
p.add(bt1);
p.add(bt2);

getContentPane().add(p);          //可直接写成 add(p);
```

4．加载数据库驱动程序、连接数据库

在类中定义一个用于存放数据库连接对象的静态成员变量:

```
static Connection con;
```

在主方法中加载数据库驱动程序,并连接数据库。
实现代码如下:

```
try
{
    //加载 JDBC - ODBC 驱动程序
    Class.forName("sun.jdbc.odbc.JdbcOdbcDriver");
    con = DriverManager.getConnection("jdbc:odbc:driver = {Microsoft Access Driver ( * .mdb)};DBQ = suser.mdb","","");
}
catch(ClassNotFoundException e1)
{
    JOptionPane.showMessageDialog(null,"不能识别类:" + e1);
}
catch(SQLException e2)
{
    JOptionPane.showMessageDialog(null,"连接数据库出错:" + e2);
}
catch(Exception e3)
{
    JOptionPane.showMessageDialog(null,"其他错误:" + e3);
}
```

Class 类的 forName()方法在执行时可能产生的异常类型为 ClassNotFoundException,DriverManager 的 getConnection()方法在执行者时可能产生的异常类型为 SQLException,因此在程序中对可能产生的异常进行了捕捉处理。

任务二 编制事件处理代码

【任务描述】

编制事件处理框架代码,编制"确定"、"取消"按钮代码,运行并测试程序。

【任务分析】

窗体中的两个按钮会产生动作事件,按事件处理的三个步骤分别编制代码,在"确定"按钮中要编写合法性检查及查询数据库的代码。

【任务实施】

1. 编制事件处理框架代码

① 使 login 类实现动作事件监听接口 ActionListener,主要实现代码如下:

```
public class login extends JFrame implements ActionListener
```

② 用 addActionListener()方法对两个按钮的动作事件进行注册监听,注册监听一般在构造方法中完成。主要实现代码如下:

```
bt1.addActionListener(this);
bt2.addActionListener(this);
```

③ 定义动作事件处理方法 actionPerformed(),编写代码框架。

```
public void actionPerformed(ActionEvent e)
{
    Object obj = e.getSource();
    if(obj == bt1)
    {
    else if(obj == bt2)
    {
    }
    ...
}
```

程序中先通过 getSource()方法获取事件源,再将"确定"、"取消"按钮的事件分开处理。

2. 编写"确定"、"取消"按钮动作事件代码

在"确定"按钮中,先取出输入的用户名、密码、两次密码,判断其合法性,即用户名、密码不能为空,并且两次输入的密码必须一致;如果不合法,则提示重新输入,合法则根据用户名及密码到用户表查找该用户是否存在,不存在则提示"数据库中不存在该用户!"的信息,存在则提示"登录成功!"的信息。实现代码如下:

```
if(tf1.getText().equals(""))
{
    JOptionPane.showMessageDialog(this,"用户名必须输入,不能为空!");
    tf1.requestFocus();
    return;
}
if((String.valueOf(tf2.getPassword())).equals(""))
{
    JOptionPane.showMessageDialog(this,"密码必须输入,不能为空!");
```

```
        tf2.requestFocus();
        return;
}
if((!(String.valueOf(tf2.getPassword())).equals(String.valueOf(tf3.getPassword()))))
{
        JOptionPane.showMessageDialog(this,"两次输入的密码不一致,请重新输入!");
        tf3.requestFocus();
        return;
}
try
{
        Statement stat = con.createStatement();
        ResultSet result = stat.executeQuery("select * from suser where username = '" + tf1.getText() + "' and password = '" + (String.valueOf(tf2.getPassword())) + "'" );

        if(result.next())
        {
                JOptionPane.showMessageDialog(this,"登录成功!");
        }
        else
        {
                JOptionPane.showMessageDialog(this,"数据库中不存在该用户!");
        }

        result.close();
        stat.close();

}
catch(SQLException e2)
{
        JOptionPane.showMessageDialog(this,"操作数据库出错:" + e2);
}
catch(Exception e3)
{
        JOptionPane.showMessageDialog(this,"其他错误:" + e3);
}
```

在"取消"按钮中,用如下代码退出程序:

```
System.exit(0);
```

3. 运行程序并进行测试

分别输入数据库表中存在和不存在的用户进行测试。

项目小结

用户登录程序是一个简单的Java桌面程序,通过本项目的实践,主要是为了让读者了解和掌握使用Java开发桌面程序的一些基本方法及步骤,掌握图形界面编制的基本方法、数据库常用的连接及查询方法、事件处理的基本步骤,为以后的开发打下基础。

项目二　简单的列表框程序

项目描述

列表框程序是一个简单的Java桌面应用程序，它主要用来展示不采用任何布局方式的情况下界面的编制及对列表框组件的具体操作。

项目预览

程序运行时，出现如图2-1所示窗体。

插入一些项及使用左移、右移后，窗体如图2-2所示。

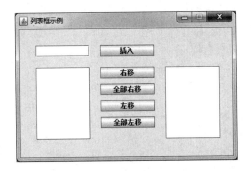 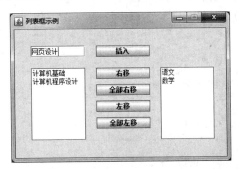

图2-1　列表框示例窗体(1)　　　　　图2-2　列表框示例窗体(2)

项目分析

可以用一个面板来组织其上的组件，列表框中项目的插入、移动主要涉及到对列表框相关方法的使用等。

环境需求

① 操作系统：Windows XP 或 Windows 7；
② JDK 版本：JDK Version1.6；
③ 集成开发环境：JCreator Pro 3.5。

项目知识点

① Swing 中常用容器 JFrame、JPanel 的使用方法；
② null 布局(空布局)下组件的定位方法；
③ 按钮组件 JButton、列表组件 List 的使用方法；
④ 事件处理机制及动作事件的处理方法。

任务一　定义窗体类，编制界面

【任务描述】

定义一个窗体类，类名与主文件名同名。定义与生成相关组件来形成界面。

【任务分析】

从 JFrame 类继承出子类创建一个窗体,用一个面板来组织其上的组件面板,不采用任何布局方式,直接设置文本框、按钮及列表框组件的大小及位置生成界面。

【任务实施】

1. 编写程序框架代码

① 在桌面上新建一个名称为 ListTest 的文件夹;
② 新建一个名为 ListTest.java 的源程序文件,保存在 ListTest 文件夹中;
③ 定义一个公共类,类名为 ListTest,该类从 JFrame 窗体类继承;
④ 定义构造方法,并设置窗体的大小及可见性;
⑤ 编写一个主方法(main()方法)。

实现代码如下:

```java
//导入包
import java.awt.*;
import javax.swing.*;
//定义公共类
public class ListTest extends JFrame
{
    //定义构造方法
    public ListTest (String title)
    {
        super(title);

        setDefaultCloseOperation(JFrame.EXIT_ON_CLOSE);
        setSize(400,260);
        Dimension ss = Toolkit.getDefaultToolkit().getScreenSize();
        setLocation((ss.width-this.getWidth())/2,(ss.height-this.getHeight())/2);
        setResizable(false);
        setVisible(true);
    }
    //定义主方法
    public static void main(String []args)
    {
        new ListTest ("列表框示例");
    }
}
```

2. 编制界面

(1) 定义一个面板对象及相关组件对象为类的成员变量

组件包括一个文本框组件(JTextField)、五个按钮组件(JButton)、两个列表框组件(List)。

实现代码如下:

```java
JPanel p;
JTextField tfitem;
JButton btins,bt1,bt2,bt3,bt4;
List lleft,lright;
```

(2) 在构造方法中创建组件并加入至窗体中

先创建面板将框架的布局方式改为不采用任何布局方式,可用如下语句实现:

```java
p.setLayout(null);
```

再用组件的 setBounds()方法设置组件在面板中的位置及大小,将组件先放置在面板上,再将面板放置在窗体的内容窗格中。

主要实现代码如下：

```
p = new JPanel();
p.setLayout(null);

tfitem = new JTextField();
…
tfitem.setBounds(30,30,100,20);
…
p.add(tfitem);
…
getContentPane().add(p);            //可直接写成 add(p);
```

任务二　编制事件处理代码

【任务描述】

编制事件处理代码，运行并测试程序。

【任务分析】

窗体中的五个按钮会产生动作事件，按事件处理的三个步骤分别编制代码，"左移"、"全部左移"按钮代码与"右移"、"全部右移"按钮代码基本一致，仅数据移动方向相反。

【任务实施】

1. 编制事件处理框架代码

① 使 ListTest 类实现动作事件监听接口 ActionListener，实现代码如下：

```
public class ListTest extends JFrame implements ActionListener
```

② 用 addActionListener()方法对五个按钮的动作事件进行注册监听，注册监听一般在构造方法中完成。实现代码如下：

```
btins.addActionListener(this);
bt1.addActionListener(this);
bt2.addActionListener(this);
bt3.addActionListener(this);
bt4.addActionListener(this);
```

③ 定义动作事件处理方法 actionPerformed()，编写代码框架。

```
public void actionPerformed(ActionEvent e)
{
    Object obj = e.getSource();
    if(obj == btins)
    {
    else if(obj == bt1)
    {
    
    }
    …
}
```

程序中先通过 getSource()方法获取事件源，再将五个按钮的事件分开处理。

2. 编写五个按钮动作事件代码

在"插入"按钮中,先取出文本框中输入的文本,判断其是否为空,不为空,即可插入左边列表框,实现代码如下:

```
String sitem = tfitem.getText();
if( sitem.equals(""))
    return;
lleft.add(sitem);
```

在"右移"按钮中,先取出左边列表框中选中的项,如果有选中项,则将其插入至右边列表框中,同时删除左边列表框中的项,实现代码如下:

```
String sitem = lleft.getSelectedItem();
if(   sitem! = null )
{
    lright.add(sitem);
    lleft.remove(sitem);
}
```

在"全部右移"按钮中,按顺序取出左边列表框中的项,将其插入至右边列表框中,完成后删除左边列表框中的项,实现代码如下:

```
int cnt = lleft.getItemCount();
for(int i = 0;i < cnt;i++)
{
    lright.add(lleft.getItem(i));
}
for(int i = cnt - 1;i >= 0;i-- )
{
    lleft.remove(i);
}
```

"左移"、"全部左移"按钮的功能与"右移"、"全部右移"类似,其代码可参照编写。

3. 运行程序并进行测试

在文本框中输入后插入列表框,进行测试。

项目小结

列表框程序也是一个简单的Java桌面程序,通过本项目的实践,主要是为了让读者进一步了解和掌握使用Java开发桌面程序的一些基本方法及步骤,掌握列表框的常见处理方法、事件处理方法等,为以后的开发打下基础。

项目三　学生成绩管理系统

项目描述

学生成绩管理系统是一个 Java 桌面数据库应用程序。它能够实现对教师、课程、班级、学生基本信息的维护,班级考试课程的选择,以及学生成绩的录入、查询、打印等功能。

项目预览

程序运行时,在用户成功登录后出现如图 3-1 所示多文档主窗体。

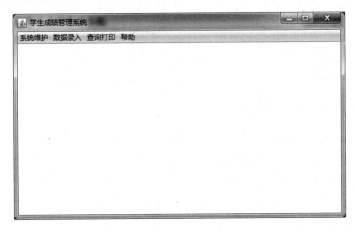

图 3-1　学生成绩管理系统主窗体

可以通过单击菜单项后打开对应的模块窗体进行操作。

项目分析

前面两个项目均是比较简单的 Java 桌面程序,主要涉及窗体、事件、数据库的查询及一些常用组件的操作。学生成绩管理系统是一个比较复杂和综合性的桌面程序,也是比较常见的数据库系统,主要功能是对一些基本实体如教师、课程、班级、学生数据的浏览及维护,班级课程的选择及学生成绩的录入、查询操作。具体开发时,将涉及到许多利用 Java 开发桌面程序的知识与技巧,包括多文档窗体、文档窗体、菜单、对话框、多种组件的编制,事件的处理,数据库的查询、插入、修改、删改操作,表格的生成及表格数据的处理等,是学习 Java 开发的理想项目。

项目设计

1. 多文档窗体的设计

应用程序界面采用 Swing 的多文档窗体界面。主窗体从 JFrame 类继承过来,上面挂有菜单,每一个菜单项打开一个文档窗体,或者称为内部框架窗体,它是从 JInternalFrame 继承下来的,它们被组织在一个被称为 JDesktopPane 类型的虚拟桌面对象上。

2. 数据库的设计

为了实现系统所要求的功能,数据库中设计了 7 张表,其中教师、课程、班级、学生、用户 5 个基本实体分别对应一张表,班级所上的课程和学生考试的成绩数据分别组织在对应的表中,有列名相同的两张表之间均存在关联关系。

教师情况表,表名为 steacher,表结构见表 3-1。

表 3-1 教师情况表 steacher 的表结构

列名	数据类型	长度	允许为空	主键	含义
stno	文本	3	否	是	教师代码
stname	文本	10	否		教师姓名
stnote	文本	50	是		备注

课程情况表,表名为 scourse,表结构见表 3-2。

表 3-2 课程情况表 scourse 的表结构

列名	数据类型	长度	允许为空	主键	含义
sono	文本	3	否	是	课程代码
soname	文本	30	否		课程名称
sonote	文本	50	是		备注

班级情况表,表名为 sclass,表结构见表 3-3。

表 3-3 班级情况表 sclass 的表结构

列名	数据类型	长度	允许为空	主键	含义
scno	文本	4	否	是	班级号
scname	文本	30	否		专业名称
stno	文本	3	否		班主任代码
sdname	文本	20	是		所在系部
scpnum	数字		是		班级人数
scnote	文本	50	是		备注

学生情况表,表名为 sstudent,表结构见表 3-4。

表 3-4 学生情况表 sstudent 的表结构

列名	数据类型	长度	允许为空	主键	含义
ssno	文本	6	否	是	学号
ssname	文本	10	否		学生姓名
ssex	文本	2	是		性别
ssnote	文本	50	是		备注

班级选课表,表名为 sccourse,表结构见表 3-5。

表 3-5 班级选课表 sccourse 的表结构

列名	数据类型	长度	允许为空	主键	含义
scno	文本	4	否	是	班级号
sono	文本	3	否	是	所选课程
stno	文本	3	否		任课教师

学生成绩表,表名为 sscore,表结构见表 3-6。

表 3-6 学生成绩表 sscore 的表结构

列名	数据类型	长度	允许为空	主键	含义
ssno	文本	6	否	是	学号
sono	文本	3	否	是	课程
ssscore	数字		是		成绩

系统用户表,表名为 suser,表结构见表 3-7。

表 3-7 系统用户表 suser 的表结构

列名	数据类型	长度	允许为空	主键	含义
id	自动编号		否	是	id 号
username	文本	20	否		用户名
password	文本	20	否		密码

3．系统功能模块的设计

系统功能模块图如图 3-2 所示。

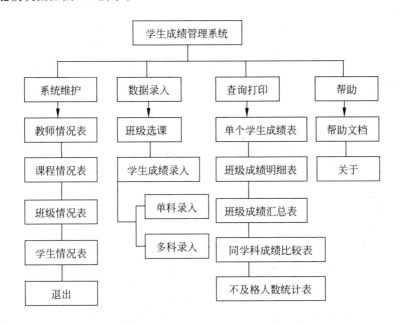

图 3-2 系统功能模块图

环境需求

① 操作系统：Windows XP 或 Windows 7；

② 数据库平台：Microsoft Access；

③ JDK 版本：JDK Version1.6；

④ 集成开发环境：JCreator Pro 3.5。

项目知识点

① Swing 中常用容器 JFrame、JInternalFrame、JDialog、JPanel 的使用方法；

② 布局管理器及 GridLayout、BorderLayout、FlowLayout 及 null 布局(空布局)方式的使用方法；
③ Swing 及 AWT 中各种常用组件的使用方法；
④ 事件处理机制及各种常用事件的处理方法；
⑤ SQL 语言及常用查询、插入、更新、删除语句的使用方法；
⑥ JDBC-ODBC 桥接方式连接数据库、查询及操纵数据库的方法；
⑦ 表格模型的基本概念及表格模型的定义、生成，表格组件的使用方法；
⑧ 组件的动态生成及使用方法。

任务一 准备数据库，编制主窗体

【任务描述】

设计并创建一个 Access 数据库。编制多文档窗体的主窗体，创建菜单并关联在主窗体上，连接数据库。

【任务分析】

先创建数据库，再在数据库中创建 7 张表。多文档窗体的主窗体从 JFrame 类继承过来，将包含菜单项的菜单条关联上去。利用 JDBC-ODBC 桥接方式访问数据库。

【任务实施】

1. 准备数据库

① 在桌面上新建一个名称为 scoremis 的文件夹；
② 在 Microsoft Access 中创建一个数据库，名称为 smis.mdb，保存在 scoremis 文件夹中；
③ 在数据库中创建 7 张表，表结构见表 3-1 至表 3-7。

2. 编制主窗体

① 新建一个名为 ScoreMis.java 的源程序文件，保存在 scoremis 文件夹中；
② 定义一个窗体类，类名为 ScoreMis，该类从 JFrame 窗体类继承；
③ 定义构造方法，将主窗体设置成屏幕大小及可见；
④ 在 ScoreMis 类中，定义一个虚拟桌面 JDesktopPane 类的对象作为类的成员变量，生成并加入至主窗体中；
⑤ 编写一个主方法(main()方法)。

实现代码如下：

```
//导入包
import java.awt.*;
import java.awt.event.*;
import javax.swing.*;
import java.sql.*;
import java.beans.PropertyVetoException;
//定义公共类
public class ScoreMis extends JFrame
{
    JDesktopPane desktopPane;                              //定义虚拟桌面对象

    //定义构造方法
```

```java
public ScoreMis (String title)
{
    super(title);

    desktopPane = new JDesktopPane();              //生成虚拟桌面对象
    getContentPane().add(desktopPane);             //将虚拟桌面对象加入窗体

    setDefaultCloseOperation(JFrame.EXIT_ON_CLOSE);

    setLocation(0,0);
    Dimension ss = Toolkit.getDefaultToolkit().getScreenSize();   //获取屏幕大小
    setSize(ss.width,ss.height);                   //窗体设成最大尺寸
    setVisible(true);
}
//定义主方法
public static void main(String []args)
{
    new ScoreMis ("学生成绩管理系统");
}
}
```

3．编制菜单

定义一个菜单条类(JMenuBar)对象；定义五个菜单类(JMenu)对象：系统维护、数据录入、学生成绩录入、查询打印、帮助；定义五个菜单下对应的菜单项类(JMenuItem)对象：教师情况维护等。

主要实现代码如下：

```java
JMenuBar menuBar;                                  //定义菜单条类对象
JMenu m1,m2,m3,m4,m5;                              //定义菜单类对象
JMenuItem m11,m12,…;                               //定义菜单项对象
```

在构造方法中，生成一个菜单条类(JMenuBar)对象，生成五个菜单类(JMenu)对象，生成五个菜单下对应的菜单项类(JMenuItem)对象，将菜单加入菜单条，将菜单项加入菜单中，将菜单条与主窗体关联起来。

主要实现代码如下：

```java
menuBar = new JMenuBar();                          //生成菜单条类对象
m1 = new JMenu("系统维护");                         //生成菜单类对象
…
m11 = new JMenuItem("教师情况维护");                //生成菜单项对象
…
menuBar.add(m1);                                   //将菜单加入菜单条
…
m1.add(m11);                                       //将菜单项加入菜单中
…
setJMenuBar(menuBar);                              //将菜单条与主窗体关联起来
```

要使两个菜单项之间有分隔线，可调用菜单(JMenu)类的addSeparator()方法。

4．加载数据库驱动程序、连接数据库

在类中定义一个用于存放数据库连接对象的静态成员变量：

```java
static Connection con;
```

在主方法中加载数据库驱动程序，并连接数据库：

```java
Class.forName("sun.jdbc.odbc.JdbcOdbcDriver");
```

con = DriverManager.getConnection("jdbc:odbc:driver = {Microsoft Access Driver (* .mdb)};DBQ = smis.mdb","","");

上述代码应放在 try…catch 语句块中,异常的处理可参考项目一中的相关代码。

5．编制菜单事件处理框架代码

① 使 ScoreMis 类实现动作事件监听接口 ActionListener,实现代码如下:

```
public class ScoreMis extends JFrame implements ActionListener
```

② 用 addActionListener 方法对所有菜单项的动作事件进行注册监听,主要实现代码如下:

```
m11.addActionListener(this);
m12.addActionListener(this);
…
```

③ 定义菜单动作事件处理方法 actionPerformed(),编写代码框架,主要实现代码如下:

```
public void actionPerformed(ActionEvent e)
{
    Object obj = e.getSource();
    if(obj == m11)
    {
    else if(obj == m12)
    {

    }
    …
}
```

程序中先通过 getSource()方法获取事件源,再将所有菜单产生的事件分开处理。

6．应用程序外观设置

在 ScoreMis 类的结束处,可以使用以下代码把外观设置成自己所使用的平台的外观:

```
static
{
    try
    {
        UIManager.setLookAndFeel(UIManager.getSystemLookAndFeelClassName());
    }
    catch (Exception e)
    {
        e.printStackTrace();
    }
}
```

7．运行程序并进行测试

(略)。

任务二　编制教师基本情况维护模块

【任务描述】

本模块用来维护教师的基本情况,主要是对教师情况表 steacher 中数据进行查询、插入、更新及删除操作。运行示意图如图 3-3、图 3-4、图 3-5 所示。

项目三　学生成绩管理系统

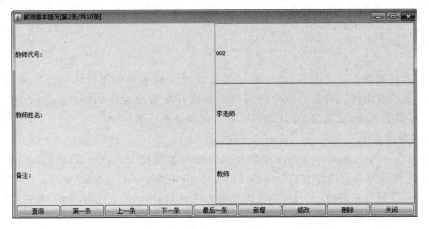

图 3-3　教师基本情况维护窗体

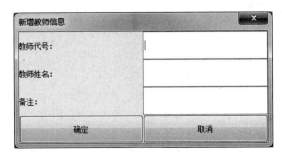

图 3-4　"新增教师信息"对话框

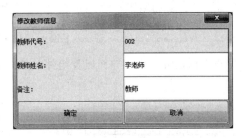

图 3-5　"修改教师信息"对话框

【任务分析】

对教师情况表 steacher 中数据进行查询、记录的前后滚动浏览、删改操作可以通过如图 3-3 所示窗体完成，对记录的新增、修改可以通过如图 3-4、图 3-5 所示对话框实现。

【任务实施】

1. 编制教师基本情况维护文档窗体

（1）新建教师基本情况维护文档窗体类并编制其基本代码

新建一个名为 TeacherFrame.java 的源程序文件，保存在 scoremis 文件夹中，在该文件中定义一个公共的文档窗体类 TeacherFrame，该类从 JInternalFrame 类继承，JInternalFrame 类相当于多文档窗体中的一个文档窗体，定义该类的构造方法，并设置窗体的大小及可见性。实现代码如下：

```java
import java.awt.*;
import java.awt.event.*;
import javax.swing.*;
import java.sql.*;
//定义教师基本情况维护窗体类
public class TeacherFrame extends JInternalFrame
{
    //定义构造方法
    public TeacherFrame(String title,boolean resizable,boolean closable,boolean maximizable,boolean iconifiable)
    {
        super(title,resizable,closable,maximizable,iconifiable);

        setSize(800,400);
```

```
        setLocation(20,20);
        setVisible(true);
    }
}
```

实际上,上述窗体类定义完后,即可通过教师情况维护菜单中用代码打开窗体,如图3-10所示,不过其上还没有组件。构造方法中的第一个参数 title 表示窗体的标题,后4个布尔型参数表示窗体是否可调整大小、是否可关闭、是否可最大化及是否可图标化。

(2) 编制窗体界面

在窗体中放置两个面板 p1、p2,面板的布局方式均为网格方式(GridLayout),而两个面板 p1、p2 在窗体中的布局方式为边界方式(BorderLayout),p1 在"Center",p2 在"South";在面板 p1 中加入控件,包括三个标签控件(JLabel),三个文本框控件(JTextField);在面板 p2 中加入控件,包括九个按钮控件(JButton)。

先将面板及组件定义为窗体类的成员变量,实现代码如下:

```
JPanel p1,p2;
JLabel lstno,lstname,lstnote;
JTextField tfstno,tfstname,tfstnote;
JButton btsearch,btfrist,btprev,btnext,btlast,btnew,btmodi,btdel,btclose;
```

再在构造方法中创建面板及组件并加入至窗体中,实现代码如下:

```
p1 = new JPanel();
p2 = new JPanel();
p1.setLayout(new  GridLayout(3,2));
p2.setLayout(new  GridLayout(1,9));

lstno = new JLabel("教师代号: ");
lstname = new JLabel("教师姓名: ");
lstnote = new JLabel("备注: ");

tfstno = new JTextField();
tfstname = new JTextField();
tfstnote = new JTextField();
tfstno.setEditable(false);
tfstname.setEditable(false);
tfstnote.setEditable(false);

btsearch = new JButton("查询");
btfrist = new JButton("第一条");
btprev = new JButton("上一条");
btnext = new JButton("下一条");
btlast = new JButton("最后一条");
btnew = new JButton("新增");
btmodi = new JButton("修改");
btdel = new JButton("删除");
btclose = new JButton("关闭");

p1.add(lstno);
p1.add(tfstno);
p1.add(lstname);
p1.add(tfstname);
p1.add(lstnote);
p1.add(tfstnote);

p2.add(btsearch);
p2.add(btfrist);
p2.add(btprev);
p2.add(btnext);
```

项目三 学生成绩管理系统

```
p2.add(btlast);
p2.add(btnew);
p2.add(btmodi);
p2.add(btdel);
p2.add(btclose);

getContentPane().add(p1,"Center");     //可直接写成 add (p1,"Center");
getContentPane().add(p2,"South");      //可直接写成 add (p2,"South");
```

(3) 编制窗体事件处理框架代码

参考项目一任务二中所述事件处理的三个步骤，修改类的定义，先使 TeacherFrame 类实现动作事件监听接口 ActionListener，实现代码如下：

```
public class TeacherFrame extends JInternalFrame implements ActionListener
```

再在构造方法中用 addActionListener()方法对三个按钮的动作事件进行注册监听，主要实现代码如下：

```
btsearch.addActionListener(this);
btfrist.addActionListener(this);
btprev.addActionListener(this);
…
```

最后定义动作事件处理方法 actionPerformed()，实现代码如下：

```
public void actionPerformed(ActionEvent e)
{
    Object obj = e.getSource();
    if(obj == btsearch)              //"查询"按钮
    { … }
    else if(obj == btfrist)          //"第一条"按钮
    { … }
    else if(obj == btprev)           //"上一条"按钮
    { … }
    else if(obj == btnext)           //"下一条"按钮
    { … }
    else if(obj == btlast)           //"最后一条"按钮
    { … }
    else if(obj == btadd)            //"新增"按钮
    { … }
    else if(obj == btmodi)           //"修改"按钮
    { … }
    else if(obj == btdel)            //"删除"按钮
    { … }
    else if(obj == btclose)          //"关闭"按钮
    { … }
}
```

(4) 编写前五个按钮动作事件代码

在类中定义一个用于存放数据库记录结果集的成员变量：

```
ResultSet result;
```

在类中定义两个用于存放数据库记录结果集中记录总数及当前记录号的成员变量：

```
int rowCnt,curRow;
```

在"查询"按钮中，它调用 search()方法，并设置窗体标题。实现代码如下：

```
search();
setTitle("教师基本情况[第" + curRow + "条/共" + rowCnt + "条]");
```

search()方法是TeacherFrame类中自定义的私有方法,在"查询"、"新增"、"修改"、"删除"事件处理中均要被调用。在search()方法中,将steacher表中的数据按教师代码顺序全部取出,并显示第一条记录。search()方法定义如下:

```java
private void search()
{
    try
    {
        Statement stat = con.createStatement(ResultSet.TYPE_SCROLL_INSENSITIVE,ResultSet.CONCUR_READ_ONLY);
        result = stat.executeQuery("select * from steacher order by stno");
        if(result.last())           //将记录指针移至记录结果集末尾
        {
            rowCnt = result.getRow();   //获取记录数
        }
        else
        {
            rowCnt = 0;             //无记录
        }
        if(result.first())          //将记录指针移至记录结果集第一条
        {
            tfstno.setText(result.getString(1));
            tfstname.setText(result.getString(2));
            tfstnote.setText(result.getString(3));

            curRow = result.getRow();
        }
        else                        //无记录
        {
            tfstno.setText("");
            tfstname.setText("");
            tfstnote.setText("");

            curRow = 0;
        }
    }
    catch(SQLException e2)
    {
        JOptionPane.showMessageDialog(this,"操作数据库出错:" + e2);
    }
    catch(Exception e3)
    {
        JOptionPane.showMessageDialog(this,"其他错误:" + e3);
    }
}
```

请注意createStatement方法中参数的写法,参数TYPE_SCROLL_INSENSITIVE指示记录可前后滚动,参数CONCUR_READ_ONLY指示记录不可以更新。

在"第一条"按钮中,使用ResultSet类的first()方法将记录记针移至第一条,显示记录数据,并设置标题。实现代码如下:

```java
if(result!= null && result.first())
{
    tfstno.setText(result.getString("stno")); //result.getString("stno")等同于result.getString(1)
    tfstname.setText(result.getString("stname"));
    tfstnote.setText(result.getString("stnote"));

    curRow = result.getRow();
    setTitle("教师基本情况[第" + curRow + "条/共" + rowCnt + "条]");
}
```

上述代码中的语句可能会产生类型为 SQLException 的异常，这在其他几个按钮中也会出现，因此，可以将整个 if…else…语法放在 try…catch…语句中进行异常处理。

在"上一条"按钮中，使用 ResultSet 类的 previous()方法将记录记针移至上一条，显示记录数据，并设置标题，可用 ResultSet 类的 isFirst()方法判断记录记针是否已指在第一条上。实现代码如下：

```
if(result!= null && !result.isFirst() && result.previous())
{
    tfstno.setText(result.getString("stno"));
    tfstname.setText(result.getString("stname"));
    tfstnote.setText(result.getString("stnote"));

    curRow = result.getRow();
    setTitle("教师基本情况[第" + curRow + "条/共" + rowCnt + "条]");
}
```

在"下一条"按钮中，使用 ResultSet 类的 next()方法将记录记针移至下一条，显示记录数据，并设置标题，可用 ResultSet 类的 isLast()方法判断记录记针是否已指在最后一条上。实现代码如下：

```
if(result!= null && !result.isLast() && result.next())
{
    tfstno.setText(result.getString("stno"));
    tfstname.setText(result.getString("stname"));
    tfstnote.setText(result.getString("stnote"));

    curRow = result.getRow();
    setTitle("教师基本情况[第" + curRow + "条/共" + rowCnt + "条]");
}
```

在"最后一条"按钮中，使用 ResultSet 类的 last()方法将记录记针移至最后一条，显示记录数据，并设置标题。实现代码如下：

```
if(result!= null && result.last())
{
    tfstno.setText(result.getString("stno"));
    tfstname.setText(result.getString("stname"));
    tfstnote.setText(result.getString("stnote"));

    curRow = result.getRow();
    setTitle("教师基本情况[第" + curRow + "条/共" + rowCnt + "条]");
}
```

（5）编写"教师基本情况维护"菜单项的动作事件代码

定义并生成一个教师基本情况维护文档窗体类的对象，并将其加入桌面面板中，并显示在最前面。实现代码如下：

```
TeacherFrame teacherFrame = new TeacherFrame("教师基本情况维护",false,true,true,true);
desktopPane.add(teacherFrame);
try
{
    teacherFrame.setSelected(true);
}
catch (PropertyVetoException e1)
{
    e1.printStackTrace();
}
```

调用构造方法时，构造方法中的后4个布尔型参数表示窗体不可调整大小、可以关闭、可以最大化及可以在状态栏上缩成一个小图标。

（6）运行程序并进行测试

观察窗体的运行情况，看是否能正确地查询并前后滚动浏览记录。

2．编制"新增"、"修改"对话框类

（1）新建对话框类并编制其基本代码

新建一个名为TeacherNewModyDialog.java的源程序文件，保存在scoremis文件夹中，在该文件中定义一个公共类TeacherNewModyDialog，该类从JDialog类继承，因此它是一个对话框。定义该类的构造方法，并设置对话框的大小及可见性。实现代码如下：

```java
import java.awt.*;
import java.awt.event.*;
import javax.swing.*;
import java.sql.*;
//定义教师基本情况新增修改对话框类
public class TeacherNewModyDialog extends JDialog
{
    //定义构造方法
    public TeacherNewModyDialog (JFrame parent, String title, boolean modal, int newmody, String no, String name, String note)
    {
        super(parent,title,modal);

        setSize(400,200);
        Dimension ss = Toolkit.getDefaultToolkit().getScreenSize();                //获取屏幕大小
        setLocation((ss.width-this.getWidth())/2,(ss.height-this.getHeight())/2);  //使对话框居中
        setResizable(false);
        setVisible(true);
    }
}
```

实际上，上述对话框类定义完成后，即可通过"新增"、"修改"按钮用代码打开对话框如图3-4和图3-5所示对话框，不过其上还没有组件。构造方法中的前4个参数的含义如下。

parent：表示对话框的所有者，即父窗体或父对话框。

title：表示对话框的标题。

modal：表示对话框是否是模式对话框，值为true时表示是模式对话框，值为false时表示是无模式对话框。模式对话框能阻塞用户向其他顶层窗体输入，即用户必须在关闭模式对话框后才能进行其他操作，无模式对话框则没有这种限制。

newmody：表示对话框是进行新增还是修改操作。值1表示是新增操作，值2表示是修改操作，此值是开发人员自定的，用于在打开对话框时区别新增还是修改操作。

构造方法中的后3个参数分别用来接收传递过来的教师代码、姓名及备注数据。

（2）编制对话框界面

在对话框中放置一个面板p，面板的布局方式为网格方式（GridLayout），有4行2列，在面板p中加入控件，包括三个标签控件（JLabel）、三个文本框控件（JTextField）、两个按钮控件（JButton）。

先将面板及组件定义为对话框类的成员变量，实现代码如下：

```java
JPanel p;
JLabel lstno,lstname,lstnote;
JTextField tfstno,tfstname,tfstnote;
JButton btok,btcancel;
```

再在构造方法中创建面板及组件并加入至对话框中，实现代码如下：

项目三 学生成绩管理系统

```
p = new JPanel();
p.setLayout(new GridLayout(4,2));
lstno = new JLabel("教师代号：");
lstname = new JLabel("教师姓名：");
lstnote = new JLabel("备注：");

tfstno = new JTextField();
tfstname = new JTextField();
tfstnote = new JTextField();

btok = new JButton("确定");
btcancel = new JButton("取消");

p.add(lstno);
p.add(tfstno);
p.add(lstname);
p.add(tfstname);
p.add(lstnote);
p.add(tfstnote);
p.add(btok);
p.add(btcancel);

getContentPane().add(p);
```

在构造方法中设置对话框大小的语句之前，将要修改的数据设置至文本框中，实现代码如下：

```
if(newmody == 1)
{
    tfsono.requestFocus();
}
else if(newmody == 2)
{
    tfsono.setText(no);
    tfsoname.setText(name);
    tfsonote.setText(note);
    tfsono.setEditable(false);
    tfsono.setFocusable(false);
    tfsoname.requestFocus();
}
```

（3）编制对话框事件处理框架代码

参考项目一任务二中所述事件处理的三个步骤，先使 TeacherNewModyDialog 类实现动作事件监听接口 ActionListener，再在构造方法中用 addActionListener()方法对两个按钮的动作事件进行注册监听，最后定义动作事件处理方法 actionPerformed()，编写代码框架。第三步的主要实现代码如下：

```
public void actionPerformed(ActionEvent e)
{
    Object obj = e.getSource();
    if(obj == btok)                    //"确定"按钮
    { … }
    else if(obj == btcancel)           //"取消"按钮
    { … }
}
```

（4）编写"确定"、"取消"按钮动作事件代码

在类中定义一个用于接收构造方法中与参数 newmody 同名的成员变量 newmody：

```
int newmody;
```

在构造方法中通过以下语句接收传递过来的值：

```
this.newmody = newmody;
```

在"确定"按钮中，先取出输入教师代码、教师姓名，判断其合法性，即教师代码、教师姓名不能为空，且教师代码必须为 3 位数字；如果不合法，则提示重新输入，合法则根据 newmody 的值分别进行插入记录或更新记录的操作。实现代码如下：

```
String stno = tfstno.getText();
if(stno.equals("") || stno.length()!= 3 || !isNumeric(stno) )
{
    JOptionPane.showMessageDialog(this,"教师代码必须输入,不能为空,并且只能是 3 位数字!");
    tfstno.requestFocus();
    return;
}
String stname = tfstname.getText();
if(stname.equals(""))
{
    JOptionPane.showMessageDialog(this,"教师姓名必须输入,不能为空!");
    tfstname.requestFocus();
    return;
}
int row = 0;
try
{
    if(newmody == 1)                          //新增操作
    {
        PreparedStatement stat = ScoreMis.con.prepareStatement("insert into steacher(stno,stname,stnote) values(?,?,?)");
        stat.setString(1,stno);
        stat.setString(2,stname);
        stat.setString(3,tfstnote.getText());

        row = stat.executeUpdate();           //执行插入
    }
    else if(newmody == 2)                     //修改操作
    {
        PreparedStatement stat = ScoreMis.con.prepareStatement("update steacher set stname = ?,stnote = ? where stno = ?");
        stat.setString(1,stname);
        stat.setString(2,tfstnote.getText());
        stat.setString(3,tfstno.getText());

        row = stat.executeUpdate();           //执行更新
    }
}
catch(SQLException e2)
{
    JOptionPane.showMessageDialog(this,"操作数据库出错:" + e2);
}
catch(Exception e3)
{
    JOptionPane.showMessageDialog(this,"其他错误:" + e3);
}
if(row == 1)
{
    rtn = 1;
    dispose();                                //关闭对话框
}
else
    JOptionPane.showMessageDialog(this,"数据保存失败!");
```

上述代码中,executeUpdate()方法在执行后通过其返回值是否为 1 判断是否成功的插入或更新记录。isNumeric()方法用来判断一个字符串是否由数字构成。方法定义如下:

```
private boolean isNumeric(String str)
{
    for (int i = 0; i < str.length(); i++)
    {
        if (!Character.isDigit(str.charAt(i)))
        {
            return false;
        }
    }
    return true;
}
```

在"取消"按钮中,直接通过调用 dispose()方法关闭对话框:

```
dispose();
```

(5) 编写输入数据时用回车键代替制表键切换焦点的代码

在构造方法中用 addActionListener()方法对 3 个文本框的动作事件进行注册监听,然后在动作事件处理方法 actionPerformed()中修改 if…else 语句,加入对 3 个文本框动作事件的处理。实现代码如下:

```
else if(obj == tfstno)
{
    FocusManager.getCurrentManager().focusNextComponent((Component)e.getSource());
}
else if(obj == tfstname)
{
    FocusManager.getCurrentManager().focusNextComponent((Component)e.getSource());
}
else if(obj == tfstnote)
{
    FocusManager.getCurrentManager().focusNextComponent((Component)e.getSource());
    btok.doClick();                          //程序触发保存
}
```

3. 编写"新增"、"修改"按钮动作事件代码

为了区分对话框是通过执行"确定"关闭的还是"取消"关闭的,可以先在 TeacherNewModyDialog 对话框类定义一个成员变量:

```
int rtn = 0;
```

变量 rtn 的初始值为 0,在"确定"按钮的代码结束处调用 dispose()方法前将其值改为 1:

```
rtn = 1;
```

方法 getReturn()的定义如下:

```
public int getReturn()
{
    return rtn;
}
```

在"新增"按钮中,先定义并生成一个对话框类的对象,构造方法中后 3 个实参均为空字符串。待对话框关闭后通过调用其 getReturn()方法获取返回值,根据返回值是 1 或 0 来区分对话框是通

过"确定"关闭的还是"取消"关闭的,如果是通过"确定"关闭,即成功插入后退出,则要重新调用 search()方法查询记录,并刷新标题。实现代码如下:

```
TeacherNewModyDialog dlg = new TeacherNewModyDialog(null,"新增教师信息",true,1,"","","");
if(dlg.getReturn() == 0)
    return;
search();
setTitle("教师基本情况[第" + curRow + "条/共" + rowCnt + "条]");
```

在"修改"按钮中,先检查 curRow 变量的值看是否有记录修改,无则不能打开对话框。构造方法中后 3 个实参均为待修改的教师代码、姓名及备注。其余处理类同"新增"按钮。实现代码如下:

```
if(curRow == 0)
    return;
TeacherNewModyDialog dlg = new TeacherNewModyDialog(null,"修改教师信息",true,2,tfstno.getText(),tfstname.getText(),tfstnote.getText());
    if(dlg.getReturn() == 0)
        return;
search();
setTitle("教师基本情况[第" + curRow + "条/共" + rowCnt + "条]");
```

4.编写"删除"、"关闭"按钮动作事件代码

在"删除"按钮中,先检查 curRow 变量的值看是否有记录删除,没有则直接返回;如果有弹出一确认删除的对话框,按"是"后从数据库中删除当前记录数据,重新调用 search()方法查询记录,并刷新标题。实现代码如下:

```
if(curRow == 0)
    return;
if(JOptionPane.showConfirmDialog(this,"你是否要删除当前记录数据?","删除确认",JOptionPane.YES_NO_OPTION ) == JOptionPane.NO_OPTION)
    return;
PreparedStatement stat = ScoreMis.con.prepareStatement("delete from steacher where stno = ?");
stat.setString(1,tfstno.getText());
int row = stat.executeUpdate();
if(row == 0)
    JOptionPane.showMessageDialog(this,"数据删除失败!");
search();
setTitle("教师基本情况[第" + curRow + "条/共" + rowCnt + "条]");
```

在"关闭"按钮中,直接调用 dispose()方法关闭文档窗体:

```
dispose();
```

5.运行程序并进行测试

输入规范的教师信息数据,对整个模块的功能进行测试。

任务三 编制课程基本情况维护模块

【任务描述】

本模块用来维护课程的基本情况,主要是对课程情况表 scourse 中数据进行查询、插入、更新及删除操作。运行示意图如图 3-6、图 3-7、图 3-8 所示。

项目三　学生成绩管理系统

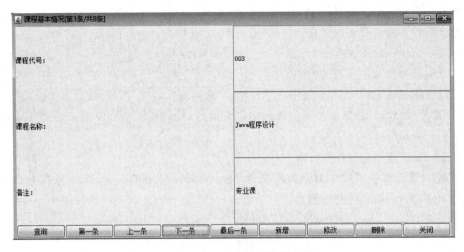

图 3-6　课程基本情况维护窗体

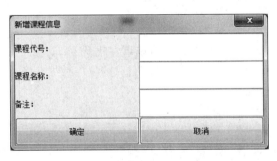

图 3-7　"新增课程信息"对话框

图 3-8　"修改课程信息"对话框

【任务分析】

本任务的实施方法与教师基本情况维护模块的任务实施方法完全一致,由于所维护的数据库表不同,代码中所涉及的数据库操作及相关组件对象的命名有所不同,在具体实施时完全可以参考前述步骤。

【任务实施】

1. 将教师情况维护模块文档窗体类文件 TeacherFrame.java 另存为 CourseFrame.java 后进行修改

① 将公共类名由 TeacherFrame 改为 CourseFrame。

② 将构造方法名由 TeacherFrame() 改为 CourseFrame()。
③ 修改组件对象名及文本信息。
④ 修改 search() 方法中的 select 语句。
⑤ 修改代码中获取记录数据的语句。
⑥ 将设置标题的 setTitle() 方法中的字符串由"教师基本情况"改为"课程基本情况"。
⑦ 将"新增"、"修改"按钮动作事件代码中打开的对话框的类名由 TeacherNewModyDialog 改为 CourseNewModyDialog。
⑧ 修改"删除"按钮动作事件代码中 delete 语句。

2．**将教师情况维护模块中对话框类文件 TeacherNewModyDialog.java 另存为 CourseNewModyDialog.java 后进行修改**

① 将公共类名由 TeacherNewModyDialog 改为 CourseNewModyDialog。
② 将构造方法名由 TeacherNewModyDialog() 改为 CourseNewModyDialog()。
③ 修改组件对象名及文本信息。
④ 修改构造方法中的 if…else…语句。
⑤ 修改"确定"按钮动作事件代码。

3．**编写"课程基本情况维护"菜单项的动作事件代码**

可以将"教师基本情况维护"菜单项的动作事件代码复制后进行修改。

4．**运行程序并进行测试**

输入规范的课程信息数据，对整个模块的功能进行测试。

任务四　编制班级基本情况维护模块

【任务描述】

本模块用来维护班级的基本情况，主要是对班级情况表 sclass 中数据进行查询、插入、更新及删除操作。运行示意图如图 3-9、图 3-10、图 3-11 所示。

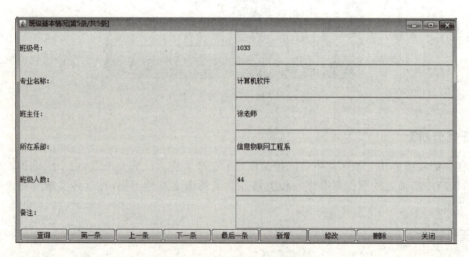

图 3-9　班级基本情况维护窗体

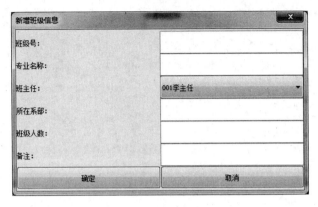

图 3-10 "新增班级信息"对话框

图 3-11 "修改班级信息"对话框

【任务分析】

本任务的实施方法与教师基本情况维护模块的任务实施方法基本一致,由于所维护的数据库表的不同,以及班级情况表 sclass 与教师情况表 steacher 之间有关联关系,代码中所涉及的数据库操作及相关组件对象的命名有所不同,在具体实施时完全可以参考前述步骤。

【任务实施】

1. **将教师情况维护模块文档窗体类文件 TeacherFrame.java 另存为 ClassFrame.java 后进行修改**

① 将公共类名由 TeacherFrame 改为 ClassFrame；

② 将构造方法名由 TeacherFrame() 改为 ClassFrame()；

③ 修改组件对象名及文本信息,将原来存放教师信息的标签控件类(JLabel)对象名、文本控件类(JTextField)对象名及标签中的文本改为对应于班级信息,p1 面板网格布局由原来的 3 行 2 列改为 6 行 2 列,将组件加入 p1 面板的语句作相应调整,窗体的大小可设为宽 800、高 400。

将组件对象的定义由:

```
JLabel lstno,lstname,lstnote;
JTextField tfstno,tfstname,tfstnote;
```

改为：

```
JLabel lscno, lscname, lstname, lsdname, lscpnum, lscnote;
JTextField tfscno, tfscname, tfstno, tfstname, tfsdname, tfscpnum, tfscnote;
```

将组件对象的创建改为:

```
lscno = new JLabel("班级号: ");
lscname = new JLabel("专业名称: ");
lstname = new JLabel("班主任: ");
lsdname = new JLabel("所在系部: ");
lscpnum = new JLabel("班级人数: ");
lscnote = new JLabel("备注: ");

tfscno = new JTextField();
tfscname = new JTextField();
tfstname = new JTextField();
tfsdname = new JTextField();
tfscpnum = new JTextField();
tfscnote = new JTextField();
tfstno = new JTextField();

tfscno.setEditable(false);
tfscname.setEditable(false);
tfstname.setEditable(false);
tfsdname.setEditable(false);
tfscpnum.setEditable(false);
tfscnote.setEditable(false);
```

注意,这里的组件 tfstno 是用来存放数据库表中取出的教师代码,不需要加入至面板中,界面上显示的是教师姓名。

④ 修改 search()方法中的 select 语句。

由原来的:

```
"select * from steacher order by stno"
```

改为:

```
" select scno, scname, sclass. stno, stname, sdname, scpnum, scnote from sclass, steacher where sclass. stno = steacher. stno order by scno"
```

两张表之间通过教师代码 stno 产生关联,将教师情况表中的姓名取了出来。

⑤ 修改获取记录数据的代码。

调整为:

```
tfscno.setText(result.getString("scno"));
tfscname.setText(result.getString("scname"));
tfstno.setText(result.getString("stno"));
tfstname.setText(result.getString("stname"));
tfsdname.setText(result.getString("sdname"));
tfscpnum.setText(String.valueOf(result.getInt("scpnum")));
tfscnote.setText(result.getString("scnote"));
```

⑥ 将设置标题的 setTitle()方法中的字符串由"教师基本情况"改为"班级基本情况"。

⑦ 将"新增"、"修改"按钮动作事件代码中打开的对话框的类名由 TeacherNewModyDialog 改为 ClassNewModyDialog,同时修改构造方法中的实际参数,将存放在 tfscno 等 7 个文本框中的班级信息传递过去。

⑧ 修改"删除"按钮动作事件代码中 delete 语句。

由原来的:

```
delete from steacher where stno = ?
```

改为:

```
delete from sclass where scno = ?
```

将存放在 tfscno 文本框中的班级号设置至 delete 语句参数中。

2. 将教师情况维护模块中对话框类文件 TeacherNewModyDialog.java 另存为 ClassNewModyDialog.java 后进行修改

① 将公共类名由 TeacherNewModyDialog 改为 ClassNewModyDialog;
② 修改构造方法名及参数。

将原来的:

```
public TeacherNewModyDialog(JFrame parent, String title, boolean modal, int newmody, String no, String name, String note)
```

改为:

```
public ClassNewModyDialog(JFrame parent, String title, boolean modal, int newmody, String scno, String scname, String stno, String stname, String sdname, String scpnum, String scnote)
```

构造方法中原来接收教师信息的 3 个参数改为接收班级信息的 7 个参数。

③ 修改组件对象名及文本信息,面板网格布局由原来的 4 行 2 列改为 7 行 2 列,将组件加入 p1 面板的语句作相应调整。

将标签及文本组件对象的定义改为:

```
JLabel lscno,lscname,lstname,lsdname,lscpnum,lscnote;
JTextField tfscno,tfscname,tfstno,tfstname,tfsdname,tfscpnum,tfscnote;
```

还要增加一个下拉列表框的定义,用于班主任的选择:

```
JComboBox csc;
```

将组件对象的创建改为:

```
lscno = new JLabel("班级号: ");
lscname = new JLabel("专业名称: ");
lstname = new JLabel("班主任: ");
lsdname = new JLabel("所在系部: ");
lscpnum = new JLabel("班级人数: ");
lscnote = new JLabel("备注: ");

tfscno = new JTextField();
tfscname = new JTextField();
tfstname = new JTextField();
tfsdname = new JTextField();
tfscpnum = new JTextField();
tfscnote = new JTextField();
tfstno = new JTextField();

csc = new JComboBox();
```

④ 在构造方法中将所有教师信息取至该下拉列表框中,以便选择班级的班主任。
实现代码如下:

```
try
{
    Statement stat = ScoreMis.con.createStatement();
```

```
            ResultSet result = stat.executeQuery("select * from steacher order by stno");
            while(result.next())
            {
                csc.addItem(result.getString(1) + result.getString(2));
            }
        }
        catch(SQLException e2)
        {
            JOptionPane.showMessageDialog(this,"操作数据库出错:" + e2);
        }
        catch(Exception e3)
        {
            JOptionPane.showMessageDialog(this,"其他错误:" + e3);
        }
```

⑤ 修改构造方法中的 if…else…语句。

改为：

```
if(newmody == 1)
{
    tfscno.requestFocus();
}
else if(newmody == 2)
{
    tfscno.setText(scno);
    tfscname.setText(scname);
    tfstno.setText(stno);
    tfstname.setText(stname);
    csc.setSelectedItem(stno + stname);      //根据传递过来的班主任代码及姓名选中指定项
    tfsdname.setText(sdname);
    tfscpnum.setText(scpnum);
    tfscnote.setText(scnote);

    tfscno.setEditable(false);
    tfscno.setFocusable(false);
    tfscname.requestFocus();
}
```

⑥ 修改"确定"按钮动作事件代码。

参考教师基本情况维护模块，对班级号、专业名称、班主任进行合法性检查，如果不合法，则提示重新输入，合法则根据 newmody 的值分别进行插入记录或更新记录的操作。其中的 SQL 语句作相应调整。

将 insert 语句由原来的：

insert into steacher(stno,stname,stnote) values(?,?,?)

改为：

insert into sclass(scno,scname,stno,sdname,scpnum,scnote) values(?,?,?,?,?,?)

将 update 语句由原来的：

update steacher set stname = ?, stnote = ? where stno = ?

改为：

update sclass set scname = ?, stno = ?, sdname = ?, scpnum = ?, scnote = ? where scno = ?

同时调整一下相应的 setXXX 方法，以便正确地设置 SQL 语句的参数值。

3．编写"班级基本情况维护"菜单项的动作事件代码

可以将"教师基本情况维护"菜单项的动作事件代码复制后进行修改。

4．运行程序并进行测试

输入规范的班级信息数据，对整个模块的功能进行测试。

任务五　表格及表格模型

【任务描述】

用两个实例说明 Swing 中表格组件 JTable 的基本使用方法。自定义一个表格模型，用实例来说明表格模型的使用方法。

【任务分析】

当对表格的操作比较简单时，可以直接使用 JTable 组件来创建表格。如果要在程序界面中显示大量数据，并且对表格中数据操作比较复杂时，直接使用 JTable 构造的表格组件往往不能满足要求，我们可以在创建 JTable 组件前，先创建一个表格模型，通过表格模型来管理和操作表格中的数据。本任务中自定义了一个表格模型类，其中包含了对多个表格数据的操作方法，这在本项目后面的任务中要被大量使用。

【任务实施】

1．直接使用 JTable 组件来创建表格和使用表格

例程 1　TableTest_1.java

程序中将表格标题存放在 String 类型的数组中，表格数据存放在 String 类型的二维数组中，将两个 String 类型的数组作为 JTable 构造方法的参数，直接构造表格。运行示意图如图 3-12 所示。实现代码如下：

```java
import java.awt.*;
import javax.swing.*;
public class TableTest_1 extends JFrame
{
    String[] heads = {"学号","姓名","性别","爱好","住址"};
    String[][] data = { {"103301","AAA","女","唱歌","清凉新村"},{"103302","BBB","男","计算机","丽华新村"} };
    JTable table;                                    //定义表格组件对象
    public TableTest_1(String title)
    {
        super(title);
        table = new JTable(data,heads);              //创建表格组件对象
        getContentPane().add(new JScrollPane(table));  //将组件对象放入滚动面板中

        setSize(500,300);
        Dimension ss = Toolkit.getDefaultToolkit().getScreenSize();
        setLocation((ss.width-this.getWidth())/2,(ss.height-this.getHeight())/2);
        setResizable(false);
        setVisible(true);
    }

    public static void main(String[] args)
    {
        new TableTest_1("表格示例 1");
    }
```

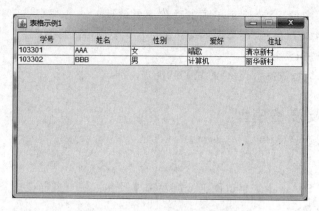

图 3-12　表格示例程 1 运行界面

例程 2　TableTest_2.java

程序中将表格标题存放在 Vector 类型的对象中，表格数据也存放在 Vector 类型的对象中，Vector 类型的对象 row 代表表格的一行数据，Vector 类型的对象 rows 代表表格的多行数据，将两个 Vector 类型的对象作为 JTable 构造方法的参数，直接构造表格。运行示意图如图 3-13 所示。实现代码如下：

```java
import java.util.*;
import java.awt.*;
import javax.swing.*;
public class TableTest_2 extends JFrame
{
    String[] heads = {"学号","姓名","性别","爱好","住址"};
    String[][] data = { {"103301","AAA","女","唱歌","清凉新村"},{"103305","BBB","男","计算机","丽华新村"} };
    Vector head;
    Vector row,rows;
    JTable table;                                          //定义表格组件对象
    public TableTest_2(String title)
    {
        super(title);

        head = new Vector();
        for(int i = 0;i < 5;i++)
            head.add(heads[i]);

        rows = new Vector();
        for(int j = 0;j < 2;j++)
        {
            row = new Vector();
            for(int k = 0;k < 5;k++)
                row.add(data[j][k]);
            rows.add(row);
        }
        table = new JTable(rows,head);                     //创建表格组件对象
        getContentPane().add(new JScrollPane(table));      //将组件对象放入滚动面板中

        setSize(500,300);
        Dimension ss = Toolkit.getDefaultToolkit().getScreenSize();
        setLocation((ss.width - this.getWidth())/2,(ss.height - this.getHeight())/2);
        setResizable(false);
        setVisible(true);
    }

    public static void main(String[] args)
```

```
        {
            new TableTest_2("表格示例2");
        }
}
```

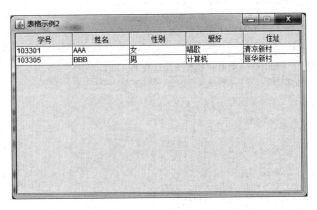

图 3-13　表格示例程 2 运行界面

2. 使用自定义模型类来构造表格及使用表格

(1) 编写一个自定义表格模型类 MyTableModel.java

可以将 Swing 中提供的抽象表格模型类 AbstractTableModel 作为基类，派生出一个表格模型类，在该类中，重新定义了构造方法以及向表格中添加数据行的方法、删除表格中数据行的方法、设置和读取表格单元格中数据的方法及获取表格中行数的方法等，通过设计这些方法，大大方便了对表格中数据的操作。自定义表格模型类 MyTableModel.java 的定义如下：

```java
import java.util.*;
import java.sql.*;
import javax.swing.table.*;
import javax.swing.*;
public class MyTableModel extends AbstractTableModel
{
    //存放表格数据的 Vector
    private Vector content = null;
    //存放表格标题的数组
    private String[] title_name;
    //表格中的列数
    private int colCnt;
    //可以编辑列的起止索引号,各列均不能编辑时,colSt,colEnd 的值均为 -1
    private int colSt,colEnd;
    //构造方法
    public MyTableModel(String[] title_name,int n)
    {
        this.title_name = new String[n];
        colCnt = n;
        for(int i = 0;i < colCnt;i++)
        {
            this.title_name[i] = new String();
            this.title_name[i] = title_name[i];
        }
        content = new Vector();
    }
    //添加一行,数据行存放在 Vector 类型的 row
    public void addRow(Vector row)
    {
        content.add(row);
    }
```

```java
//添加多行,数据行存放在 rows 中,rows 中每一个对象是一个 Vector 类型的 row
public void addRows(Vector rows)
{
    for(int i = 0;i < rows.size();i++)
    {
        content.add(rows.get(i));
    }
}
//添加多行,数据由 ResultSet 类型的对象 result 传来
public void addRows(ResultSet result) throws SQLException
{
    Vector row;
    Vector rows = new Vector();
    int i = 1;
    while(result.next())
    {
        row = new Vector();
        row.add(String.valueOf(i++));
        for(int j = 1;j <= colCnt - 1;j++)
        {//JOptionPane.showMessageDialog(null,"其他错误:" + j);
            row.add(result.getString(j));
        }
        rows.add(i - 2,row);
    }
    addRows(rows);
}
//删除一行
public void removeRow(int rowIndex)
{
    content.remove(rowIndex);
}
//删除所有行,全部删除
public void removeRows(int rowIndex, int count)
{
    for(int i = 0;i < count;i++)
    {
        if(content.size()> rowIndex)
        {
            content.remove(rowIndex);
        }
    }
}
//获取行数
public int getRowCount()
{
    return content.size();
}
//获取某一单元格的值
public Object getValueAt(int rowIndex, int columnIndex)
{
    return ((Vector)content.get(rowIndex)).get(columnIndex);
}
//更新表格中某一单元格的值
public void setValueAt(Object value, int rowIndex, int columnIndex)
{
    ((Vector)content.get(rowIndex)).remove(columnIndex);
    ((Vector)content.get(rowIndex)).add(columnIndex,value);
    
    this.fireTableCellUpdated(rowIndex,columnIndex);
}
//设定可编辑列的起止范围,从 colSt 到 colEnd
```

```java
    public void setColumnEditable(int colSt,int colEnd)
    {
        this.colSt = colSt;
        this.colEnd = colEnd;
    }
    //决定表格中哪些单元格的值可以修改,返回 false 表示不能修改
    public boolean isCellEditable(int rowIndex,int columnIndex)
    {
        //序列号不能修改
        if(columnIndex >= colSt && columnIndex <= colEnd)
        {
            return true;
        }
        return false;
    }

    //获取列数
    public int getColumnCount()
    {
        return title_name.length;
    }

    //获取列名
    public String getColumnName(int columnIndex)
    {
        return title_name[columnIndex];
    }

    //返回某列的数据类型
    public Class<?> getColumnClass(int columnIndex)
    {
        if( getValueAt(0,columnIndex) == null )
        {
            return (new String("")).getClass();
        }
        return getValueAt(0,columnIndex).getClass();
    }

}
```

（2）编写一个测试该模型类的简单例子程序 MyTableModelTest.java

程序中先定义存放表格标题的 String 类型数组、表格对象及表格模型对象,在构造方法中创建模型对象,根据模型对象创建表格对象。在按钮的动作事件中,调用模型对象的方法对表格中的数据进行操纵。运行示意图如图 3-14 所示。实现代码如下:

```java
import java.util.*;
import java.awt.*;
import java.awt.event.*;
import javax.swing.*;
public class MyTableModelTest extends JFrame implements ActionListener
{
    JPanel p1,p2;
    JButton bt1,bt2,bt3,bt4,bt5;
    String[] heads = {"学号","姓名","性别","年龄"};    //表格标题
    JTable table;                                       //表格对象
    MyTableModel model;                                 //表格模型对象
    public MyTableModelTest(String title)
    {
        super(title);

        p1 = new JPanel();
```

```java
            p2 = new JPanel();
            p1.setLayout(new GridLayout(1,5));
            p2.setLayout(new BorderLayout());

            bt1 = new JButton("增加一行");
            bt2 = new JButton("增加二行");
            bt3 = new JButton("删除末行");
            bt4 = new JButton("全部删除");
            bt5 = new JButton("关闭");

            p1.add(bt1);
            p1.add(bt2);
            p1.add(bt3);
            p1.add(bt4);
            p1.add(bt5);

            model = new MyTableModel(heads,4);         //创建模型对象
            table = new JTable(model);                 //创建表格对象
            p2.add(new JScrollPane(table));

            bt1.addActionListener(this);
            bt2.addActionListener(this);
            bt3.addActionListener(this);
            bt4.addActionListener(this);
            bt5.addActionListener(this);

            getContentPane().add(p1,"South");
            getContentPane().add(p2,"Center");

            setDefaultCloseOperation(JFrame.EXIT_ON_CLOSE);

            setSize(650,500);
            Dimension ss = Toolkit.getDefaultToolkit().getScreenSize();
            setLocation((ss.width-this.getWidth())/2,(ss.height-this.getHeight())/2);
            setResizable(false);
            setVisible(true);
        }

        public void actionPerformed(ActionEvent e)
        {
            Object obj = e.getSource();

            if(obj == bt1)                             //增加一行
            {
                Vector row = new Vector();
                row.add("066101");
                row.add("小王");
                row.add("女");
                row.add(19);

                model.addRow(row);
                table.updateUI();
            }
            else if(obj == bt2)                        //增加二行
            {
                Vector rows = new Vector();

                Vector row1 = new Vector();
                row1.add("066101");
                row1.add("小王");
                row1.add("女");
                row1.add(19);
                rows.add(row1);
```

```
            Vector row2 = new Vector();
            row2.add("066105");
            row2.add("小李");
            row2.add("男");
            row2.add(20);
            rows.add(row2);

            model.addRows(rows);
            table.updateUI();
        }
        else if(obj == bt3)                          //删除末行
        {
            int rowcnt = table.getRowCount();
            if(rowcnt == 0)
                return;
            model.removeRow(rowcnt - 1);

            table.updateUI();
        }
        else if(obj == bt4)                          //全部删除
        {
            int rowcnt = table.getRowCount();
            if(rowcnt == 0)
                return;
            model.removeRows(0,rowcnt);
            table.updateUI();
        }
        else if(obj == bt5)
        {
            System.exit(0);
        }
    }
    public static void main(String args[])
    {
        new MyTableModelTest("表格模型示例");
    }
}
```

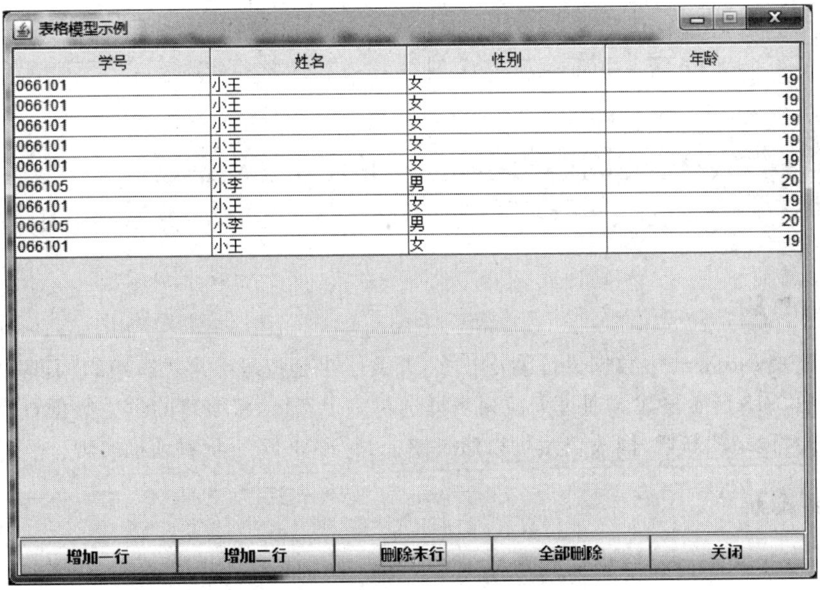

图 3-14 表格模型示例运行界面

任务六 编制学生基本情况维护模块

【任务描述】

本模块用来维护学生的基本情况,主要是对学生情况表 sstudent 中数据进行查询、插入、更新及删除操作。运行示意图如图 3-15、图 3-16、图 3-17 所示。

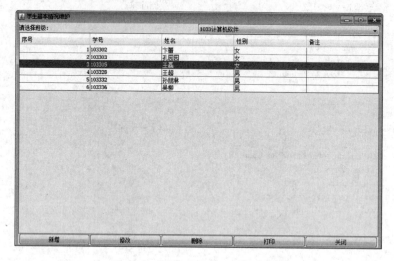

图 3-15 "学生基本情况维护"窗体

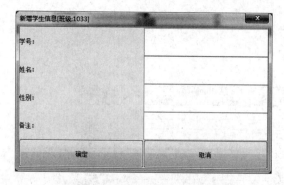

图 3-16 "新增学生信息"对话框

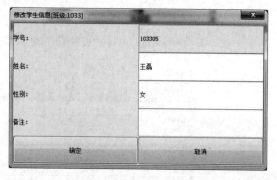

图 3-17 "修改学生信息"对话框

【任务分析】

学生情况表 sstudent 中数据量可能比较多,并且学生是按班级来进行组织的,因此可以将文档窗体设计成图 3-15 所示格式。通过下拉列表框选择一个班级,将该班的学生数据查询至一个表格中进行维护,对记录的新增、修改仍然可以通过图 3-16、图 3-17 所示对话框实现。

【任务实施】

1. 编制学生基本情况维护文档窗体

(1) 新建学生基本情况维护文档窗体类并编制其基本代码

新建一个名为 StudentFrame.java 的源程序文件,保存在 scoremis 文件夹中,在该文件中定义一个公共的文档窗体类 StudentFrame，该类从 JInternalFrame 类继承，JInternalFrame 类相当于多文档窗体中的一个文档窗体，定义该类的构造方法，并设置窗体的大小及可见性。实现代码如下：

```java
import java.awt.*;
import java.awt.event.*;
import javax.swing.*;
import java.sql.*;
//定义教师基本情况维护窗体类
public class StudentFrame extends JInternalFrame
{
    //定义构造方法
    public StudentFrame(String title,boolean resizable,boolean closable,boolean maximizable,boolean iconifiable)
    {
        super(title,resizable,closable,maximizable,iconifiable);

        setSize(800,500);
        setLocation(30,30);
        setVisible(true);
    }
}
```

（2）编制窗体界面

在窗体中放置三个面板 p1、p2、p3，三个面板 p1、p2、p3 在框架中的布局方式为边界方式（BorderLayout），p1 在"North"，p2 在"Center"，p3 在"South"；p1 面板的布局方式为网格方式（GridLayout），有 1 行 2 列，面板中一个标签控件（JLabel），一个选择班级的下拉列表框（JComboBox）控件；p2 面板的布局方式为边界方式，面板中有一个表格控件（JTable）；p3 面板的布局方式为网格方式（GridLayout），有 1 行 5 列，包括五个按钮控件（JButton）。

先将面板及组件定义为窗体类的成员变量，实现代码如下：

```java
JPanel p1,p2,p3;
JLabel lsc;
JComboBox csc;
JButton btnew,btmodi,btdel,btprn,btclose;
String[] tableHeads = {"序号","学号","姓名","性别","备注"};  //表格标题
JTable table;
MyTableModel model;
```

再在构造方法中创建面板及组件并加入至窗体中，实现代码如下：

```java
p1 = new JPanel();
p2 = new JPanel();
p3 = new JPanel();

p1.setLayout(new GridLayout(1,2));
p2.setLayout(new BorderLayout());
p3.setLayout(new GridLayout(1,5));

lsc = new JLabel("请选择班级：");
csc = new JComboBox();

btnew = new JButton("新增");
btmodi = new JButton("修改");
btdel = new JButton("删除");
btprn = new JButton("打印");
btclose = new JButton("关闭");

p1.add(lsc);
p1.add(csc);
```

```
model = new MyTableModel(tableHeads,5);
model.setColumnEditable(-1,-1);
table = new JTable(model);
table.setSelectionMode(ListSelectionModel.SINGLE_SELECTION);   //只能单行选择
p2.add(new JScrollPane(table));

p3.add(btnew);
p3.add(btmodi);
p3.add(btdel);
p3.add(btprn);
p3.add(btclose);

getContentPane().add(p1,"North");                    //可直接写成 add (p1," North ");
getContentPane().add(p2,"Center");                   //可直接写成 add (p2," Center ");
getContentPane().add(p3,"South");                    //可直接写成 add (p1," South ");
```

(3) 编写填充班级下拉列表框中数据项的代码

定义一个存放当前班级号的窗体类的成员变量 cursc,实现代码如下：

```
String cursc;
```

在类的构造方法中,填充选择班级的下拉列表框中数据项,并将当前班级号取出存放在变量 cursc 中,实现代码如下：

```
try
{
    Statement stat = ScoreMis.con.createStatement();
    ResultSet result = stat.executeQuery("select scno,sname from sclass order by scno");
    while(result.next())
    {
        csc.addItem(result.getString("scno") + result.getString("scname"));   //组合字符串
    }
}
catch(SQLException e2)
{
    JOptionPane.showMessageDialog(this,"操作数据库出错:" + e2);
}
catch(Exception e3)
{
    JOptionPane.showMessageDialog(this,"其他错误:" + e3);
}

if(csc.getItemCount() == 0)
{
    JOptionPane.showMessageDialog(this,"不存在班级信息,请先输入班级!");
}
else
{
    cursc = String.valueOf(csc.getItemAt(0)).substring(0,4);                  //取字符串中前4位
}
```

(4) 编写查询表格中数据的方法

在类中定义一个方法,方法名为 addDataToTable(),该方法能根据当前班级号将该班学生数据取出并填充至表格中,方法定义如下：

```
private void addDataToTable()
{
    Vector rows = new Vector();
    Vector row;
    try
```

```
        {
            Statement stat = ScoreMis.con.createStatement();
            ResultSet result = stat.executeQuery( "select * from sstudent where left(ssno,4) = '" + cursc + "' order by ssno" );

            int i = 1;
            while(result.next())
            {
                row = new Vector();
                row.add(i);
                row.add(result.getString(1));
                row.add(result.getString(2));
                row.add(result.getString(3));
                row.add(result.getString(4));

                rows.add(row);
                i++;
            }
            model.addRows(rows);
            model.fireTableDataChanged();
            if(table.getRowCount()>0)
            {
                table.setRowSelectionInterval(0,0);        //选择第一行
            }
        }
        catch(SQLException e2)
        {
            JOptionPane.showMessageDialog(null,"操纵数据库出错:" + e2);
        }
        catch(Exception e3)
        {
            JOptionPane.showMessageDialog(null,"其他错误:" + e3);
        }
    }
```

在该方法中，先将当前班级的学生数据取出，并存放至 Vector 类型的变量 rows 中，使用表格模型类的 addRows() 方法将数据填充至表格中，并选择第一行。

（5）编写调用 addDataToTable() 方法的代码

在构造方法中设置窗体大小语句的前面，编写调用 addDataToTable() 方法的代码，实现代码如下：

```
addDataToTable();
```

（6）编写"学生基本情况维护"菜单项的动作事件代码

可以将"教师基本情况维护"菜单项的动作事件代码复制后进行修改。

（7）运行程序并进行测试

观察窗体的运行情况，看下拉列表框中是否正确地填充了班级数据，表格中是否正确地查询到了当前班级的学生数据。

2. **将教师情况维护模块中对话框类文件 TeacherNewModyDialog.java 另存为 StudentNewModyDialog.java 后进行修改**

① 将公共类名由 TeacherNewModyDialog 改为 StudentNewModyDialog。

② 修改构造方法名及参数。

将原来的：

```
public TeacherNewModyDialog(JFrame parent, String title, boolean modal, int newmody, String no, String name, String note)
```

改为：

```
public StudentNewModyDialog(JFrame parent, String title, boolean modal, int newmody, String scno, String no,
String name, String sex, String note)
```

构造方法中原来接收教师信息的 3 个参数改为接收学生信息的 4 个参数，还增加了一个接收班级号的参数 scno。

同时，将调用 super() 方法的代码改为：

```
super(parent, title + "[班级:" + scno + "]", modal);
```

以便在对话框标题上能显示出当前班级号。

③ 在构造方法编写接收班级号的代码。

定义一个 String 类型的变量为类的成员变量，用于存放当前班级号：

```
String scno;
```

在构造方法中原接收 newmody 值的语句的下面，编写以下代码：

```
this.scno = scno;
```

④ 修改组件对象名及文本信息，面板网格布局由原来的 4 行 2 列改为 5 行 2 列，将组件加入 p1 面板的语句作相应调整。

将标签及文本组件对象的定义改为：

```
JLabel lssno, lssname, lssex, lssnote;
JTextField tfssno, tfssname, tfssex, tfssnote;
```

将组件对象的创建改为：

```
lssno = new JLabel("学号：");
lssname = new JLabel("姓名：");
lssex = new JLabel("性别：");
lssnote = new JLabel("备注：");

tfssno = new JTextField();
tfssname = new JTextField();
tfssex = new JTextField();
tfssnote = new JTextField();
```

⑤ 修改构造方法中的 if…else…语句。

改为：

```
if(newmody == 1)
{
    tfssno.requestFocus();
}
else if(newmody == 2)
{
    tfssno.setText(no);
    tfssname.setText(name);
    tfssex.setText(sex);
    tfssnote.setText(note);

    tfssno.setEditable(false);
    tfssno.setFocusable(false);
    tfssname.requestFocus();
}
```

⑥ 修改"确定"按钮动作事件代码。

参考教师基本情况维护模块,对学生学号、姓名进行合法性检查,要求学号的前 4 位必须是当前班级号,如果不合法,则提示重新输入,合法则根据 newmody 的值分别进行插入记录或更新记录的操作。其中的 SQL 语句作相应调整。

将 insert 语句改为:

insert into sstudent(ssno,ssname,ssex,ssnote) values(?,?,?,?)

将 update 语句改为:

update sstudent set ssname = ?,ssex = ?,ssnote = ? where ssno = ?

同时调整一下相应的 setXXX 方法,以便正确地设置 SQL 语句的参数值。

3．编写窗体的事件处理框架代码

在本窗体中会产生两种事件,一是单击按钮产生的动作事件,另一是选择下拉列表框项目时产生的项目变化事件。如前面项目所述事件处理的三个步骤,先使 TeacherFrame 类实现动作事件监听接口 ActionListener 及项目选择变化事件监听接口 ItemListener；再在构造方法中用 addActionListener()方法对五个按钮的动作事件进行注册监听,用 addItemListener 方法对下拉列表框项目时产生的项目变化事件进行注册监听；最后定义动作事件处理方法 actionPerformed(),定义下拉列表框项目时产生的项目变化事件处理方法 itemStateChanged()。

(1) 修改窗体类的定义,使其实现两种事件监听接口

实现代码如下:

```
public class StudentFrame extends JInternalFrame implements ActionListener,ItemListener
```

(2) 在构造方法中,对产生的两种事件进行注册监听

实现代码如下:

```
btnew.addActionListener(this);
btmodi.addActionListener(this);
btdel.addActionListener(this);
btprn.addActionListener(this);
btclose.addActionListener(this);

csc.addItemListener(this);                    //对下拉列表框事件注册监听
```

(3) 定义动作事件处理方法 actionPerformed()

主要实现代码如下:

```
public void actionPerformed(ActionEvent e)
{
    Object obj = e.getSource();
    if(obj == btadd)                          //"新增"按钮
    { … }
    else if(obj == btmodi)                    //"修改"按钮
    { … }
    else if(obj == btdel)                     //"删除"按钮
    { … }
    else if(obj == btprn)                     //"打印"按钮
    { … }
    else if(obj == btclose)                   //"关闭"按钮
    { … }
}
```

(4) 定义下拉列表框项目时产生的项目变化事件处理方法 itemStateChanged()

主要实现代码如下:

```
public void itemStateChanged(ItemEvent e)
{
…
}
```

4．编写五个按钮动作事件代码

（1）"新增"按钮动作事件代码

先定义并生成一个对话框类的对象,构造方法中后 3 个实参均为空字符串。待对话框关闭后通过调用其 getReturn()方法获取返回值,根据返回值是 1 或 0 来区分对话框是通过"确定"关闭的还是"取消"关闭的。如果是通过"确定"关闭,即成功插入后退出,则要删除表格中原有数据,并重取数据刷新至表格。实现代码如下：

```
if(cursc.equals("") )
{
    JOptionPane.showMessageDialog(this,"请先输入班级信息!");
    return;
}
StudentNewModyDialog dlg = new StudentNewModyDialog(null,"新增学生信息",true,1,cursc,"","","","");
if(dlg.getReturn() == 0)
    return;

model.removeRows(0,model.getRowCount());
addDataToTable();
```

（2）"修改"按钮动作事件代码

先检查表格中是否有数据行可以修改,无则不能打开对话框,有则先取出表格中当前数据行各列的值,打开对话框,构造方法中后 5 个实参为当前班级号及待修改的学生数据。其余处理类同"新增"按钮。实现代码如下：

```
int row = -1;
if(table!= null)
{
    row = table.getSelectedRow();
    if(row == -1)
        return;
}

String ssno = String.valueOf(table.getValueAt(row,1));
String ssname = String.valueOf(table.getValueAt(row,2));
String ssex = String.valueOf(table.getValueAt(row,3));
String ssnote = String.valueOf(table.getValueAt(row,4));

StudentNewModyDialog dlg = new StudentNewModyDialog(null,"修改学生信息",true,2,cursc,ssno,ssname,ssex,ssnote);
if(dlg.getReturn() == 0)
    return;

model.removeRows(0,model.getRowCount());
addDataToTable();
```

（3）"删除"按钮动作事件代码

将表格中当前行中数据从数据库中删除,并刷新表格。实现代码如下：

```
int row = -1;
if(table!= null)
{
    row = table.getSelectedRow();
    if(row == -1)
```

```
            return;
        }
if(JOptionPane.showConfirmDialog(this,"你是否要删除当前记录数据?","删除确认",JOptionPane.YES_NO_OPTION ) ==
JOptionPane.NO_OPTION)
            return;
int rowcnt = 0;
try
{
    PreparedStatement stat = ScoreMis.con.prepareStatement("delete from sstudent where ssno = ?");
    stat.setString(1,String.valueOf(table.getValueAt(row,1)));
    rowcnt = stat.executeUpdate();
}
catch(SQLException e2)
{
    JOptionPane.showMessageDialog(null,"操纵数据库出错:" + e2);
}
catch(Exception e3)
{
    JOptionPane.showMessageDialog(null,"其他错误:" + e3);
}
if(rowcnt == 0)
    JOptionPane.showMessageDialog(this,"数据删除失败!");
model.removeRows(0,model.getRowCount());
addDataToTable();
```

(4)"打印"按钮动作事件代码

暂无实现。

(5)"关闭"按钮动作事件代码

在"关闭"按钮中,直接调用 dispose()方法关闭文档窗体:

```
dispose();
```

5. 编写下拉列表框项目变化事件代码

先获取下拉列表框中当前所选班级的班级号,清除表格中原有数据,再根据当前班级号重新检索数据并刷新表格。实现代码如下:

```
public void itemStateChanged(ItemEvent e)
{
    cursc = String.valueOf(csc.getSelectedItem()).substring(0,4);   //获取当前班级号
    model.removeRows(0,model.getRowCount());
    addDataToTable();
}
```

6. 运行程序并进行测试

输入规范的学生数据,对整个模块的功能进行测试。

任务七 编制班级选课模块

【任务描述】

本模块用来选择班级考试课程及对应任课教师,选课数据进入班级选课表 sccourse,同时向学生成绩表 sscore 中插入相应的学生成绩记录,学生成绩初始值设为 0。运行示意图如图 3-18 和图 3-19 所示。

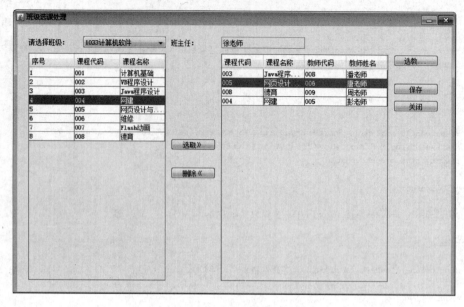

图 3-18 "班级选课处理"窗体

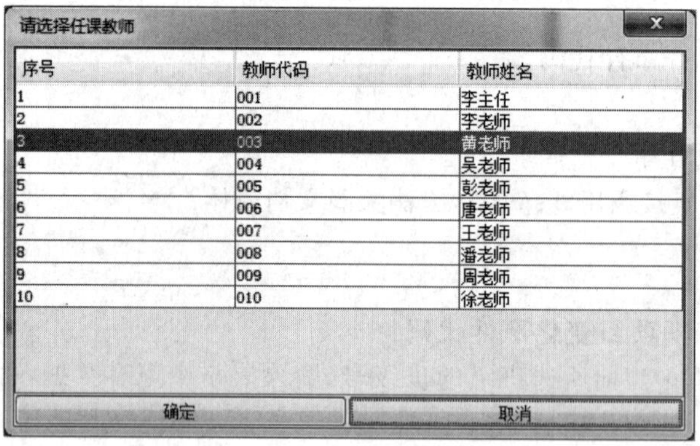

图 3-19 "教师选择"对话框

【任务分析】

打开窗体时,将全部班级取出并插入下拉列表框中,将全部课程取出并填充至表格一(左边表格)中,将第一个班级的已选课及任课教师取出并填充至表格二(右边表格)中;重新选择班级时,将当前班级的已选课及任课教师取出并填充至表格二(右边表格)中;单击"选取"按钮时,将表格一(左边表格)已选择行对应课程插入表格二(右边表格)中,重复的不插入,单击"删除"按钮时,将表格二(右边表格)已选择行对应课程删除。单击"选教师…"按钮时,打开选择教师对话框,选择表格二(右边表格)已选择行的任课教师;单击"保存"按钮时,将当前班级选择的考试课程数据及相应学生数据保存至数据库中,保存前先将该班原先的选课数据及学生成绩表中数据删除。

【任务实施】

1. 新建班级选课文档窗体类并编写其基本代码

新建一个名为 ScCourseFrame.java 的源程序文件,保存在 scoremis 文件夹中,在该文件中定义一个公共的文档窗体类 ScCourseFrame,该类从 JInternalFrame 类继承,JInternalFrame 类相当

于多文档窗体中的一个文档窗体,定义该类的构造方法,并设置窗体的大小及可见性。实现代码如下:

```
import java.util.*;
import java.awt.*;
import java.awt.event.*;
import javax.swing.*;
import java.sql.*;
import javax.swing.table.*;
public class ScCourseFrame extends JInternalFrame                //定义构造方法
{
     public ScCourseFrame(String title,boolean resizable,boolean closable,boolean maximizable,boolean iconifiable)
    {
        super(title,resizable,closable,maximizable,iconifiable);

        setSize(800,500);
        setLocation(20,20);
        setVisible(true);
    }
}
```

2. 编制窗体界面

窗体上有如下几个组件。

两个标签组件(JLabel):"请选择班级:"、"班主任:";

一个文本组件(JTextField):用于存放当前班级的班主任的姓名;

两个下拉列表框组件(JComboBox):一个存放班级信息、一个存放班主任的姓名;

两个表格组件(JTable):表格一(左边表格)存放所有课程,表格二(右边表格)存放当前班级所选课程;

五个按钮组件(JButton):"选取》"、"删除《"、"选教师"、"保存"、"关闭"。

窗体上放置三个面板 p、p1、p2,其布局方式及作用分别如下。

p1:边界布局(BorderLayout),其上只有一个表格(JTabel)控件,为表格一;

p2:边界布局(BorderLayout),其上只有一个表格(JTabel)控件,为表格二;

p:不采用任何布局方式(null),所有组件及 p1、p2 面板都放在其上,直接采用 setBounds()方法确定控件及面板 p1、p2 在其上的位置及大小,最后将面板 p 放到框架中。

先将面板及组件定义为窗体类的成员变量,实现代码如下:

```
JPanel p,p1,p2;
JLabel l1,l2;
JTextField tf1;
JComboBox csc,csstname;

String[] tableHeads1 = {"序号","课程代码","课程名称"};
String[] tableHeads2 = {"课程代码","课程名称","教师代码","教师姓名"};

JTable table1,table2;
MyTableModel  model1,model2;

JButton btadd,btdel,btst,btsave,btclose;
```

再在构造方法中创建面板及组件并加入至窗体中,实现代码如下:

```
p = new JPanel();
p1 = new JPanel();
p2 = new JPanel();
```

```
p.setLayout(null);
p1.setLayout(new BorderLayout());
p2.setLayout(new BorderLayout());

l1 = new JLabel("请选择班级：");
csc = new JComboBox();
l2 = new JLabel("班主任：");
tf1 = new JTextField();
csstname = new JComboBox();
l1.setBounds(20,20,85,20);
csc.setBounds(120,20,150,20);
l2.setBounds(280,20,85,20);
tf1.setBounds(370,20,150,20);
tf1.setEditable(false);

p.add(l1);
p.add(csc);
p.add(l2);
p.add(tf1);

model1 = new MyTableModel(tableHeads1,3);
table1 = new JTable(model1);
table1.setSelectionMode(ListSelectionModel.SINGLE_SELECTION);   //只能单行选择
p1.setBounds(20,50,250,400);
p1.add(new JScrollPane(table1));
p.add(p1);

model2 = new MyTableModel(tableHeads2,4);
table2 = new JTable(model2);
table2.setSelectionMode(ListSelectionModel.SINGLE_SELECTION);   //只能单行选择
p2.setBounds(370,50,300,400);
p2.add(new JScrollPane(table2));
p.add(p2);

btadd = new JButton("选取»");
btdel = new JButton("删除«");
btst = new JButton("选教师...");
btsave = new JButton("保存");
btclose = new JButton("关闭");
btadd.setBounds(280,200,80,20);
btdel.setBounds(280,250,80,20);
btst.setBounds(680,50,80,20);
btsave.setBounds(680,100,80,20);
btclose.setBounds(680,130,80,20);

p.add(btadd);
p.add(btdel);
p.add(btst);
p.add(btsave);
p.add(btclose);

getContentPane().add(p);
```

3．编写"班级选课处理"菜单项的动作事件代码

可以将"教师基本情况维护"菜单项的动作事件代码复制后进行修改。

4．编写窗体类中填充班级下拉列表框中数据项的方法

先定义一个存放当前班级号的窗体类的成员变量cursc，实现代码如下：

```
String cursc;
```

再在类中定义一个方法，方法名为addDataToSCList()。在该方法，从班级情况表、教师情况表中取出班级号、专业名称、班主任姓名,分别填充至两个下拉列表框中,并取出第一个班级号,存放至变量 cursc。实现代码如下:

```java
private void addDataToSCList()
{
    try
    {
        Statement stat = ScoreMis.con.createStatement();
        ResultSet result = stat.executeQuery("select scno,scname,stname from sclass,steacher where sclass.stno = steacher.stno order by scno");
        while(result.next())
        {
            csc.addItem(result.getString(1) + result.getString(2));    //班级号 + 专业名称
            csstname.addItem(result.getString(3));                     //班主任姓名
        }
    }
    catch(SQLException e2)
    {
        JOptionPane.showMessageDialog(this,"操作数据库出错:" + e2);
    }
    catch(Exception e3)
    {
        JOptionPane.showMessageDialog(this,"其他错误:" + e3);
    }
    if(csc.getItemCount() == 0)
    {
        JOptionPane.showMessageDialog(this,"不存在班级信息,请先输入班级!");
    }
    else
    {
        cursc = String.valueOf(csc.getItemAt(0)).substring(0,4);
        tf1.setText(String.valueOf(csstname.getItemAt(0)));
    }
}
```

该方法在构造方法中被调用。在构造方法中将面板 p 加入窗体中的 getContentPane().add(p)语句后面,编写以下调用代码:

```
addDataToSCList();
```

这样,窗体被打开后,即能看到班级数据显示在下拉列表框中。

5. 编写窗体类中填充表格一(左边表格)中数据的方法

在类中定义一个方法,方法名为 addDataToTable1()。在该方法中,从课程情况表中取出所有课程,利用表格模型,将数据填充至左边表格中。该方法的具体实现代码可参考本项目任务六中的 addDateTable()方法。在构造方法中调用 addDataToSCList()方法的语句后面,编写以下调用代码:

```
addDataToTable1();
```

这样,窗体被打开后,即能看到课程数据显示在表格一(左边表格)中。

6. 编制窗体事件处理框架代码

在本窗体中会产生两种事件,一是单击按钮产生的动作事件,另一是选择下拉列表框项目时产生的项目变化事件。如前面项目所述事件处理的三个步骤,先使 ScCourseFrame 类实现动作事件监听接口 ActionListener 及项目选择变化事件监听接口 ItemListener;再在构造方法中用 addActionListener()方法对五个按钮的动作事件进行注册监听,用 addItemListener 方法对下拉列表框项目时产生的项目变化事件进行注册监听;最后定义动作事件处理方法 actionPerformed(),定义下拉列表框项目时产生的项目变化事件处理方法 itemStateChanged()。

(1) 修改窗体类的定义,使其实现两种事件监听接口
实现代码如下:

```java
public class ScCourseFrame extends JInternalFrame implements ActionListener,ItemListener
```

(2) 在构造方法中,对产生的两种事件进行注册监听
实现代码如下:

```java
btadd.addActionListener(this);
btdel.addActionListener(this);
btst.addActionListener(this);
btsave.addActionListener(this);
btclose.addActionListener(this);

csc.addItemListener(this);                          //对下拉列表框事件注册监听
```

(3) 定义动作事件处理方法 actionPerformed()
主要实现代码如下:

```java
public void actionPerformed(ActionEvent e)
{
    Object obj = e.getSource();
    if(obj == btadd)                                //选取
    { … }
    else if(obj == btdel)                           //删除
    { … }
    else if(obj == btst)                            //选择任课教师
    { … }
    else if(obj == btsave)                          //保存
    { … }
    else if(obj == btclose)                         //关闭
    { … }
}
```

(4) 定义下拉列表框项目时产生的项目变化事件处理方法 itemStateChanged()
主要实现代码如下:

```java
public void itemStateChanged(ItemEvent e)
{
    …
}
```

7. 编写窗体类中"选取》"、"删除《"按钮动作事件代码

在"选取》"按钮动作事件中,将表格一中当前所选课程添加至表格二中,重复的不插入。实现代码如下:

```java
int row = -1;
if(table1!= null)
{
    row = table1.getSelectedRow();
    if(row == -1)
        return;
}

String sono = String.valueOf(table1.getValueAt(row,1));
String soname = String.valueOf(table1.getValueAt(row,2));

int rowCnt = table2.getRowCount();
for(int i = 0;i < rowCnt;i++)
{
```

```
        if(String.valueOf(table2.getValueAt(i,0)).equals(sono))
            return;
}

Vector rowdata = new Vector();
rowdata.add(sono);
rowdata.add(soname);
rowdata.add("");
rowdata.add("");

model2.addRow(rowdata);
//model2.fireTableDataChanged();          //会取消原选中的行
table2.updateUI();                        //不影响选中的行

if(table2.getRowCount() == 1)
    table2.setRowSelectionInterval(0,0);  //选择第一行
```

在"删除《"按钮动作事件中,将表格二中当前所选课程删除。实现代码如下:

```
int row = -1;
if(table2!= null)
{
    row = table2.getSelectedRow();
    if(row == -1)
        return;
}

model2.removeRow(row);
model2.fireTableDataChanged();

if(table2.getRowCount()>0)
{
    table2.setRowSelectionInterval(0,0);  //选择第一行
}
```

8. 编制选择教师的对话框类

(1) 新建对话框类并编制其基本代码

新建一个名为 SelectTeacherDialog.java 的源程序文件,保存在 scoremis 文件夹中,在该文件中定义一个公共类 SelectTeacherDialog,该类从 JDialog 类继承,因此它是一个对话框。定义该类的构造方法,并设置对话框的大小及可见性。

(2) 编制对话框界面

在对话框中放置两个面板 p1、p2,面板 p1 的布局方式为边界(BorderLayout)方式,其上有一个表格控件;面板 p2 的布局方式为网格(GridLayout)方式,有 1 行 2 列,其上有两个按钮控件(JButton)。两个面板 p1、p2 在窗体的边界(BorderLayout)布局方式中分别处于"Center"、"South"位置。

(3) 编写对话框类中填充表格中数据的方法

在类中定义一个方法,方法名为 addDataToTable()。在该方法中,从教师情况表中取出所有教师,利用表格模型,将数据填充至表格中。具体实现及调试代码可参考本任务步骤 5 中编制的代码。

(4) 编写两个用于获取返回值的方法

定义两个 String 类型的变量为对话框类的成员变量。实现代码如下:

```
String rtnStno,rtnStname;
```

它们分别用于存放表格中所选取教师的教师代码和教师姓名。

再定义两个分别用于获取返回值的对话框类的方法。实现代码如下:

```
public String getStno()
{
    return rtnStno;
}
public String getStname()
{
    return rtnStname;
}
```

(5) 编制对话框事件处理框架代码

如前面项目所述事件处理的三个步骤相一致。

(6) 编写"确定"、"取消"按钮动作事件代码

在"确定"按钮中，先取出输入教师代码、教师姓名，判断其合法性，即教师代码、教师姓名不能为空，且教师代码必须为3位数字；如果不合法，则提示重新输入，合法则根据newmody的值分别进行插入记录或更新记录的操作。实现代码如下：

```
int row = -1;
if(table!= null)
{
    row = table.getSelectedRow();
    if(row == -1)
        return;
}
rtnStno = String.valueOf(table.getValueAt(row,1));
rtnStname = String.valueOf(table.getValueAt(row,2));

dispose();
```

在"取消"按钮中，直接通过调用dispose()方法关闭对话框：

```
dispose();
```

9. 编写窗体类中"选教师…"按钮动作事件代码

打开选择教师对话框，选择一教师返回后，取出所选择的教师的代码及名称，并设置到表格二中选中行中。实现代码如下：

```
int row = -1;
if(table2!= null)
{
    row = table2.getSelectedRow();
    if(row == -1)
        return;
}
SelectTeacherDialog selectTeacherDialog = new SelectTeacherDialog(null," 请选择任课教师",true);

String rtnStno = selectTeacherDialog.getStno();
if(rtnStno == null || rtnStno.equals(""))
        return;
String rtnStname = selectTeacherDialog.getStname();
if(rtnStname == null || rtnStname.equals(""))
        return;
model2.setValueAt(rtnStno,row,2);
model2.setValueAt(rtnStname,row,3);
table2.updateUI();
```

10. 编写窗体类中"保存"、"关闭"按钮动作事件代码

在"保存"按钮中，先删除班级选课表sccourse中当前班级的已选课程记录及学生成绩表sscore中相应学生成绩记录(如果记录存在)，再将当前所选择的课程情况及相应学生成绩记录插

入至两个表中,学生成绩表中的成绩初始值设为 0。实现代码如下:

```java
try
{
    //删除当前班级全部学生成绩记录
    PreparedStatement stat1 = ScoreMis.con.prepareStatement("delete from sscore where left(ssno,4) = ?");
    stat1.setString(1,cursc);
    stat1.executeUpdate();

    //删除当前班级全部选课记录
    PreparedStatement stat2 = ScoreMis.con.prepareStatement("delete from sccourse where scno = ?");
    stat2.setString(1,cursc);
    stat2.executeUpdate();

    Statement stat3 = ScoreMis.con.createStatement(ResultSet.TYPE_SCROLL_INSENSITIVE,ResultSet.CONCUR_READ_ONLY);
    ResultSet result = stat3.executeQuery("select ssno from sstudent where left(ssno,4) = '" + cursc + "' order by ssno");

    //插入班级选课记录及学生成绩记录
    int rowCnt = table2.getRowCount();
    for(int i = 0;i < rowCnt;i++)
    {
        String sono = String.valueOf(table2.getValueAt(i,0));
        String stno = String.valueOf(table2.getValueAt(i,2));
        if(stno == null || stno.equals(""))
        {
            JOptionPane.showMessageDialog(null,"第" + String.valueOf(i+1) + "行的任课教师未选择,请选择!");
            table2.setRowSelectionInterval(i,i);
            return;
        }

        PreparedStatement stat4 = ScoreMis.con.prepareStatement("insert into sccourse values(?,?,?)");
        stat4.setString(1,cursc);
        stat4.setString(2,sono);
        stat4.setString(3,stno);
        stat4.executeUpdate();

        if( result!= null && result.first() )
        {
            result.previous();
            while(result.next())
            {
                String ssno = result.getString(1);
                PreparedStatement stat5 = ScoreMis.con.prepareStatement("insert into sscore values(?,?,?)");
                stat5.setString(1,ssno);
                stat5.setString(2,sono);
                stat5.setInt(3,0);
                stat5.executeUpdate();
            }
        }
    }
}
catch(SQLException e2)
{
    JOptionPane.showMessageDialog(null,"操纵数据库出错:" + e2);
}
catch(Exception e3)
{
    JOptionPane.showMessageDialog(null,"其他错误:" + e3);
}
```

在"关闭"按钮中,直接调用 dispose()方法关闭文档窗体:

```java
dispose();
```

11. 编写窗体类中填充表格二(右边表格)中数据的方法

在类中定义一个方法，方法名为 addDataToTable2()。在该方法中，从班级选课表、教师情况表、课程情况表取出当前班级的已选课程及任课教师数据并填充至表格二(右边表格)中。具体实现代码可参考本任务步骤 5 中编制的 addDataToTable()方法。该方法在构造方法及下拉列表框项目变化事件方法中被调用。

12. 编制下拉列表框项目变化事件(ItemEvent)处理代码

在第 6 步中已完成事件处理的三步框架代码，这里只要完成项目变化事件处理方法 itemStateChanged()。实现代码如下：

```java
public void itemStateChanged(ItemEvent e)
{
    cursc = String.valueOf(csc.getSelectedItem()).substring(0,4);
    tf1.setText(String.valueOf(csstname.getItemAt(csc.getSelectedIndex())));

    //清除表二中原班级选课数据
    model2.removeRows(0,model2.getRowCount());

    //取出当前班级已选课程
    addDataToTable2();
}
```

13. 运行程序并进行测试

对本模块进行测试时，也可以分步进行。完成步骤 1 至步骤 3 后可测试模块界面，完成步骤 4 后可测试班级下拉列表框中数据，完成步骤 5 后可测试表格一(左边表格)中数据，完成步骤 7 后可测试"选取》"、"删除《"按钮功能，完成步骤 9 后可测试"选教师…"按钮功能，完成步骤 10 后可测试"保存"按钮功能，完成步骤 11 后可测试表格二(右边表格)中数据，完成步骤 12 后可测试重选班级后表格二中数据。

任务八　编制单科方式录入学生成绩模块

【任务描述】

本模块根据班级选课情况，按单科方式录入学生成绩，更新至学生成绩表 sscore 中。单科方式是指选择班级的一门考试课程，输入该班级所有学生该门课程的成绩。运行示意图如图 3-20 所示。

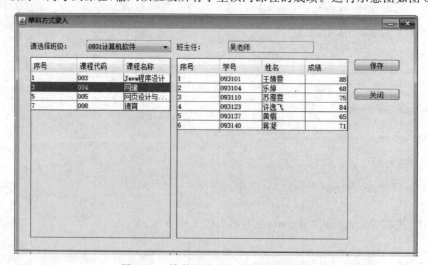

图 3-20　单科方式录入学生成绩窗体

项目三 学生成绩管理系统

【任务分析】

打开窗体时,将所有班级从数据库表中取出插入至下拉列表框中,将当前班级的已选课程全部取出并填充至表格一(左边表格)中,根据当前班级及已选课程中的第一门课取出学生成绩表中记录数据并填充至表格二(右边表格)中;选择另外一门课程时,重新取出学生成绩表中记录数据并填充至表格二(右边表格)中;选择班级时,重取当前班级的已选课程并填充至表格一(左边表格)中,根据当前班级及已选课程中的第一门取出学生成绩表中记录数据并填充至表格二(右边表格)中;输入或修改成绩后,单击"保存"按钮时,学生成绩数据更新至数据库表中。本任务与任务七有很多类似之处,具体实施时可参考任务七。

【任务实施】

1. 新建单科方式录入学生成绩文档窗体类并编写其基本代码

新建一个名为 SingleScoreFrame.java 的源程序文件,保存在 scoremis 文件夹中,在该文件中定义一个公共的文档窗体类 SingleScoreFrame,该类从 JInternalFrame 类继承,JInternalFrame 类相当于多文档窗体中的一个文档窗体,定义该类的构造方法,并设置窗体的大小及可见性。实现代码如下:

```java
import java.util.*;
import java.awt.*;
import java.awt.event.*;
import javax.swing.*;
import java.sql.*;
import javax.swing.table.*;

public class SingleScoreFrame extends JInternalFrame                    //定义构造方法
{
    public SingleScoreFrame (String title, boolean resizable, boolean closable, boolean maximizable, boolean iconifiable)
    {
        super(title,resizable,closable,maximizable,iconifiable);

        setSize(710,500);
        setLocation(20,20);
        setVisible(true);
    }
}
```

2. 编制窗体界面

窗体上有以下几个组件。

两个标签组件(JLabel):"请选择班级:"、"班主任:"。

一个文本组件(JTextField):用于存放当前班级的班主任的姓名。

两个下拉列表框组件(JComboBox):一个存放班级信息、一个存放班主任的姓名。

两个表格组件(JTable):表格一(左边表格)存放当前班级的已选课程,表格二(右边表格)存放当前班级某一门课的学生成绩。

二个按钮组件(JButton):"保存"、"关闭"。

窗体上放置三个面板 p、p1、p2,其布局方式及作用分别如下。

p1:边界布局(BorderLayout),其上只有一个表格(JTabel)控件,为表格一。

p2:边界布局(BorderLayout),其上只有一个表格(JTabel)控件,为表格二。

p:不采用任何布局方式(null),所有组件及 p1、p2 面板都放在其上,直接采用 setBounds()方法确定控件及面板 p1、p2 在其上的位置及大小,最后将面板 p 放到框架中。

先将面板及组件定义为窗体类的成员变量,实现代码如下:

```
JPanel p,p1,p2;
JLabel l1,l2;
JTextField tf1;
JComboBox csc,csstname;

String[] tableHeads1 = {"序号","课程代码","课程名称"};
String[] tableHeads2 = {"序号","学号","姓名","成绩"};
JTable table1,table2;
MyTableModel model1,model2;

JButton btsave,btclose;
```

再在构造方法中创建面板及组件并加入至窗体中,实现代码如下:

```
p = new JPanel();
p1 = new JPanel();
p2 = new JPanel();

p.setLayout(null);
p1.setLayout(new BorderLayout());
p2.setLayout(new BorderLayout());

l1 = new JLabel("请选择班级:");
csc = new JComboBox();
l2 = new JLabel("班主任:");
tf1 = new JTextField();
csstname = new JComboBox();
l1.setBounds(20,20,85,20);
csc.setBounds(120,20,150,20);
l2.setBounds(280,20,85,20);
tf1.setBounds(370,20,150,20);
tf1.setEditable(false);

p.add(l1);
p.add(csc);
p.add(l2);
p.add(tf1);

model1 = new MyTableModel(tableHeads1,3);
model1.setColumnEditable(-1,-1);                              //表格中所有列不能编辑
table1 = new JTable(model1);
table1.setSelectionMode(ListSelectionModel.SINGLE_SELECTION); //只能单行选择
p1.setBounds(20,50,250,400);
p1.add(new JScrollPane(table1));
p.add(p1);

model2 = new MyTableModel(tableHeads2,4);
model2.setColumnEditable(3,3);                                //只有第4列能编辑
table2 = new JTable(model2);
p2.setBounds(280,50,300,400);
p2.add(new JScrollPane(table2));
p.add(p2);

btsave = new JButton("保存");
btclose = new JButton("关闭");
btsave.setBounds(590,50,80,20);
btclose.setBounds(590,100,80,20);

p.add(btsave);
```

```
p.add(btclose);

getContentPane().add(p);
```

3. 编写"单科方式录入"菜单项的动作事件代码

可以将"教师基本情况维护"菜单项的动作事件代码复制后进行修改。

4. 编写窗体类中填充班级下拉列表框中数据项的方法

先定义一个存放当前班级号的窗体类的成员变量cursc,实现代码如下:

```
String cursc;
```

再将班级选课模块中 addDataToSCList()方法复制至本类中。该方法在构造方法中被调用。在构造方法中将面板 p 加入窗体中的 getContentPane().add(p)语句后面,编写以下调用代码:

```
addDataToSCList();
```

这样,窗体被打开后,即能看到班级数据显示在下拉列表框中。

5. 编写窗体类中填充表格一(左边表格)中数据的方法

在类中定义一个方法,方法名为 addDataToTable1()。在该方法中,从班级选取课表及课程情况表中取出当前班级的所选课程,利用表格模型,将数据填充至左边表格中。该方法在构造方法及下拉列表框项目变化事件方法中被调用。具体实现代码可参考本项目任务七中编制的 addDataToTable2()方法。

6. 编制窗体类中表格一行选择发生变化事件(ListSelectionEvent)代码

在本窗体中会产生三种事件,一是单击按钮产生的动作事件,二是选择下拉列表框项目时产生的项目变化事件,三是表格一行选择发生变化事件(ListSelectionEvent)。本步骤编制第三种事件代码,其余两种在后面步骤中说明。

如前面项目所述事件处理的三个步骤,先使 SingleScoreFrame 类实现行选择发生变化事件监听接口 ListSelectionListener;再在构造方法中用 addListSelectionListener()方法对行选择发生变化事件进行注册监听,最后定义行选择发生变化事件处理方法 valueChanged()。

(1) 修改窗体类的定义,使其实现事件监听接口

实现代码如下:

```
public class SingleScoreFrame extends JInternalFrame implements ListSelectionListener
```

(2) 在构造方法中,对行选择发生变化事件进行注册监听

先定义一个存放行选择状态的窗体类的成员变量 selectionMode,实现代码如下:

```
ListSelectionModel selectionMode;
```

再进行注册监听,实现代码如下:

```
selectionMode = table1.getSelectionModel();
selectionMode.addListSelectionListener(this);          //对行选择发生变化事件注册监听
```

(3) 定义行选择发生变化处理方法 valueChanged()

先定义一个存放表格一中所选当前课程代码的窗体类的成员变量 curso,实现代码如下:

```
String curso;
```

方法的实现代码如下:

```
public void valueChanged(ListSelectionEvent e)
{
```

```
        int row = table1.getSelectedRow();
        if(row == -1)
            return;
        curso = String.valueOf(table1.getValueAt(row,1));    //获取当前所选课程代码
        model2.removeRows(0,model2.getRowCount());           //清除表二原班级成绩数据
        //查询填充表格二中当前班级当前课程的学生成绩表中记录
        addDataToTable2();                                    //该方法在步骤7中实现,可先注释
    }
```

在该方法中,先获取表格一中所选择课程的课程代码,再调用 addDataToTable2()方法,根据当前班级号 cursc 及当前课程代码 curso 查询表格二中数据,即学生成绩记录。

7. 编写窗体类中填充表格二(右边表格)中数据的方法

在类中定义一个方法,方法名为 addDataToTable2()。在该方法中,根据当前班级号 cursc 及当前课程代码 curso 查询学生成绩记录,并填充至表格二(右边表格)中的方法。具体实现代码可参考本任务步骤 5 中编制的 addDataToTable1()方法。该方法在行选择发生变化事件方法 ValueChanged 中被调用,见步骤 6。

8. 编制动作事件及下拉列表框项目变化事件处理框架代码

按步骤 6 所述,再使 SingleScoreFrame 类实现动作事件监听接口 ActionListener 及项目选择变化事件监听接口 ItemListener;在构造方法中用 addActionListener()方法对两个按钮的动作事件进行注册监听,用 addItemListener 方法对下拉列表框项目时产生的项目变化事件进行注册监听;最后定义动作事件处理方法 actionPerformed(),定义下拉列表框项目时产生的项目变化事件处理方法 itemStateChanged()。

(1) 修改窗体类的定义,使其实现两种事件监听接口

实现代码如下:

```
public class ScCourseFrame extends JInternalFrame implements ListSelectionListener,ActionListener,ItemListener
```

(2) 在构造方法中,对产生的两种事件进行注册监听

实现代码如下:

```
btsave.addActionListener(this);
btclose.addActionListener(this);
csc.addItemListener(this);                    //对下拉列表框事件注册监听
```

(3) 定义动作事件处理方法 actionPerformed()

主要实现代码如下:

```
public void actionPerformed(ActionEvent e)
{
    Object obj = e.getSource();
    if(obj == btsave)                         //保存
    { ... }
    else if(obj == btclose)                   //关闭
    { ... }
}
```

(4) 定义下拉列表框项目时产生的项目变化事件处理方法 itemStateChanged()

主要实现代码如下:

```
public void itemStateChanged(ItemEvent e)
{
    ...
}
```

9. 编写窗体类中"保存"、"关闭"按钮动作事件代码

在"保存"按钮中,只需将学生成绩数据更新至数据库中。实现代码如下:

```
try
{
    int rowCnt = table2.getRowCount();
    for(int i = 0;i < rowCnt;i++)
    {
        String ssno = String.valueOf(table2.getValueAt(i,1));
        int ssscore = Integer.parseInt(String.valueOf(table2.getValueAt(i,3)));

        PreparedStatement stat = ScoreMis.con.prepareStatement("update sscore set ssscore = ? where ssno = ? and sono = ?");
        stat.setInt(1,ssscore);
        stat.setString(2,ssno);
        stat.setString(3,curso);

        stat.executeUpdate();
    }
}
catch(SQLException e2)
{
    JOptionPane.showMessageDialog(null,"操纵数据库出错:" + e2);
}
catch(Exception e3)
{
    JOptionPane.showMessageDialog(null,"其他错误:" + e3);
}
```

在"关闭"按钮中,直接调用 dispose()方法关闭文档窗体:

```
dispose();
```

10. 编制下拉列表框项目变化事件(ItemEvent)处理代码

在步骤 8 中已完成事件处理的三步框架代码,这里只要完成项目变化事件处理方法 itemStateChanged()中的代码。实现代码如下:

itemStateChanged()中的代码.实现代码如下:
```
public void itemStateChanged(ItemEvent e)
{
    cursc = String.valueOf(csc.getSelectedItem()).substring(0,4);
    tf1.setText(String.valueOf(csstname.getItemAt(csc.getSelectedIndex())));

    //清除表一原班级选课数据
    model1.removeRows(0,model1.getRowCount());
    model1.fireTableDataChanged();

    //清除表二原班级成绩数据
    model2.removeRows(0,model2.getRowCount());
    model2.fireTableDataChanged();

    //填充表一中当前班级已选课程
    addDataToTable1();
}
```

11. 运行程序并进行测试

对本模块进行测试时,也可以分步进行。完成步骤 1 至步骤 3 后可测试模块界面,完成步骤 4 后可测试班级下拉列表框中数据,完成步骤 5 后可测试表格一(左边表格)中数据,完成步骤 7 后可测试表格二中数据,完成步骤 9 后可测试"保存"按钮功能,完成步骤 10 后可测试重选班级后二个表格中数据。

任务九 编制多科方式录入学生成绩模块

【任务描述】

本模块根据班级选课情况,按多科方式录入学生成绩,更新至学生成绩表 sscore 中。多科方式是指选择班级的一个学生,输入该学生各门课程的成绩。运行示意图如图 3-21 所示。

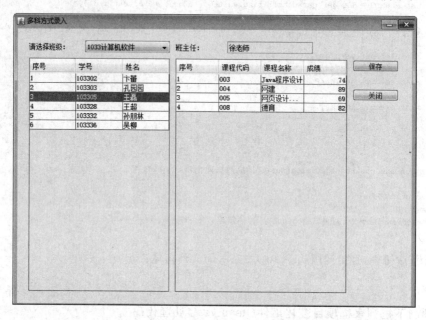

图 3-21 多科方式录入学生成绩窗体

【任务分析】

多科方式录入学生成绩模块与前面的单科方式录入学生成绩模块界面完全一致,其处理过程略有不同。在单科方式下,表格一(左边表格中)数据是当前班级的已选课程,表格二(右边表格)中数据是根据当前班级及表格一中所选课程取出的学生成绩数据;在多科方式下,表格一(左边表格)中数据是当前班级的学生,表格二(右边表格)中数据是根据当前班级及表格一选择的学生取出的该学生各门课的成绩数据;"保存"按钮的功能相同。具体实施时可在任务七的基础上修改完成。

【任务实施】

① 将单科方式成绩录入模块源程序文件 SingleScoreFrame.java,另存为多科方式成绩录入模块源程序文件 MultipleScoreFrame.java。

② 将窗体类类名及构造方法名由 SingleScoreFrame 改为 MultipleScoreFrame。

③ 修改二个表格的标题定义。

将表格的标题定义改为:

String[] tableHeads1 = {"序号","学号","姓名"};
String[] tableHeads2 = {"序号","课程代码","课程名称","成绩"};

④ 将类中定义的成员变量 curso 改为 curss。

curso 用于存放单科方式下表格一中当前选择的一门课的课程号;

curss 用于存放多科方式下表格一中当前选择的一个学生的学号。
⑤ 将方法 valueChanged() 中取当前课程号改为取当前学生学号。
⑥ 将方法 addDataToTable1() 中取当前班级的已选课程改为取当前班级的学生。
⑦ 将方法 addDataToTable2() 中取当前课程所有学生的成绩改为取当前学生各门课的成绩。
⑧ 修改"保存"按钮按钮动作事件代码。
⑨ 编写"多科方式录入"菜单项的动作事件代码。可以将"教师基本情况维护"菜单项的动作事件代码复制后进行修改。
⑩ 运行程序并进行测试。对本模块进行测试时,也可以分步进行,以便查看各步骤的功能是否已正确实现。

任务十 编制单个学生成绩查询模块

【任务描述】

本模块的功能是根据学生学号查询该学生的基本信息及各门课程的成绩,并且计算总分、平均分。运行示意图如图 3-22 所示。

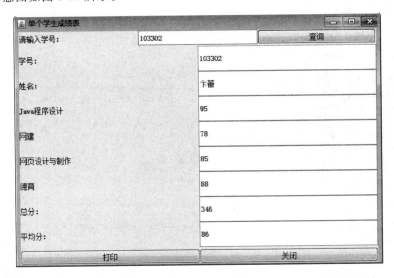

图 3-22 单个学生成绩查询窗体

【任务分析】

根据输入的学生学号,从学生基本情况表 sstudent 中可查询出学生的基本信息,从所在班级的选课情况表 sccourse 中可以取出考试课程,从学生成绩表 sscore 中可以查询出学生各门课的成绩。由于不同班级的考试课程可能不同,每个班学生的成绩记录数可能不一样,因此用来显示数据的组件个数必须是动态变化的。

【任务实施】

1. 新建单个学生成绩查询模块文档窗体类并编制其基本代码

新建一个名为 SearchStudentScore.java 的源程序文件,保存在 scoremis 文件夹中,在该文件中定义一个公共的文档窗体类 SearchStudentScore,该类从 JInternalFrame 类继承,JInternalFrame 类

相当于多文档窗体中的一个文档窗体,定义该类的构造方法,并设置窗体的大小及可见性。具体实现代码可参考本项目任务六中编制的代码。

2. 编制窗体界面

在窗体中放置三个面板 p1、p2、p3,三个面板 p1、p2、p3 在框架中的布局方式为边界方式(BorderLayout),一个在"North",一个在"Center",一个在"South";p1 面板的布局方式为网格方式(GridLayout),有 1 行 3 列,p1 面板中一个标签组件(JLabel),一个输入学号的文本组件(JTextField),一个按钮组件(JButton);p2 面板的布局方式为网格方式(GridLayout),其行、列数及上面的组件是动态生成的;p3 面板的布局方式为网格方式(GridLayout),有 1 行 2 列,包括两个按钮组件(JButton)。

先将面板及组件定义为窗体类的成员变量,实现代码如下:

```
JPanel p1,p2,p3;
JLabel l1;
JTextField tf1;
JButtonbtsearch,btprn,btclose;
```

再在构造方法中创建面板及组件并加入至窗体中,实现代码如下:

```
p1 = new JPanel();
p2 = new JPanel();
p3 = new JPanel();
p1.setLayout(new GridLayout(1,3));
p3.setLayout(new GridLayout(1,2));

l1 = new JLabel("请输入学号:");
tf1 = new JTextField();
btsearch = new JButton("查询");

btprn = new JButton("打印");
btclose = new JButton("关闭");

p1.add(l1);
p1.add(tf1);
p1.add(btsearch);

p3.add(btprn);
p3.add(btclose);

getContentPane().add(p1,"North");
getContentPane().add(p2,"Center");
getContentPane().add(p3,"South");
```

3. 编写"单个学生成绩表"菜单项的动作事件代码

可以将"教师基本情况维护"菜单项的动作事件代码复制后进行修改。

4. 编写窗体动作事件处理框架代码

如前面项目所述事件处理的三个步骤一致。

5. 编写三个按钮动作事件代码

"查询"按钮主要实现步骤如下:

① 先判断学号的合法性(是否为空,是否由数字组成,长度是否是 6 位),如不合法,弹出一对话框,显示出错信息后返回;

② 如果学号合法,根据学号到学生情况表中取学生的姓名,如果取不到,说明该学生不存在,弹出一对话框,显示出错信息后返回;

③ 如果学生存在,根据学号到学生成绩表和课程表中将该学生的考试课程及成绩取出,如果

取不到考试课程,弹出一对话框,显示信息后返回;

④ 根据考试课程门数设置 p2 面板的网格布局方式的行数(考试课程门数+4),列数为 2 列,并定义 JLable、JTextField 控件数组,其长度与行数一致,并用一循环将数组元素(控件对象)生成出来,加入至 p2 面板中;

⑤ 最后对学生成绩进行循环,将学号、姓名、总分、平均分、课程名称及成绩值设置至控件中。实现代码如下:

```java
String ssno = tf1.getText();
if(ssno.equals("") || ssno.length()!= 6 || !isNumeric(ssno))   //判断学号的合法性
{
        JOptionPane.showMessageDialog(this,"学号必须输入,不能为空,并且只能是数字!");
        tf1.requestFocus();
        return;
}
try
{
        Statement stat1 = ScoreMis.con.createStatement();
        ResultSet result1 = stat1.executeQuery( "select ssname from sstudent where ssno = '" + ssno + "'" );
          String ssname = "";
          if(result1.next())                                    //学生存在
          {
              ssname = result1.getString(1);
          }
          else
          {
              JOptionPane.showMessageDialog(this,"不存在该学生!");
              tf1.requestFocus();
              return;
          }
                                                                //获取学生的考试课程及成绩
        Statement stat2 = ScoreMis.con.createStatement(ResultSet.TYPE_SCROLL_INSENSITIVE, ResultSet.CONCUR_READ_ONLY);
            ResultSet result2 = stat2.executeQuery( "select sscore.sono, soname, ssscore from sscore, scourse where sscore.ssno = '" + ssno + "' and sscore.sono = scourse.sono order by sscore.sono" );

          int num = 0;
          if(result2.last())
          {
              num = result2.getRow();                           //考试门数
          }
          if(num == 0)
          {
              JOptionPane.showMessageDialog(this,"该学生还没有成绩!");
              tf1.requestFocus();
              return;
          }

          p2.removeAll();
          p2.setLayout(new GridLayout(num + 4,2));

          //定义组件数组
          JLabel lb[] = new JLabel[num + 4];
          JTextField tf[] = new JTextField[num + 4];
                                                                //创建组件对象并加入面板中
          for(int i = 0;i < num + 4;i++)
          {
```

```
            lb[i] = new JLabel();
            tf[i] = new JTextField();
            p2.add(lb[i]);
            p2.add(tf[i]);
            tf[i].setEditable(false);
            tf[i].setBackground(Color.WHITE);
        }
        lb[0].setText("学号：");
        tf[0].setText(ssno);

        lb[1].setText("姓名：");
        tf[1].setText(ssname);

        lb[num + 2].setText("总分：");
        lb[num + 3].setText("平均分：");

        String soname;
        int ssscore;
        int sum = 0,avg;

        result2.first();
        result2.previous();
        int j = 0;
        while(result2.next())                    //循环取出课程及成绩并设置至面板组件中
        {
            soname = result2.getString("soname");
            ssscore = result2.getInt("ssscore");
            sum = sum + ssscore;

            lb[j + 2].setText(soname);
                tf[j + 2].setText(String.valueOf(ssscore));
                j++;
        }
        avg = sum/num;
        tf[num + 2].setText(String.valueOf(sum));
        tf[num + 3].setText(String.valueOf(avg));
        p2.updateUI();
}
catch(SQLException e2)
{
    JOptionPane.showMessageDialog(null,"操纵数据库出错:" + e2);
}
catch(Exception e3)
{
    JOptionPane.showMessageDialog(null,"其他错误:" + e3);
}
```

代码中的 isNumeric() 方法用来判断一个字符串是否由数字构成。这在前面的教师、课程等情况维护模块的新增、修改对话框类已定义过，只要直接复制到本类中即可。"打印"按钮事件代码，暂无实现。

在"关闭"按钮中，直接调用 dispose() 方法关闭文档窗体：

```
dispose();
```

6. 运行程序并进行测试

输入几个不同班级的学生学号，观察运行情况。

任务十一　编制班级成绩明细查询模块

【任务描述】

本模块的功能是根据班级号查询该班级所有学生的基本信息及各门课程的成绩，并且计算总分、平均分。运行示意图如图 3-23 所示。

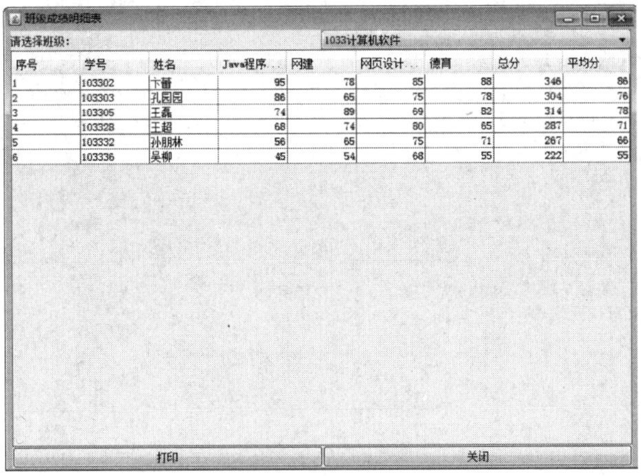

图 3-23　班级成绩明细查询窗体

【任务分析】

根据所选择的班级号，从学生基本情况表 sstudent 中可查询出学生的基本信息，从所在班级的选课情况表 sccourse 中可以取出考试课程，从学生成绩表 sscore 中可以查询出所有学生各门课的成绩。由于不同班级的考试课程可能不同，每个班学生的成绩记录数可能不一样，因此用来显示数据的表格必须是动态变化的。

【任务实施】

1. 新建班级成绩明细查询模块文档窗体类并编制其基本代码

新建一个名为 SearchClassScore.java 的源程序文件，保存在 scoremis 文件夹中，在该文件中定义一个公共的文档窗体类 SearchClassScore，该类从 JInternalFrame 类继承，JInternalFrame 类相当于多文档窗体中的一个文档窗体，定义该类的构造方法，并设置窗体的大小及可见性。具体代码可参考本项目任务十中编制的代码。

2. 编制窗体界面

在窗体中放置三个面板 p1、p2、p3，三个面板 p1、p2、p3 在框架中的布局方式为边界方式（BorderLayout），一个在"North"，一个在"Center"，一个在"South"；p1 面板的布局方式为网格方式（GridLayout），有 1 行 2 列，p1 面板中有一个标签组件（JLabel），一个选择班级的下拉列表框组件（JComboBox）；p2 面板的布局方式为边界方式（BorderLayout），面板中有一个表格组件，表格控件暂不生成；p3 面板的布局方式为网格方式（GridLayout），有 1 行 2 列，包括两个按钮组件（JButton）。

先将面板及组件定义为窗体类的成员变量，实现代码如下：

```
JPanel p1,p2,p3;
```

```
JLabel l1;
JComboBox csc;
JButton btprn,btclose;

JTable table;
MyTableModel model;
```

再在构造方法中创建面板及组件并加入至窗体中,实现代码如下:

```
p1 = new JPanel();
p2 = new JPanel();
p3 = new JPanel();
p1.setLayout(new GridLayout(1,2));
p2.setLayout(new BorderLayout());
p3.setLayout(new GridLayout(1,2));

l1 = new JLabel("请选择班级: ");
csc = new JComboBox();

btprn = new JButton("打印");
btclose = new JButton("关闭");

p1.add(l1);
p1.add(csc);

p3.add(btprn);
p3.add(btclose);

getContentPane().add(p1,"North");
getContentPane().add(p2,"Center");
getContentPane().add(p3,"South");
```

3. 编写"班级成绩明细表"菜单项的动作事件代码

可以将"教师基本情况维护"菜单项的动作事件代码复制后进行修改。

4. 编写窗体类中填充班级下拉列表框中数据项的方法

先定义一个存放当前班级号的窗体类的成员变量 cursc,实现代码如下:

```
String cursc;
```

再在类中定义一个方法,方法名为 addDataToSCList()。在该方法,从班级情况表中取出班级号、专业名称,填充至下拉列表框中,并取出第一个班级号,存放至变量 cursc。具体实现代码可参考本项目任务七中编制的 addDataToSCList()方法。该方法在构造方法中被调用。

5. 编写窗体类中填充表格中数据的方法

在类中定义一个方法,方法名为 addDataToTable()。在该方法中,根据当前班级号将该班所有学生的基本信息及各门课程的成绩数据取出,利用表格模型,将数据填充至动态生成的表格中。主要实现步骤如下:

① 定义并创建一个用于存放班级考试课程名称的 Vector 类型的对象 colname。

② 根据当前班级号将班级的考试课程名称取出,存放至 colname 对象中,并用一个 int 型变量 num 统计班级考试课程的门数。

③ 如果该班级尚未选课,则弹出一对话框并将 p2 面板中内容清空后返回。

④ 根据班级考试课程的门数 num 定义一个 String 类型的表格标题数组,数组的长度为考试课程门数 num+5,并将表格标题(固定字符串及 colname 中字符串)构造出来。

⑤ 将表格模型、表格对象创建出来并加入到面板中。

⑥ 定义用于存放一行和多行数据的 Vector 类型的对象 row、rows,并将 rows 创建出来。

⑦ 根据当前班级号从学生情况表中将当前班级的所有学生取出，并逐个循环，取出学号及姓名，根据学号从学生成绩表中按课程号排序取出每个学生的成绩，计算总分及平均分，并存放至 Vector 类型的变量 row 中，将 row 中的数据加入至 rows 中并使用表格模型类的 addRows() 方法将数据填充至表格中。

⑧ 刷新界面上的数据。

实现代码如下：

```java
private void addDataToTable()
{
    Vector colname = new Vector();                    //定义存放班级考试课程名称的对象
    int num = 0;                                      //考试课程的门数
    try
    {
        //取出考试课程名称
        Statement stat1 = ScoreMis.con.createStatement();
        ResultSet result1 = stat1.executeQuery("select soname from sccourse,scourse where scourse.sono = sccourse.sono and scno = '" + cursc + "' order by sccourse.sono");
        while(result1.next())
        {
            colname.add(result1.getString("soname"));
            num++;
        }
    }
    catch(SQLException e2)
    {
        JOptionPane.showMessageDialog(null,"操纵数据库出错:" + e2);
    }
    catch(Exception e3)
    {
        JOptionPane.showMessageDialog(null,"其他错误:" + e3);
    }
    //班级无考试课程
    if(num == 0)
    {
        JOptionPane.showMessageDialog(null,"该班级尚未选择考试课程,无成绩数据!");
        p2.updateUI();
        return;
    }
    //定义表格标题数组并构造标题
    String[] heads = new String[num + 5];
    heads[0] = "序号";
    heads[1] = "学号";
    heads[2] = "姓名";
    for(int j = 0;j < num;j++)
    {
        heads[j + 3] = String.valueOf(colname.get(j));
    }
    heads[num + 3] = "总分";
    heads[num + 4] = "平均分";
    //创建表格
    model = new MyTableModel(heads,num + 5);
    table = new JTable(model);
    p2.add(new JScrollPane(table));

    Vector row;
    Vector rows = new Vector();

    try
    {
        //取班级学生学号及姓名
        Statement stat2 = ScoreMis.con.createStatement();
        ResultSet result2 = stat2.executeQuery("select ssno,ssname from sstudent where left(ssno,4) = '" +
```

```
            cursc + "'order by ssno");
                int k = 1;
                while(result2.next())                       //对学生循环
                {
                    String ssno = result2.getString(1);
                    String ssname = result2.getString(2);

                    row = new Vector();
                    row.add(String.valueOf(k++));
                    row.add(ssno);
                    row.add(ssname);
                    //取学生各门课的成绩
                    Statement stat3 = ScoreMis.con.createStatement();
                    ResultSet result3 = stat3.executeQuery( "select ssscore from sscore where ssno = '" + ssno + "'
order by sono" );
                    int sum = 0,avg;
                    while(result3.next())                   //对成绩循环
                    {
                        int ssscore = result3.getInt("ssscore");
                        sum = sum + ssscore;
                        row.add(new Integer(ssscore));
                    }
                    avg = sum/num;
                    row.add(new Integer(sum));
                    row.add(new Integer(avg));

                    rows.add(row);
                }
                model.addRows(rows);                        //形成表格数据
                p2.updateUI();
            }
            catch(SQLException e2)
            {
                JOptionPane.showMessageDialog(null,"操纵数据库出错:" + e2);
            }
            catch(Exception e3)
            {
                JOptionPane.showMessageDialog(null,"其他错误:" + e3);
            }
        }
```

该方法在构造方法中及下拉列表框项目变化事件方法中被调用。在构造方法中调用 addDataToSCList()方法的语句后面,编写以下调用代码:

```
addDataToTable();
```

这样,窗体被打开后,即能看到班级成绩明细数据显示在表格中。在下拉列表框项目变化事件方法中被调用,请参见步骤 6 中代码。

6. 编制窗体事件处理框架代码

在本窗体中会产生两种事件,一是单击按钮产生的动作事件,另一是选择下拉列表框项目时产生的项目变化事件。如前面项目所述事件处理的三个步骤,先使 SearchClassScore 类实现动作事件监听接口 ActionListener 及项目选择变化事件监听接口 ItemListener;再在构造方法中用 addActionListener()方法对两个按钮的动作事件进行注册监听,用 addItemListener 方法对下拉列表框项目时产生的项目变化事件进行注册监听;最后定义动作事件处理方法 actionPerformed(),定义下拉列表框项目时产生的项目变化事件处理方法 itemStateChanged()。

(1) 修改窗体类的定义,使其实现二种事件监听接口

实现代码如下:

```
public class SearchClassScore extends JInternalFrame implements ActionListener,ItemListener
```

(2) 在构造方法中,对产生的两种事件进行注册监听

实现代码如下:

btprn.addActionListener(this);
btclose.addActionListener(this);
csc.addItemListener(this);   //对下拉列表框事件注册监听

(3) 定义动作事件处理方法 actionPerformed()

实现代码如下:

```
public void actionPerformed(ActionEvent e)
{
    Object obj = e.getSource();
    if(obj == btprn)                                    //打印
    {
        //暂无实现
    }
    else if(obj == btclose)                             //关闭
    {
        dispose();
    }
}
```

(4) 定义下拉列表框项目时产生的项目变化事件处理方法 itemStateChanged()

实现代码如下:

```
public void itemStateChanged(ItemEvent e)
{
    cursc = String.valueOf(csc.getSelectedItem()).substring(0,4);
    p2.removeAll();
    addDataToTable();
}
```

7. 运行程序并进行测试

从下拉列表框中选择不同的班级,观察表格中查询出的数据情况。

任务十二　编制班级成绩汇总数据查询模块

【任务描述】

本模块的功能是根据班级号查询该班级所考各门考试课程的汇总成绩,即平均分、及格率、优秀率、低分率。运行示意图如图 3-24 所示。

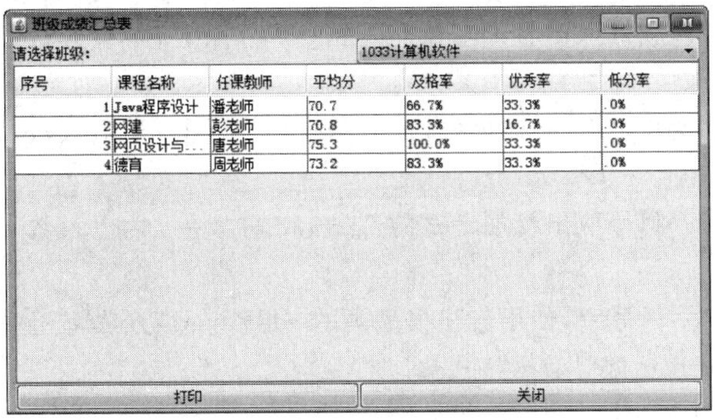

图 3-24　班级成绩汇总数据查询窗体

【任务分析】

根据所选择的班级号,从班级选课表 sscourse 中可查询出班级的考试课程及任课教师,对每一门课程,从学生成绩表 sscore 中可以查询出所有学生该门课的汇总成绩。本任务与任务十一类似,不过表格是固定的,具体实施时可参考任务十一。

【任务实施】

1. 新建班级成绩汇总数据查询模块文档窗体类并编制其基本代码

新建一个名为 SearchScAll.java 的源程序文件,保存在 scoremis 文件夹中,在该文件中定义一个公共的文档窗体类 SearchScAll,该类从 JInternalFrame 类继承,JInternalFrame 类相当于多文档窗体中的一个文档窗体,定义该类的构造方法,并设置窗体的大小及可见性。具体实现代码可参考本项目任务十一中编制的代码。

2. 编制窗体界面

在窗体中放置三个面板 p1、p2、p3,三个面板 p1、p2、p3 在框架中的布局方式为边界方式(BorderLayout),一个在"North",一个在"Center",一个在"South";p1 面板的布局方式为网格方式(GridLayout),有 1 行 2 列,p1 面板中有一个标签组件(JLabel),一个选择班级的下拉列表框组件(JComboBox);p2 面板的布局方式为边界方式(BorderLayout),面板中有一个表格组件,表格控件暂不生成;p3 面板的布局方式为网格方式(GridLayout),有 1 行 2 列,包括两个按钮组件(JButton)。具体实现代码可参考本项目任务十一。

3. 编写"班级成绩汇总表"菜单项的动作事件代码

可以将"教师基本情况维护"菜单项的动作事件代码复制后进行修改。

4. 编写窗体类中填充班级下拉列表框中数据项的方法

具体实现代码基本同本项目任务十一。

5. 编写窗体类中填充表格中数据的方法

在类中定义一个方法,方法名为 addDataToTable()。在该方法中,根据当前班级号将该班所考各门课程的平均分、及格率(60 分以上)、优秀率(80 分以上)、低分率(30 分以下)汇总,利用表格模型,将数据填充至表格中。主要步骤如下:

① 定义用于存放一行和多行数据的 Vector 类型的对象 row、rows,并将 rows 创建出来。

② 根据当前班级号从班级选课表、课程情况表、教师情况表中将课程号、课程名、教师姓名取出,并逐个循环。

③ 在循环中取出课程号、课程名、教师姓名,分别存放至 String 类型的变量中;

根据班级号及当前课程号取出学生的平均成绩,存放至 double 类型的变量 avg 中;

根据班级号及当前课程号取出参加考试的学生人数,存放至 int 类型的变量 pnumb 中;

根据班级号及当前课程号取出参加考试的及格(60 分以上)学生人数,存放至 int 类型的变量 pnumb1 中;

根据班级号及当前课程号取出参加考试的优秀(80 分以上)学生人数,存放至 int 类型的变量 pnumb2 中;

根据班级号及当前课程号取出参加考试的低分(30 分以下)学生人数,存放至 int 类型的变量 pnumb3 中;

计算及格率、优秀率、低分率,使用表格模型类的 addRows() 方法将相关数据填充至表格中。

④ 使用 DecimalFormat 类对数据格式化。

⑤ 刷新表格中的数据。

实现代码如下:

```java
private void addDataToTable()
{
    //定义格式化对象
    DecimalFormat df1 = new DecimalFormat("###.0");
    DecimalFormat df2 = new DecimalFormat("###.0%");
    //定义存放行数据的 Vector 类对象
    Vector row,rows;
    rows = new Vector();
    try
    {
        //根据当前班级号取出课程号、课程名、教师姓名
        Statement stat = ScoreMis.con.createStatement();
        ResultSet result = stat.executeQuery("select sccourse.sono, soname, stname from sccourse, scourse, steacher where sccourse.sono = scourse.sono and sccourse.stno = steacher.stno and scno = '" + cursc + "' order by sccourse.sono");
        int i = 1;
        while(result.next())                    //对课程进行循环
        {
            String sono = result.getString("sono");
            String soname = result.getString("soname");
            String stname = result.getString("stname");
            //根据班级号及当前课程号取出学生的平均成绩
            Statement stat1 = ScoreMis.con.createStatement();
            ResultSet result1 = stat1.executeQuery("select avg(ssscore) from sscore where left(ssno, 4) = '" + cursc + "' and sono = '" + sono + "'");
            double avg = 0;
            if(result1.next())
            {
                avg = result1.getDouble(1);
            }
            result1.close();
            stat1.close();
            result1 = null;
            stat1 = null;
            //根据班级号及当前课程号取出参加考试的学生人数
            Statement stat2 = ScoreMis.con.createStatement();
            ResultSet result2 = stat2.executeQuery("select count(ssscore) from sscore where left(ssno, 4) = '" + cursc + "' and sono = '" + sono + "'");
            int pnumb = 0;
            if(result2.next())
            {
                pnumb = result2.getInt(1);
            }
            result2.close();
            stat2.close();
            result2 = null;
            stat2 = null;
            //根据班级号及当前课程号取出参加考试的及格学生人数
            Statement stat3 = ScoreMis.con.createStatement();
            ResultSet result3 = stat3.executeQuery("select count(ssscore) from sscore where left(ssno, 4) = '" + cursc + "' and sono = '" + sono + "' and ssscore >= 60");
            int pnumb1 = 0;
            if(result3.next())
            {
                pnumb1 = result3.getInt(1);
            }
            result3.close();
            stat3.close();
            result3 = null;
            stat3 = null;
            //根据班级号及当前课程号取出参加考试的优秀学生人数
```

```java
            Statement stat4 = ScoreMis.con.createStatement();
            ResultSet result4 = stat4.executeQuery("select count(ssscore) from sscore where left(ssno,
4) = '" + cursc + "' and sono = '" + sono + "' and ssscore >= 80");
            int pnumb2 = 0;
            if(result4.next())
            {
                pnumb2 = result4.getInt(1);
            }
            result4.close();
            stat4.close();
            result4 = null;
            stat4 = null;

            //根据班级号及当前课程号取出参加考试的低分学生人数
            Statement stat5 = ScoreMis.con.createStatement();
            ResultSet result5 = stat5.executeQuery("select count(ssscore) from sscore where left(ssno,
4) = '" + cursc + "' and sono = '" + sono + "' and ssscore < 30");
            int pnumb3 = 0;
            if(result5.next())
            {
                pnumb3 = result5.getInt(1);
            }
            result5.close();
            stat5.close();
            result5 = null;
            stat5 = null;
            //计算及格率、优秀率、低分率
            row = new Vector();
            row.add(i);
            row.add(soname);
            row.add(stname);
            //使用DecimalFormat类对数据格式化.
            row.add( df1.format( avg ) );
            row.add( df2.format( pnumb1 * 1.0/pnumb) );
            row.add( df2.format( pnumb2 * 1.0/pnumb) );
            row.add( df2.format( pnumb3 * 1.0/pnumb) );
            rows.add(row);
            i++;
        }
        //刷新表格中的数据
        model.addRows(rows);
        model.fireTableDataChanged();

        result.close();
        stat.close();
        result = null;
        stat = null;
    }
    catch(SQLException e2)
    {
        JOptionPane.showMessageDialog(this,"操作数据库出错:" + e2);
    }
    catch(Exception e3)
    {
        JOptionPane.showMessageDialog(this,"出现其他错误:" + e3);
    }
}
```

该方法在构造方法中及下拉列表框项目变化事件方法中被调用。在构造方法中调用addDataToSCList()方法的语句后面,编写以下调用代码:

```java
addDataToTable();
```

这样,窗体被打开后,即能看到班级成绩汇总数据显示在表格中。在下拉列表框项目变化事件方法中被调用,请参见本任务步骤 6 中代码。

6. 编制窗体事件处理框架代码

与前面模块中的基本一致,具体实现代码可参考本项目任务十一。

7. 运行程序并进行测试

从下拉列表框中选择不同的班级,观察表格中查询出的汇总数据情况。

任务十三　编制同学科成绩比较数据查询模块

【任务描述】

本模块的功能是根据所选择的课程对所有考核该门课程的班级的成绩平均分、及格率、优秀率、低分率进行汇总比较。运行示意图如图 3-25 所示。

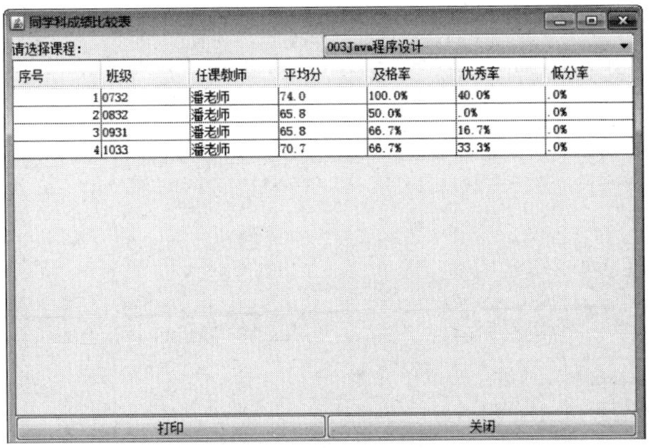

图 3-25　同学科成绩比较数据查询窗体

【任务分析】

本任务是根据所选择的课程汇总比较各班级的成绩数据,而本项目任务十二是根据所选班级汇总比较各门课程的成绩数据,二者正好相反,在具体实施时完全可以参考前述步骤。

【任务实施】

1. **将班级成绩汇总数据查询模块窗体类文件 SearchScAll.java 另存为 SearchCompScore.java 后进行修改**

① 将公共类名由 SearchScAll 改为 SearchCompScore;
② 将构造方法名由 SearchScAll() 改为 SearchCompScore();
③ 修改组件对象名及文本信息,并相应修改其 add 语句。

将标签组件(JLabel)中的文本信息改为:

l1 = new JLabel("请选择课程: ");

将下拉列表框组件(JComboBox)的定义改为:

JComboBox cso;

将存放表格标题的 String 数组的定义改为:

```
String[] heads = {"序号","班级","任课教师","平均分","及格率","优秀率","低分率"};
```

将原来存放当前班级号的成员变量名 cursc 的定义改为:

```
String curso;
```

2. 编写"同学科成绩比较表"菜单项的动作事件代码

可以将"教师基本情况维护"菜单项的动作事件代码复制后进行修改。

3. 修改窗体类中填充下拉列表框中数据项的方法

任务十二中,方法 addDataToSCList() 向下拉列表框中填充的是班级数据;在本任务中,填充的是考试课程,方法名改为 addDataToSOList(),并且课程不能重复。实现代码如下:

```
private void addDataToSOList()
{
    try
    {
        Statement stat = ScoreMis.con.createStatement();
        ResultSet result = stat.executeQuery("select distinct sccourse.sono, soname from sccourse, scourse where sccourse.sono = scourse.sono order by sccourse.sono");
        while(result.next())
        {
            cso.addItem(result.getString(1) + result.getString(2));
        }
    }
    catch(SQLException e2)
    {
        JOptionPane.showMessageDialog(this,"操作数据库出错:" + e2);
    }
    catch(Exception e3)
    {
        JOptionPane.showMessageDialog(this,"其他错误:" + e3);
    }
    if(cso.getItemCount() == 0)
    {
        JOptionPane.showMessageDialog(this,"不存在选课信息,请先选择班级的考试课程!");
        return;
    }
    else
    {
        curso = String.valueOf(cso.getItemAt(0)).substring(0,3);
    }
}
```

将构造方法中原来的调用代码:

```
addDataToSCList();
```

改为:

```
addDataToSOList();
```

4. 修改窗体类中填充表格中数据的方法

调整方法 addDataToTable() 中的代码,改为根据课程号取出所有考该门课程的班级的平均分、及格率(60分以上)、优秀率(80分以上)、低分率(30分以下)汇总,利用表格模型,将数据填充至表格中。主要步骤如下。

① 定义用于存放一行和多行数据的 Vector 类型的对象 row、rows,并将 rows 创建出来;

② 根据当前课程号从班级选课表、教师情况表中将班级号、教师姓名取出,并逐个循环;
③ 在循环中:

取出班级号、教师姓名,分别存放至 String 类型的变量中;

根据班级号及当前课程号取出学生的平均成绩,存放至 double 类型的变量 avg 中;

根据班级号及当前课程号取出参加考试的学生人数,存放至 int 类型的变量 pnumb 中;

根据班级号及当前课程号取出参加考试的及格(60 分以上)学生人数,存放至 int 类型的变量 pnumb1 中;

根据班级号及当前课程号取出参加考试的优秀(80 分以上)学生人数,存放至 int 类型的变量 pnumb2 中;

根据班级号及当前课程号取出参加考试的低分(30 分以下)学生人数,存放至 int 类型的变量 pnumb3 中;

计算及格率、优秀率、低分率,使用表格模型类的 addRows()方法将相关数据填充至表格中。
④ 使用 DecimalFormat 类对数据格式化。
⑤ 刷新表格中的数据。

实现代码如下:

```java
private void addDataToTable()
{
        //定义格式化对象
        DecimalFormat df1 = new DecimalFormat("###.0");
        DecimalFormat df2 = new DecimalFormat("###.0%");
        //定义用于存放行数据的 Vector 类型对象
        Vector row,rows;
        rows = new Vector();
        try
        {
            //根据当前课程号取出考核该课程的班级及任课教师
            Statement stat = ScoreMis.con.createStatement();
            ResultSet result = stat.executeQuery("select scno,stname from sccourse,steacher where sccourse.stno = steacher.stno and sono = '" + curso + "' order by scno");
            int i = 1;
            while(result.next())
            {
                String scno = result.getString("scno");
                String stname = result.getString("stname");
                //根据班级号及当前课程号取出学生的平均成绩
                Statement stat1 = ScoreMis.con.createStatement();
                ResultSet result1 = stat1.executeQuery("select avg(ssscore) from sscore where left(ssno,4) = '" + scno + "' and sono = '" + curso + "'");
                double avg = 0;
                if(result1.next())
                {
                    avg = result1.getDouble(1);
                }
                result1.close();
                stat1.close();
                result1 = null;
                stat1 = null;
                //根据班级号及当前课程号取出参加考试的学生人数
                Statement stat2 = ScoreMis.con.createStatement();
                ResultSet result2 = stat2.executeQuery("select count(ssscore) from sscore where left(ssno,4) = '" + scno + "' and sono = '" + curso + "'");
                int pnumb = 0;
                if(result2.next())
                {
                    pnumb = result2.getInt(1);
```

```java
                }
                result2.close();
                stat2.close();
                result2 = null;
                stat2 = null;
                //根据班级号及当前课程号取出参加考试的及格人数
                Statement stat3 = ScoreMis.con.createStatement();
                ResultSet result3 = stat3.executeQuery("select count(ssscore) from sscore where left(ssno,
4) = '" + scno + "' and sono = '" + curso + "' and ssscore >= 60");
                int pnumb1 = 0;
                if(result3.next())
                {
                    pnumb1 = result3.getInt(1);
                }
                result3.close();
                stat3.close();
                result3 = null;
                stat3 = null;
                //根据班级号及当前课程号取出参加考试的优秀人数
                Statement stat4 = ScoreMis.con.createStatement();
                ResultSet result4 = stat4.executeQuery("select count(ssscore) from sscore where left(ssno,
4) = '" + scno + "' and sono = '" + curso + "' and ssscore >= 80");
                int pnumb2 = 0;
                if(result4.next())
                {
                    pnumb2 = result4.getInt(1);
                }
                result4.close();
                stat4.close();
                result4 = null;
                stat4 = null;
                //根据班级号及当前课程号取出参加考试的低分人数
                Statement stat5 = ScoreMis.con.createStatement();
                ResultSet result5 = stat5.executeQuery("select count(ssscore) from sscore where left(ssno,
4) = '" + scno + "' and sono = '" + curso + "' and ssscore < 30");
                int pnumb3 = 0;
                if(result5.next())
                {
                    pnumb3 = result5.getInt(1);
                }
                result5.close();
                stat5.close();
                result5 = null;
                stat5 = null;
                //计算并格式化数据,将数据填充至表格中
                row = new Vector();

                row.add(i);
                row.add(scno);
                row.add(stname);
                row.add( df1.format( avg ) );
                row.add( df2.format( pnumb1 * 1.0/pnumb) );
                row.add( df2.format( pnumb2 * 1.0/pnumb) );
                row.add( df2.format( pnumb3 * 1.0/pnumb) );
                i++;
                rows.add(row);
            }
            //刷新表格中的数据
            model.addRows(rows);
            model.fireTableDataChanged();

            result.close();
            stat.close();
```

```
            result = null;
            stat = null;
        }
        catch(SQLException e2)
        {
            JOptionPane.showMessageDialog(this,"操作数据库出错:" + e2);
        }
        catch(Exception e3)
        {
            JOptionPane.showMessageDialog(this,"出现其他错误:" + e3);
        }
    }
```

5．修改下拉列表框项目变化事件处理代码

先将构造方法中对下拉列表框事件注册监听由：

```
csc.addItemListener(this);
```

改为：

```
cso.addItemListener(this);
```

再将方法 itemStateChanged()调整为：

```
public void itemStateChanged(ItemEvent e)
{
    curso = String.valueOf(cso.getSelectedItem()).substring(0,3);
    model.removeRows(0,model.getRowCount());
    model.fireTableDataChanged();
    addDataToTable();
}
```

6．运行程序并进行测试

从下拉列表框中选择不同的课程，观察表格中查询出的汇总数据情况。

任务十四　编制班级不及格人数统计模块

【任务描述】

本模块的功能是根据所选择的班级，统计班级中每一门次的不及格学生人数。运行示意图如图 3-26 所示。

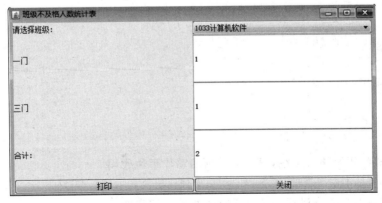

图 3-26　班级不及格人数统计窗体

【任务分析】

根据班级号,从学生基本情况表 sstudent 中可以查询出该班的学生,对每一个学生可以从学生成绩表 sscore 中查询出该学生不及格的门数,然后进行统计。由于不同班级的学生不及格的人次可能不一样,因此用来显示数据的组件个数必须是动态变化的。

【任务实施】

1. 新建班级不及格人数统计模块文档窗体类并编制其基本代码

新建一个名为 SearchBadScore.java 的源程序文件,保存在 scoremis 文件夹中,在该文件中定义一个公共的文档窗体类 SearchBadScore,该类从 JInternalFrame 类继承,JInternalFrame 类相当于多文档窗体中的一个文档窗体,定义该类的构造方法,并设置窗体的大小及可见性。具体实现代码可参考本项目任务十三中编制的代码。

2. 编制窗体界面

在窗体中放置三个面板 p1、p2、p3,三个面板 p1、p2、p3 在框架中的布局方式为边界方式(BorderLayout),一个在"North",一个在"Center",一个在"South"。p1 面板的布局方式为网格方式(GridLayout),有 1 行 2 列,p1 面板中一个标签组件(JLabel)、一个选择班级的下拉列表框(JComboBox)控件;p2 面板的布局方式为网格方式(GridLayout),其行、列数及上面的组件是动态生成的;p3 面板的布局方式为网格方式(GridLayout),有 1 行 2 列,包括两个按钮控件(JButton)。

先将面板及组件定义为窗体类的成员变量,实现代码如下:

```
JPanel p1,p2,p3;
JLabel l1;
JComboBox  csc;
JButton btprn,btclose;
```

再在构造方法中创建面板及组件并加入至窗体中,实现代码如下:

```
p1 = new JPanel();
p2 = new JPanel();
p3 = new JPanel();
p1.setLayout(new  GridLayout(1,2));
p3.setLayout(new  GridLayout(1,2));

l1 = new JLabel("请选择班级: ");
csc = new JComboBox();
p1.add(l1);
p1.add(csc);

btprn = new JButton("打印");
btclose = new JButton("关闭");

p3.add(btprn);
p3.add(btclose);

getContentPane().add(p1,"North");
getContentPane().add(p2,"Center");
getContentPane().add(p3,"South");
```

3. 编写"班级不及格人数统计表"菜单项的动作事件代码

可以将"教师基本情况维护"菜单项的动作事件代码复制后进行修改。

4. 编写窗体类中填充班级下拉列表框中数据项的方法

具体实现代码可参考本项目任务七中编制的 addDataToSCList()方法。

5. 编写窗体类中统计班级不及格人数的方法

在类中定义一个方法,方法名为 statBadNum()。在该方法中,根据当前班级号统计该班不及格人数,将数据填充至中间面板动态生成的组件中。主要步骤如下：

① 根据班级号从学生成绩表中将当前班级中成绩有不及格的学生学号取出。
② 定义并生成一个长度为10的int数组sum。
③ 对学生进行循环,循环步骤如下。
- 取出学号；
- 根据学号从学生成绩表中统计出该学生不及格的门数；
- 如果该学号有不及格科目,则将其统计(计数)至相应的sum数组元素中；

④ 定义一个 int 型变量 n,对 sum 数组进行循环,将 sum 数组中不为 0 的元素个数统计至变量 n 中。
⑤ 清空面板,重新设置面板的布局方式为网格方式,行数为 n+1,列数为 2。
⑥ 定义并生成 JLabel 及 JTextField 组件数组,长度均为 n+1。
⑦ 将组件数组中每一个元素生成出来并加入至面板中。
⑧ 如果没有不及格学生(n 为 0),则将组件上文本设置"合计："、"0"后返回。
⑨ 定义一个 String 类型数组,如下：

String titles[] = {"一门","二门","三门","四门","五门","六门","七门","八门","九门","十门"};

⑩ 定义 int 型循环变量 k 及统计总人数的变量 count。
⑪ 对 sum 数组进行循环：如果元素的值不为 0,则设置面板上组件的文本值,并统计总数。
⑫ 设置"合计："及其值,并刷新面板。

实现代码如下：

```java
private void statBadNum()
{
    try
    {
        //根据班级号取出不及格学生
        Statement stat = ScoreMis.con.createStatement();
        ResultSet result = stat.executeQuery("select distinct ssno from sscore where left(ssno,4) = '" + cursc + "' and ssscore<60 order by ssno");
        //定义数组,用于存放统计的不及格人数
        int sum[] = new int[10];
        //对不及格学生进行循环
        while(result.next())
        {
            String ssno = result.getString("ssno");
            //根据学号从学生成绩表中统计出该学生不及格的门数
            Statement stat1 = ScoreMis.con.createStatement();
            ResultSet result1 = stat1.executeQuery("select count(ssscore) from sscore where ssno = '" + ssno + "' and ssscore<60");
            int cnt = 0;
            if(result1.next())
            {
                cnt = result1.getInt(1);
            }
            //有不及格科目,则将其统计进 sum 数组中
            if(cnt>0)
            {
                sum[cnt - 1] = sum[cnt - 1]+1;
            }
        }
```

```java
//统计sum数组中不为0的元素个数
int n = 0;
for(int i = 0;i < 10;i++)
{
    if(sum[i]> 0)
    n++;
}
//清空面板,重新设置面板的布局方式为网格方式及行列数
p2.removeAll();
p2.setLayout(new GridLayout(n + 1,2));
//定义并创建JLabel及JTextField组件数组
JLabel lb[] = new JLabel[n + 1];
JTextField tf[] = new JTextField[n + 1];
//将组件数组中每一个元素创建出来并加入至面板
for(int i = 0;i < n + 1;i++)
{
    lb[i] = new JLabel();
    tf[i] = new JTextField();
    p2.add(lb[i]);
    p2.add(tf[i]);
    tf[i].setEditable(false);
    tf[i].setBackground(Color.white);
}
//没有不及格学生
if(n == 0)
{
    lb[0].setText("合计");
    tf[0].setText("0");
    p2.updateUI();
    return;
}
String titles[] = {"一门","二门","三门","四门","五门","六门","七门","八门","九门","十门"};
int k = 0,count = 0;
//对sum数组进行循环
for(int i = 0;i < 10;i++)
{
    //如果元素的值不为0,则设置面板上组件的文本值,并统计总数
    if(sum[i]> 0)
    {
        lb[k].setText(titles[i]);
        tf[k].setText(String.valueOf(sum[i]));
        count = count + sum[i];
        k++;
    }
}
//设置"合计:"及其值,并刷新面板
lb[n].setText("合计: ");
tf[n].setText(String.valueOf(count));
p2.updateUI();
}
catch(SQLException e2)
{
    JOptionPane.showMessageDialog(this,"操作数据库出错:" + e2);
}
catch(Exception e3)
{
    JOptionPane.showMessageDialog(this,"出现其他错误:" + e3);
}
}
```

该方法在构造方法中及下拉列表框项目变化事件方法中被调用。在构造方法中调用addDataToSCList()方法的语句后面,编写以下调用代码:

```
statBadNum();
```

这样,窗体被打开后,即能看到当前班级的不及格人数已经被统计出来。在下拉列表框项目变化事件方法中被调用,请参见本任务步骤 6 中代码。

6. 编制窗体事件处理框架代码

与前面模块中的基本一致,具体实现代码可参考本项目任务十一中编制的代码。

7. 运行程序并进行测试

从下拉列表框中选择不同的班级,观察统计出来的班级不及格人数情况。

任务十五 编制用户登录模块并打包

【任务描述】

本模块编制一个系统的登录对话框,用来检查登录用户的合法性。将系统编译、调试完成后,打包成 jar 文件。运行示意图如图 3-27 所示。

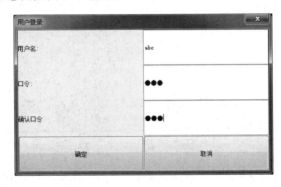

图 3-27 用户登录对话框

【任务分析】

在项目一用户登录程序中已经编制了一个独立的用于判断用户合法性的程序,只要将上述程序复制至本项目中来进行修改即可。所有程序文件编制完成后,进行编译调试,最后将所有 class 文件打包成 jar 文件。

【任务实施】

1. 复制程序

将项目一用户登录程序中编制的 login.java 源程序文件复制至本项目中。

2. 修改 login.java

① 将 login 类中的主方法 main() 删去;

② 将 login 类从 JFrame 继承改为从 JDialog 继承,并修改构造方法的参数。

主要实现代码如下:

```
public login(JFrame parent,String title,boolean model)
{
    super(parent,title,model);
    …
}
```

③ 修改 login 类中的数据库连接对象的定义和引用。
将类中原来定义类的静态成员变量 con 的语句删去：

```
static Connection con;
```

将查询时创建语句对象的代码：

```
Statement stat = con.createStatement();
```

改为：

```
Statement stat = ScoreMis.con.createStatement();
```

④ 定义用于获取返回值的方法。
在 login 类中定义一类成员变量：

```
int rtn = 0;
```

在登录成功时将其值改为 1。
再定义一返回该变量值的方法：

```
public int getRtn()
{
    return rtn;
}
```

⑤ 增加连续 3 次登录不成功退出的功能。
在 login 类中定义一类成员变量：

```
int cnt = 0;
```

该变量用于统计连续登录的次数。在查询后，如果不存在该用户，则统计次数并处理，主要实现代码如下：

```
…
else
{
    JOptionPane.showMessageDialog(this,"数据库中不存在该用户!");
    cnt++;
    if(cnt >= 3)
    {
        rtn = 0;
        dispose();
    }
}
…
```

3．编写 ScoreMis 类中打开登录对话框的代码

在 ScoreMis 类主方法 main() 中连接数据库的代码后，编写以下实现代码：

```
login logindlg = new login(null,"用户登录",true);
if (logindlg.getRtn() == 0)
System.exit(0);
```

4．运行程序并进行测试

用不同的用户进行登录，观察程序的运行情况。

5．编译源程序文件

将所有源程序文件编译成 class 类文件。

6. 用记事本编辑 manifest.mf 文件

如图 3-28 所示，请注意类文件名之前有一空格。

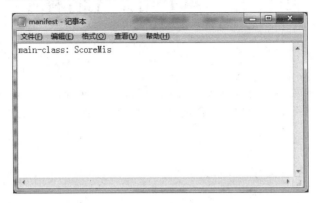

图 3-28　编辑 manifest.mf 文件示例

7. 用 jar 命令将 class 文件打包成 jar 文件

如图 3-29 所示，用 jar 命令将本系统中所有 class 文件打包成 scoremisjar 文件。

图 3-29　用 jar 命令将多个 class 文件打包

8. 运行程序

打包正确完成后，可以直接运行 scoremisjar 文件，也可以创建一个 bat 文件来运行程序。

项目小结

学生成绩管理系统是一个较为复杂的 Java 桌面程序，通过本项目的实践，读者可以比较全面地掌握利用 Java 开发数据库桌面系统的方法及技术。在项目的开发过程中，涉及到了许多如复杂的图形界面的编制，数据库中数据的查询、增、删、改操作，多种事件的处理等开发技术，也使读者对数据库的设计、业务逻辑的处理有一个深入的理解，为以后开发规模更大更复杂的系统打下基础。

项目四　网上用户注册程序

项目描述

随着互联网技术的发展，Web 应用越来越多，大多数的 Web 应用都具备用户登录和注册功能。网上用户注册程序就是一个 Java Web 应用程序，主要是实现网上用户的登录及注册。

项目预览

在浏览器地址栏中输入 http://localhost:8080/UserRegister_1/login.jsp，程序运行时，出现如图 4-1 所示登录页面。

输入用户名和密码，单击"登录"按钮，若数据库中存在该用户，则转至显示用户信息页面，如图 4-2 所示。

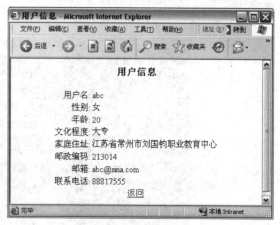

图 4-1　用户登录页面　　　　　图 4-2　用户信息显示页面

若数据库中不存在该用户，则转至登录（注册）失败页面，如图 4-3 所示。

图 4-3　登录（注册）失败页面

在单击"注册"按钮后，转至输入用户信息的页面，如图 4-4 所示。

输入信息后，单击"提交"按钮，用户信息保存至数据库中，成功则转至如图 4-2 所示显示信息的页面，失败则转至如图 4-3 所示页面。

项目四 网上用户注册程序

图 4-4 注册信息输入页面

项目分析

网上用户注册程序主要有两方面的功能，一是已经注册过的用户，可以在正确输入用户名及密码后登录服务器，并查看注册信息；二是没有注册过的用户，则可以进入注册页面进行注册。这是一个比较简单的 Web 应用程序，我们将用四种开发模式来具体实现这一项目程序，将涉及用 JSP 技术及框架技术开发 Web 应用程序的许多基础知识。

项目设计

1. 总体流程的设计

程序总体流程图如图 4-5 所示。

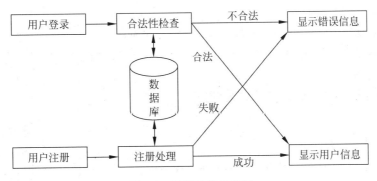

图 4-5 程序总体流程图

2. 数据库的设计

数据库的设计比较简单，只要在其中设计一张用户表即可。数据库名称为 UserRegister，表名为 suser，表结构见表 4-1。

表 4-1 用户表 suser 的表结构

列名	数据类型	长度	允许为空	主键	含义
uname	文本	16	否	是	用户名
upass	文本	16	否		密码
usex	文本	2	是		性别
uage	数字		是		年龄
ulevel	文本	10	是		文化程度
uaddress	文本	50	是		家庭住址
umail	文本	6	是		邮政编码
uemail	文本	30	是		邮箱
uphone	文本	30	是		联系电话

环境需求

操作系统：Windows XP 或 Windows 7；
数据库平台：Microsoft Access；
JDK 版本：JDK Version1.6；
集成开发环境：MyEclipse 8.5；
Web 服务器：Tomcat 6.0；
浏览器：IE6.0 及以上版本。

项目知识点

① JSP 的页面结构，常用指令、动作的使用方法；
② JSP 常用内置对象的使用方法；
③ JavaScript 脚本语言的基本使用方法；
④ JSP 三种开发模式的特点、开发的基本方法及步骤；
⑤ JavaBean 的概念及编制方法；
⑥ Servelt 的概念、编制方法及配置方法；
⑦ JSP 中使用 JDBC 连接、操纵数据库的基本方法；
⑧ Struts 框架开发模式的特点、开发的基本方法及步骤；
⑨ Struts 中表单 Bean、动作 Action 类的概念、编制方法及配置方法。

任务一 用纯 JSP 页面模式实现

【任务描述】

用纯 JSP 页面模式实现时，程序中负责页面显示、数据存储和处理、流程控制等所有代码都是编写在 JSP 页面文件中。因此，本任务中要编制该模式下所有页面文件。

【任务分析】

1. 程序运行基本流程

纯 JSP 页面模式下程序运行基本流程图如图 4-6 所示。

项目四 网上用户注册程序

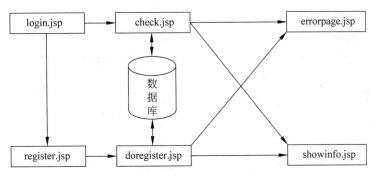

图 4-6 纯 JSP 页面模式下程序运行基本流程图

2．程序文件清单

纯 JSP 页面模式下程序文件清单见表 4-2。

表 4-2 纯 JSP 页面模式下程序文件清单

组件类别	文 件 名	功 能 描 述
JSP 页面	login.jsp	用户登录的页面
	check.jsp	检查登录用户的合法性，并转至相应的页面
	showinfo.jsp	显示用户注册信息的页面
	errorpage.jsp	登录和注册失败时显示错误信息的页面
	register.jsp	用户输入注册信息的页面
	doregister.jsp	保存用户注册信息的页面

3．项目目录结构

纯 JSP 页面模式下项目目录结构如图 4-7 所示。

```
▲ 📁 UserRegister_1
    📁 src
    ▷ 📚 JRE System Library [Sun JDK 1.6.0_13]
    ▷ 📚 J2EE 1.4 Libraries
    ▲ 📁 web
        ▲ 📁 database
            📄 UserRegister.mdb
        ▷ 📁 META-INF
        ▲ 📁 WEB-INF
            📁 lib
            📄 web.xml
        📄 check.jsp
        📄 doregister.jsp
        📄 errorpage.jsp
        📄 login.jsp
        📄 register.jsp
        📄 showinfo.jsp
```

图 4-7 纯 JSP 页面模式下项目目录结构

【任务实施】

1. 准备数据库

在 Microsoft Access 中创建一个数据库,名称为 UserRegister.mdb,创建一张名称为 suser 的表,表结构见表 4-1,并输入几条记录。

2. 新建工程

启动 MyEclipse,选择"New→Web project"(即选择 New 菜单中的"Web Project"命令,下同),新建一个名称为 UserRegister_1 的工程,如图 4-8 所示。

图 4-8 新建工程对话框

3. 将数据库复制到工程中

选中 Web 文件夹,右击,选择"New->Folder"(即选择"New"菜单中的"Folder"命令,下同),新建名称为 database 的文件夹,将步骤 1 中创建的数据库文件 UserRegister.mdb 复制至文件夹中。

4. 新建并编制 login.jsp 文件

选中 Web 文件夹,右击,选择"New->File",新建 login.jsp 文件。

如图 4-1 所示,该页面为用户登录的页面。单击"登录"按钮后,将首先通过 script 代码检查用户名及密码是否均不为空,然后提交表单,并执行 check.jsp 文件;单击"注册"超链接后,转至用户注册页面。

实现代码如下:

```
<%@ page contentType = "text/html;charset = utf-8" %>
<html>
  <head>
    <title>网上用户注册系统</title>
    <script language = "javascript" >
      function checkdata()
      {
        var uname = document.loginform.uname.value;
        if (uname == "" )
        {
          alert("用户名不能为空,必须输入!");
          document.loginform.uname.focus();
          return false;
```

```
                    }
                var upass = document.loginform.upass.value;
                if (upass == "" )
                {
                    alert("密码不能为空,必须输入!");
                    document.loginform.upass.focus();
                    return false;
                }
                return true;
            }
        </script>
    </head>
    <body onload = "document.loginform.uname.focus()">
        <center>
            <h3>用户登录</h3>
            <form name = "loginform" method = "post" action = "check.jsp">
                <table>
                    <tr>
                        <td align = "right">用户名:</td>
                        <td><input type = "text" name = "uname" style = "width:135px"></td>
                    </tr>
                    <tr>
                        <td align = "right">密    码:</td>
                        <td><input type = "password" name = "upass" style = "width:135px"></td>
                    </tr>
                    <tr>
                        <td> </td>
                        <td><input type = "submit" name = "login" value = "登 录" onclick = "return checkdata()">
                            <input type = "reset"  name = "reset"  value = "重 置">
                            <a href = "register.jsp">注册</a>
                        </td>
                    </tr>
                </table>
            </form>
        </center>
    </body>
</html>
```

5．新建并编制 check.jsp 文件

这是对用户合法性进行检查的页面。首先获取 login.jsp 页面中传递过来的用户名及密码,连接数据库,进行验证。如果是合法用户,则取出用户信息,保存在会话 session 级变量中,并转至用户信息显示页面 showinfo.jsp,将用户信息显示出来；如果不合法,则转至错误页面 errorpage.jsp。它实际上起一个控制器的作用,并不直接形成显示页面输出。

实现代码如下：

```
<%@ page   contentType = "text/html;charset = utf - 8"   %>
<%@ page import = "java.sql.*" %>
<%
    request.setCharacterEncoding("utf - 8");

    String uname = request.getParameter("uname");
    String upass = request.getParameter("upass");
    Connection con = null;
    boolean flag = false;
    try
    {
        Class.forName("sun.jdbc.odbc.JdbcOdbcDriver");
        String url = "jdbc:odbc:driver = {Microsoft Access Driver ( * .mdb)};DBQ = " + getServletContext().getRealPath("") + "/database/UserRegister.mdb";
        con = DriverManager.getConnection(url,"","");
```

```
            Statement stmt = con.createStatement();
            ResultSet result = stmt.executeQuery("select * from suser where uname = '" + uname + "' and upass = '" +
upass + "'");

            if ( result.next())
            {
                session.setAttribute("uname",result.getString("uname"));
                session.setAttribute("upass",result.getString("upass"));
                session.setAttribute("usex",result.getString("usex"));
                session.setAttribute("uage",result.getInt("uage"));
                session.setAttribute("ulevel",result.getString("ulevel"));
              session.setAttribute("uaddress",result.getString("uaddress"));
                session.setAttribute("umail",result.getString("umail"));
                session.setAttribute("uemail",result.getString("uemail"));
                session.setAttribute("uphone",result.getString("uphone"));

                flag = true;
            }
        }
        catch (ClassNotFoundException e)
        {
            e.printStackTrace();
        }
        catch(SQLException e)
        {
            e.printStackTrace();
        }
        catch(Exception e)
        {
            e.printStackTrace();
        }
        finally
        {
            if(con!= null)
            {
                try
                {
                    con.close();
                }
                catch(Exception e)
                {
                    e.printStackTrace();
                }
            }
        }
        if(flag)
        {
            response.sendRedirect("showinfo.jsp");
        }
        else
        {
            response.sendRedirect("errorpage.jsp");
        }
%>
```

6. 新建并编制 showinfo.jsp 文件

如图 4-2 所示，该页面用来显示用户的注册信息，这些信息是通过会话 session 级变量传递过来的，每一项用户注册信息均存放在对应的一个变量中。

实现代码如下：

```
<%@ page contentType = "text/html;charset = utf-8"  %>
<html>
    <head>
```

```
        <title>用户信息</title>
    </head>
    <body>
        <center>
            <h3 align="center">用户信息</h3>
            <table>
                <tr>
                    <td align="right">用户名:</td><td><%=session.getAttribute("uname")%></td>
                </tr>
                <tr>
                    <td align="right">性别:</td><td><%=session.getAttribute("usex")%></td>
                </tr>
                <tr>
                    <td align="right">年龄:</td><td><%=session.getAttribute("uage")%></td>
                </tr>
                <tr>
                    <td align="right">文化程度:</td><td><%=session.getAttribute("ulevel")%></td>
                </tr>
                <tr>
                    <td align="right">家庭住址:</td><td><%=session.getAttribute("uaddress")%></td>
                </tr>
                <tr>
                    <td align="right">邮政编码:</td><td><%=session.getAttribute("umail")%></td>
                </tr>
                <tr>
                    <td align="right">邮箱:</td><td><%=session.getAttribute("uemail")%></td>
                </tr>
                <tr>
                    <td align="right">联系电话:</td><td><%=session.getAttribute("uphone")%></td>
                </tr>
                <tr>
                    <td colspan="2" align="center"><a href="javascript:history.back()">返回</a></td>
                </tr>
            </table>
        </center>
    </body>
</html>
```

7. 新建并编制 errorpage.jsp 文件

如图 4-3 所示,当用户所输入的用户名和密码不正确或注册失败时转至该页面,并显示错误信息。

实现代码如下:

```
<%@ page contentType="text/html;charset=utf-8" %>
<html>
    <head>
        <title>错误页面</title>
    </head>
    <body>
        <p>
        登录(注册)失败!
        </p>
        <a href="javascript:history.back()">返回</a>
    </body>
</html>
```

8. 发布并测试程序

启动 Tomcat 服务器,将 UserRegister_1 应用发布至服务器;打开浏览器,在地址栏中输入 http://localhost:8080/UserRegister_1/login.jsp,运行并测试登录部分。如果在 web.xml 文件的子元素<welcome-file>对中配置 login.jsp 为默认首页,则只要在地址栏中输入 http://localhost:

8080/UserRegister_1 即可运行程序。

9. 新建并编制 register.jsp 文件

如图 4-4 所示,该页面提供用户输入注册信息。单击"提交"按钮后,将首先通过 javascript 代码检查所输信息的合法性,合法则提交表单,并执行 doregister.jsp 文件。数据输入要求:

用户名　必须输入,不能为空;
密码　　必须输入,不能为空,并且两次输入必须一致;
年龄　　必须输入,不能为空,并且只能在 16 至 21 岁之间;
邮政　　必须输入,并且只能是 6 位数字;
邮箱　　必须输入,并且要包含"@"、"."字符。

实现代码如下:

```jsp
<%@ page contentType="text/html;charset=utf-8" %>
<html>
  <head>
    <title>用户注册</title>
    <script language="javascript">
      function checkdata()
      {
          var uname=document.registerform.uname.value;
          if(uname=="")
          {
              alert("用户名不能为空,必须输入!");
              document.registerform.uname.focus();
              return false;
          }
          var upass=document.registerform.upass.value;
          if(upass=="")
          {
              alert("密码不能为空,必须输入!");
              document.registerform.upass.focus();
              return false;
          }
          var upass1=document.registerform.upass1.value;
          if(upass1!=upass)
          {
              alert("两次输入的密码不一致!");
              document.registerform.upass1.focus();
              return false;
          }
          var uage=document.registerform.uage.value;
          if(uage=="" || isNaN(uage))
          {
              alert("年龄必须输入,不能为空,并且只能是数字!");
              document.registerform.uage.focus();
              return false;
          }
          if(parseInt(uage)<16 || parseInt(uage)>21)
          {
              alert("年龄只能在 16 至 21 岁之间!");
              document.registerform.uage.focus();
              return false;
          }
          var umail=document.registerform.umail.value;
          if(umail=="" || umail.length!=6 || isNumberic(umail)==false)
          {
              alert("邮政编码必须输入,不能为空,并且必须是 6 位数字!");
              document.registerform.umail.focus();
              return false;
```

```javascript
        }
        var uemail = document.registerform.uemail.value;
        if(uemail == "" )
        {
            alert("邮箱地址必须输入,不能为空!");
            document.registerform.uemail.focus();
            return false;
        }
        if(uemail.indexOf('@',0) == -1 )
        {
            alert("邮箱地址格式有错误,应包含@字符!");
            document.registerform.uemail.focus();
            return false;
        }
        if(uemail.indexOf('.',0) == -1 )
        {
            alert("邮箱地址格式有错误,应包含.字符!");
            document.registerform.uemail.focus();
            return false;
        }
        return true;
    }

    function isNumberic( str )
    {
        var len = str.length;
        for (var i = 0;i<len;i++)
            if ( str.charAt(i)<'0' || str.charAt(i)>'9' )
                return false;
        return true;
    }
    </script>
</head>
<body onload = "document.registerform.uname.focus()">
    <center>
        <h3 align = "center">填写注册信息</h3>
        <form name = "registerform"  method = "post"  action = "doregister.jsp">
            <table>
                <tr>
                    <td>用户名:</td><td><input type = "text" name = "uname" style = "width:150px"></td>
                </tr>
                <tr>
                    <td>密码:</td><td><input type = "password"  name = "upass" style = "width:150px"></td>
                </tr>
                <tr>
                    <td>确认密码:</td><td><input type = "password" name = "upass1" style = "width:150px"></td>
                </tr>
                <tr>
                    <td>性别:</td>
                    <td><input type = "radio" name = "usex" value = "男" checked>男<input type = "radio" name = "usex" value = "女">女</td>
                </tr>
                <tr>
                    <td>年龄:</td><td><input type = "text" name = "uage"  style = "width:150px"></td>
                </tr>
                <tr>
                    <td>文化程度:</TD>
                    <td><select name = "ulevel"  style = "width:150px">
                        <option value = "中专">中专</option>
                        <option value = "大专">大专</option>
                        <option value = "本科">本科</option>
                        <option value = "研究生">研究生</option>
```

```
                </select>
            </td>
        </tr>
        <tr>
            <td>家庭住址:</td><td><input type="text" name="uaddress" style="width:150px"></td>
        </tr>
        <tr>
            <td>邮政编码:</td><td><input type="text" name="umail" style="width:150px"></td>
        </tr>
        <tr>
            <td>邮箱:</td><td><input type="text" name="uemail" style="width:150px"></td>
        </tr>
        <tr>
            <td>联系电话:</td><td><input type="text" name="uphone" style="width:150px"></td>
        </tr>
        <tr>
            <td> </td>
            <td align="center"><input type="submit" value="提 交" onclick="return checkdata()">
   <input type="reset" value="重 置"></td>
        </tr>
    </table>
    </form>
    </center>
</body>
</html>
```

10. 新建并编制 doregister.jsp 文件

这是一个对用户进行注册的页面,作为一个新用户保存至数据库中,本身并不直接形成显示页面,代码类似于 check.jsp。

实现代码如下:

```jsp
<%@ page contentType="text/html;charset=utf-8" %>
<%@ page import="java.sql.*" %>
<%
    request.setCharacterEncoding("utf-8");

    String uname = request.getParameter("uname");
    String upass = request.getParameter("upass");
    String usex = request.getParameter("usex");
    String uage = request.getParameter("uage");
    String ulevel = request.getParameter("ulevel");
    String uaddress = request.getParameter("uaddress");
    String umail = request.getParameter("umail");
    String uemail = request.getParameter("uemail");
    String uphone = request.getParameter("uphone");

    Connection con = null;
    boolean flag = false;
    try
    {
        Class.forName("sun.jdbc.odbc.JdbcOdbcDriver");
        String url = "jdbc:odbc:driver={Microsoft Access Driver (*.mdb)};DBQ=" + request.getSession().getServletContext().getRealPath("") + "/database/UserRegister.mdb";
        con = DriverManager.getConnection(url,"","");

        PreparedStatement stmt = con.prepareStatement("insert into suser values(?,?,?,?,?,?,?,?,?)");
        stmt.setString(1,uname);
        stmt.setString(2,upass);
        stmt.setString(3,usex);
```

```
                stmt.setInt(4,Integer.parseInt(uage));
                stmt.setString(5,ulevel);
                stmt.setString(6,uaddress);
                stmt.setString(7,umail);
                stmt.setString(8,uemail);
                stmt.setString(9,uphone);

                int row = stmt.executeUpdate();
                if(row == 1)
                {
                    session.setAttribute("uname",uname);
                    session.setAttribute("upass",upass);
                    session.setAttribute("usex",usex);
                    session.setAttribute("uage",uage);
                    session.setAttribute("ulevel",ulevel);
                    session.setAttribute("uaddress",uaddress);
                    session.setAttribute("umail",umail);
                    session.setAttribute("uemail",uemail);
                    session.setAttribute("uphone",uphone);

                    flag = true;
                }
        }
        catch(ClassNotFoundException e)
        {
                e.printStackTrace();
        }
        catch(SQLException e)
        {
                e.printStackTrace();
        }
        catch(Exception e)
        {
                e.printStackTrace();
        }
        finally
        {
                if(con!= null)
                    try
                    {
                        con.close();
                    }
                    catch(Exception e)
                    {
                        e.printStackTrace();
                    }
        }
        if(flag)
        {
            response.sendRedirect("showinfo.jsp");
        }
        else
        {
            response.sendRedirect("errorpage.jsp");
        }
%>
```

11. 发布并测试程序

启动 Tomcat 服务器,将 UserRegister_1 应用发布至服务器;打开浏览器,在地址栏中输入 http://localhost:8080/UserRegister_1/login.jsp,运行并测试注册部分。

任务二　用 JSP＋JavaBean 模式实现

【任务描述】

用 JSP＋JavaBean 模式实现时，程序中的数据存储和处理由 JavaBean 实现，而数据显示与输入、流程控制等仍由页面负责。因此，本任务可以在纯 JSP 页面实现模式的基础上，编制 JavaBean 组件并对相关页面文件进行修改。

【任务分析】

1．程序运行基本流程

JSP＋JavaBean 模式下程序运行基本流程图如图 4-9 所示。

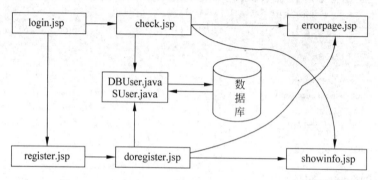

图 4-9　JSP＋JavaBean 模式下程序运行基本流程图

2．程序文件清单

JSP＋JavaBean 模式下程序文件清单见表 4-3。

表 4-3　JSP＋JavaBean 模式下程序文件清单

组件类别	文件名	功能描述
JSP 页面	login.jsp	用户登录的页面
	check.jsp	检查登录用户的合法性，并转至相应的页面
	showinfo.jsp	显示用户注册信息的页面
	errorpage.jsp	登录和注册失败时显示错误信息的页面
	register.jsp	用户输入注册信息的页面
	doregister.jsp	保存用户注册信息的页面
JavaBean(模型 Model)	DBUser.java	连接及操纵数据库的组件，一个工具类
	SUser.java	代表用户的实体类

3．项目目录结构

JSP＋JavaBean 模式下项目目录结构图如图 4-10 所示。

【任务实施】

1．新建工程

启动 MyEclipse，选择"New->Web Project"，新建一个名称为 UserRegister_2 的工程。

2．将数据库复制到工程中

选中 Web 文件夹，右击，选择"New->Folder"，新建名称为 database 的文件夹，将 UserRegister_1

项目四 网上用户注册程序

图 4-10 JSP＋JavaBean 模式下项目目录结构图

工程中的数据库文件 UserRegister.mdb 复制到文件夹中。

3．新建包结构

选中 src 文件夹,右击,选择"New->Package",在 src 文件夹下新建一个名为 com.register 的包结构,JavaBean 类文件均存放在该包中。

4．新建并编制 JavaBean 组件

（1）新建并编制 SUser.java 文件

选中 com.register 包,右击,选择"New->File",新建 SUser.java 类文件。这是一个实体 JavaBean 组件,该类的一个对象代表一个具体用户,包含该用户的详细信息。

实现代码如下：

```java
package com.register;
import java.io.Serializable;
public class SUser implements Serializable
{
    //私有成员变量
    private String uname;
    private String upass;
    private String usex;
    private int    uage;
    private String ulevel;
    private String uaddress;
    private String umail;
    private String uemail;
    private String uphone;

    public SUser()
    {
    }
    //一组 get 与 set 方法
    public String getUname()
    {
        return uname;
    }
    public void setUname(String uname)
    {
        this.uname = uname;
    }

    public String getUpass()
```

```java
{
 return upass;
}
public void setUpass(String upass)
{
 this.upass = upass;
}

public String getUsex()
{
 return usex;
}
public void setUsex(String usex)
{
 this.usex = usex;
}

public int getUage()
{
 return uage;
}

public void setUage(int uage)
{
 this.uage = uage;
}

public String getUlevel()
{
 return ulevel;
}

public void setUlevel(String ulevel)
{
 this.ulevel = ulevel;
}

public String getUaddress()
{
 return uaddress;
}

public void setUaddress(String uaddress)
{
 this.uaddress = uaddress;
}

public String getUmail()
{
 return umail;
}
public void setUmail(String umail)
{
 this.umail = umail;
}

public String getUemail()
{
 return uemail;
}
public void setUemail(String uemail)
{
 this.uemail = uemail;
```

```java
            }
        public String getUphone()
        {
            return uphone;
        }
        public void setUphone(String uphone)
        {
            this.uphone = uphone;
        }
}
```

(2) 新建并编制 DBUser.java 文件

这是一个工具 JavaBean 组件，它封装了对数据库的所有操作，包括在构造方法中加载驱动程序、根据用户名及密码获取一个用户注册信息、增加一个注册用户等方法。

实现代码如下：

```java
package com.register;
import javax.servlet.ServletContext;
import java.sql.*;
public class DBUser
{
    private ServletContext application;
    //定义构造方法,加载驱动程序,接收全局应用对象
    public DBUser(ServletContext application)
    {
        try
        {
            Class.forName("sun.jdbc.odbc.JdbcOdbcDriver");
        }
        catch (ClassNotFoundException e)
        {
            e.printStackTrace();
        }
        catch(Exception e)
        {
            e.printStackTrace();
        }
        this.application = application;
    }
    //根据用户名及密码获取一个用户信息的方法
    public SUser getSUser(String uname,String upass)
    {
        Connection con = null;
        SUser suser = null;

        try
        {
            //连接字符串
            String url = "jdbc:odbc:driver = {Microsoft Access Driver ( * .mdb)};DBQ = " + application.getRealPath("") + "/database/UserRegister.mdb";
            con = DriverManager.getConnection(url,"","");
            Statement stmt = con.createStatement();
            ResultSet result = stmt.executeQuery("select * from suser where uname = '" + uname + "' and upass = '" + upass + "'");

            if ( result.next())
            {
                suser = new SUser();

                suser.setUname(result.getString("uname"));
                suser.setUpass(result.getString("upass"));
```

```java
                    suser.setUsex(result.getString("usex"));
                    suser.setUage(result.getInt("uage"));
                    suser.setUlevel(result.getString("ulevel"));
                    suser.setUaddress(result.getString("uaddress"));
                    suser.setUmail(result.getString("umail"));
                    suser.setUemail(result.getString("uemail"));
                    suser.setUphone(result.getString("uphone"));
                }
            }
            catch(SQLException e)
            {
                e.printStackTrace();
            }
            catch(Exception e)
            {
                e.printStackTrace();
            }
            finally
            {
                if(con!=null)
                {
                    try
                    {
                        con.close();
                    }
                    catch(Exception e)
                    {
                        e.printStackTrace();
                    }
                }
            }
            return suser;
    }
    //增加一个注册用户的方法
    public boolean addSUser(SUser suser)
    {
        Connection con = null;
        boolean flag = false;
        try
        {
            String url = "jdbc:odbc:driver = {Microsoft Access Driver ( * .mdb)};DBQ = " + application.getRealPath("") + "/database/UserRegister.mdb";
            con = DriverManager.getConnection(url,"","");
            PreparedStatement stmt = con.prepareStatement("insert into suser values(?,?,?,?,?,?,?,?,?) ");

            stmt.setString(1,suser.getUname());
            stmt.setString(2,suser.getUpass());
            stmt.setString(3,suser.getUsex());
            stmt.setInt(4,suser.getUage());
            stmt.setString(5,suser.getUlevel());
            stmt.setString(6,suser.getUaddress());
            stmt.setString(7,suser.getUmail());
            stmt.setString(8,suser.getUemail());
            stmt.setString(9,suser.getUphone());

            int rowcnt = stmt.executeUpdate();
            if(rowcnt == 1)
                flag = true;
        }
        catch(SQLException e)
        {
            e.printStackTrace();
        }
```

```
        catch(Exception e)
        {
            e.printStackTrace();
        }
        finally
        {
            if(con!= null)
            {
                try
                {
                    con.close();
                }
                catch(Exception e)
                {
                    e.printStackTrace();
                }
            }
        }
        return flag;
    }
}
```

5. 编制 login.jsp 文件

login.jsp 文件与 UserRegister_1 中的 login.jsp 文件完全一致,直接将 UserRegister_1 工程中的 login.jsp 文件复制到本工程中。

6. 编制 check.jsp 文件

由于已经定义了 JavaBean 组件,因此可以利用 JavaBean 组件连接并操纵数据库,对用户进行验证。将 UserRegister_1 工程中的 check.jsp 文件复制到本工程中,对其进行修改。

实现代码如下:

```jsp
<%@ page contentType="text/html;charset=utf-8" %>
<%@ page import="com.register.*" %>

<%
    request.setCharacterEncoding("utf-8");

    String uname = request.getParameter("uname");
    String upass = request.getParameter("upass");
    //定义并创建 DBUser 类的对象
    DBUser dbuser = new DBUser(application);
    //获取一个用户
    SUser suser = dbuser.getSUser(uname,upass);
    if(suser!= null)
    {
        //保存用户数据至 session 级变量
        session.setAttribute("suser",suser);
        response.sendRedirect("showinfo.jsp");
    }
    else
    {
        response.sendRedirect("errorpage.jsp");
    }
%>
```

7. 编制 showinfo.jsp 文件

由于已经将封装用户信息的 SUser 类的一个对象 suser 保存在了 session 级变量中,因此可以使用 JSP 的 useBean、getProperty 动作来获取和显示用户的注册信息。将 UserRegister_1 工程中的 showinfo.jsp 文件复制到本工程中,对其进行修改。

实现代码如下：

```jsp
<%@ page contentType="text/html;charset=utf-8" %>
<jsp:useBean id="suser" class="com.register.SUser" scope="session" />
<html>
  <head>
    <title>用户信息</title>
  </head>
  <body>
    <center>
      <h3 align="center">用户信息</h3>
      <table>
        <tr>
          <td align="right">用户名：</td>
          <td><jsp:getProperty name="suser" property="uname"/></td>
        </tr>
        <tr>
          <td align="right">性别：</td>
          <td><jsp:getProperty name="suser" property="usex"/></td>
        </tr>
        <tr>
          <td align="right">年龄：</td>
          <td><jsp:getProperty name="suser" property="uage"/></td>
        </tr>
        <tr>
          <td align="right">文化程度：</td>
          <td><jsp:getProperty name="suser" property="ulevel"/></td>
        </tr>
        <tr>
          <td align="right">家庭住址：</td>
          <td><jsp:getProperty name="suser" property="uaddress"/></td>
        </tr>
        <tr>
          <td align="right">邮政编码：</td>
          <td><jsp:getProperty name="suser" property="umail"/></td>
        </tr>
        <tr>
          <td align="right">邮箱：</td>
          <td><jsp:getProperty name="suser" property="uemail"/></td>
        </tr>
        <tr>
          <td align="right">联系电话：</td>
          <td><jsp:getProperty name="suser" property="uphone"/></td>
        </tr>
        <tr>
          <td colspan="2" align="center"><a href="javascript:history.back()">返回</a></td>
        </tr>
      </table>
    </center>
  </body>
</html>
```

8. 编制 errorpage.jsp 文件

errorpage.jsp 文件与 UserRegister_1 中的 errorpage.jsp 文件完全一致，直接将 UserRegister_1 工程中的 errorpage.jsp 文件复制到本工程中。

9. 发布并测试程序

启动 Tomcat 服务器，将 UserRegister_2 应用发布至服务器；打开浏览器，在地址栏中输入 http://localhost:8080/UserRegister_2/login.jsp，运行并测试登录部分。

项目四 网上用户注册程序

10. 编制 register.jsp 文件

register.jsp 文件与 UserRegister_1 中的 register.jsp 文件完全一致,直接将 UserRegister_1 工程中的 register.jsp 文件复制到本工程中。

11. 编制 doregister.jsp 文件

先用 JSP 的动作 useBean、setProperty 来接收传递过来的用户注册数据,利用 JavaBean 类 DBUser 的方法将用户注册信息保存至数据库中。将 UserRegister_1 工程中的 doregister.jsp 文件复制到本工程中,对其进行修改。

实现代码如下:

```
<%@ page contentType = "text/html;charset = utf - 8" %>
<%@ page import = "com.register.*" %>

<% request.setCharacterEncoding("utf - 8"); %>

<jsp:useBean id = "suser" class = "com.register.SUser" scope = "session" />
<jsp:setProperty name = "suser" property = " * " />
<%
    DBUser dbuser = new DBUser(application);
    //增加一个注册用户
    boolean flag = dbuser.addSUser(suser);
    if(flag)
    {
        response.sendRedirect("showinfo.jsp");
    }
    else
    {
        response.sendRedirect("errorpage.jsp");
    }
%>
```

12. 启动 Tomcat 服务器

启动 Tomcat 服务器,并将 UserRegister_2 应用发布至服务器。

13. 发布并测试程序

将 UserRegister_2 应用发布至服务器,打开浏览器,在地址栏中输入 http://localhost:8080/UserRegister_2/login.jsp,运行并测试注册部分。

任务三　用 JSP+JavaBean+Servlet 模式实现

【任务描述】

用 JSP+JavaBean+Servlet 模式实现时,程序中的数据显示与输入由页面负责,数据存储和处理由 JavaBean 组件负责,流程控制由 Servlet 类负责,三者实现分离。因此,本任务可以在 JSP+JavaBean 实现模式的基础上,编制 Servlet 类负责流程控制,并对相关页面文件进行修改。

【任务分析】

1. 程序运行基本流程

JSP+JavaBean+Servlet 模式下程序运行基本流程图如图 4-11 所示。

2. 程序文件清单

JSP+JavaBean+Servlet 模式下程序文件清单见表 4-4。

Java & JSP应用程序实例开发

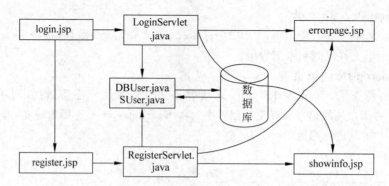

图 4-11　JSP＋JavaBean＋Servlet 模式下程序运行基本流程图

表 4-4　JSP＋JavaBean＋Servlet 模式下程序文件清单

组件类别	文件名	功能描述
视图层(View)	login.jsp	用户登录的页面
	register.jsp	用户输入注册信息的页面
	showinfo.jsp	显示用户注册信息的页面
	errorpage.jsp	登录和注册失败时显示错误信息的页面
控制器层(Controller)	LoginServlet.java	处理用户登录的 Servlet 类
	RegisterServlet.java	处理用户注册的 Servlet 类
	web.xml	应用程序的通用配置文件
模型层(Model)	DBUser.java	连接及操纵数据库的组件，一个工具类
	SUser.java	代表用户的实体类

3．项目目录结构

JSP＋JavaBean＋Servlet 模式下项目目录结构图如图 4-12 所示。

图 4-12　JSP＋JavaBean＋Servlet 模式下项目目录结构图

【任务实施】

1．新建工程

启动 MyEclipse，选择"New->Web Project"，新建一个名称为 UserRegister_3 的工程。

2. 将数据库复制到工程中

选中 Web 文件夹,右击,选择"New->Folder",新建名称为 database 的文件夹,将 UserRegister_1 工程中的数据库文件 UserRegister.mdb 复制到文件夹中。

3. 新建包结构

选中 src 文件夹,右击,选择"New->Package",在 src 文件夹下新建一个名为 com.register 的包结构,JavaBean 类文件及 Servlet 类文件均存放在该包中。

4. 编制 JavaBean 组件

DBUser.java 与 UserRegister_2 工程中的 DBUser.java 文件完全一致,可以直接将其复制到本工程中。

SUser.Java 与 UserRegister_2 工程中的 SUser.Java 略有差别,可以先将其直接复制到本工程中,再在类中增加一个如下代码所示的构造方法:

```java
public SUser(String uname, String upass, String usex, int uage, String ulevel, String uaddress, String umail, String uemail, String uphone)
{
    this.uname = uname;
    this.upass = upass;
    this.usex = usex;
    this.uage = uage;
    this.ulevel = ulevel;
    this.uaddress = uaddress;
    this.umail = umail;
    this.uemail = uemail;
    this.uphone = uphone;
}
```

5. 新建并编制用于登录的 Servlet 类文件 LoginServlet.java

LoginServlet 类的功能与 UserRegister_1、UserRegister_2 工程中的 check.jsp 程序功能一致。在该类的 doPost()方法中获取 login.jsp 页面中传递过来的用户名及密码后,利用 JavaBean 类 DBUser、SUser 的方法连接并操纵数据库,验证用户的合法性。如果是合法用户,则将用户信息存放在会话 session 级变量 suser 中,转至用户信息显示页面 showinfo.jsp;如果不合法,则转至错误页面 errorpage.jsp。该 Servlet 实际上起一个控制器的作用。

实现代码如下:

```java
package com.register;

import java.io.*;
import javax.servlet.*;
import javax.servlet.http.*;
public class LoginServlet extends HttpServlet
{
    //doGet 方法
    public void doGet(HttpServletRequest request, HttpServletResponse response) throws ServletException, IOException
    {
        doPost(request,response);
    }
    //doPost 方法
    public void doPost(HttpServletRequest request, HttpServletResponse response) throws ServletException, IOException
    {
        request.setCharacterEncoding("utf-8");
        String uname = request.getParameter("uname");
```

```
            String upass = request.getParameter("upass");

            HttpSession session = request.getSession();
            DBUser dbuser = new DBUser(this.getServletContext());
            SUser suser = dbuser.getSUser(uname,upass);
            if(suser!= null)
            {
                session.setAttribute("suser",suser);
                response.sendRedirect("showinfo.jsp");
            }
            else
            {
                response.sendRedirect("errorpage.jsp");
            }
        }
    }
```

6．配置 LoginServlet

一个 Servlet 类就是一个组件，必须在 web.xml 中配置后才能使用。对于每一个 servlet，必须创建一个 servlet 声明和一个对应的 servlet 映射。

配置代码如下：

```
<servlet>   <!-- servlet 声明 -->
    <servlet-name>LoginServlet</servlet-name>
    <servlet-class>com.register.LoginServlet</servlet-class>
</servlet>
<servlet-mapping>  <!-- servlet 映射 -->
    <servlet-name>LoginServlet</servlet-name>
    <url-pattern>/loginservlet</url-pattern>
</servlet-mapping>
```

servlet 声明即为给一个完全限定的 Java 类起一个可读的名字，如给完全限定的 Java 类 com.register.LoginServlet 起的名字为 LoginServlet，servlet 映射即为告诉服务器如何把到来的请求路径转发到这个 servlet，如将所有路径为 /loginservlet 的请求映射到 LoginServlet 这个 servlet。

7．编制 3 个页面文件 login.jsp、showinfo.jsp、errorpage.jsp

先将 UserRegister_2 工程中的 3 个页面文件 login.jsp、showinfo.jsp、errorpage.jsp 复制到本工程中，只需修改 login.jsp，在提交表单时，原先请求执行的是 check.jsp 页面文件，现在要执行的是一个名为 loginservlet 的 servlet 控制器。

修改代码如下：

```
…
<form name="loginform" method="post" action="loginservlet">
…
</form>
```

8．发布并测试程序

启动 Tomcat 服务器，将 UserRegister_3 应用发布至服务器；打开浏览器，在地址栏中输入 http://localhost:8080/UserRegister_3/login.jsp，运行并测试登录部分。

9．新建并编制用于注册的 Servlet 类文件 RegisterServlet.java

RegisterServlet 类功能与 UserRegister_1、UserRegister_2 工程中的 doregister.jsp 程序功能一致。在该类的 doPost() 方法中获取 register.jsp 页面中传递过来的用户注册数据，利用 JavaBean 类 DBUser、SUser 的方法连接并操纵数据库，向数据库中增加一个用户。成功则将用户信息存放在会话 session 级变量 suser 中，转至用户信息显示页面 showinfo.jsp；失败则转至错误页面 errorpage。

jsp。该 Servlet 实际上起一个控制器的作用。

实现代码如下：

```java
package com.register;

import java.io.*;
import javax.servlet.*;
import javax.servlet.http.*;
public class RegisterServlet extends HttpServlet
{
    public void doGet(HttpServletRequest request, HttpServletResponse response) throws ServletException, IOException
    {
        doPost(request,response);
    }
    public void doPost(HttpServletRequest request, HttpServletResponse response) throws ServletException, IOException
    {
        request.setCharacterEncoding("utf-8");

        String uname = request.getParameter("uname");
        String upass = request.getParameter("upass");
        String upass2 = request.getParameter("upass2");
        String usex = request.getParameter("usex");
        int uage = Integer.parseInt(request.getParameter("uage"));
        String ulevel = request.getParameter("ulevel");
        String uaddress = request.getParameter("uaddress");
        String umail = request.getParameter("umail");
        String uemail = request.getParameter("uemail");
        String uphone = request.getParameter("uphone");

        HttpSession session = request.getSession();
        DBUser dbuser = new DBUser(this.getServletContext());
        SUser suser = new SUser(uname,upass,usex,uage,ulevel,uaddress,umail,uemail,uphone);
        boolean flag = dbuser.addSUser(suser);
        if(flag)
        {
            session.setAttribute("suser",suser);
            response.sendRedirect("showinfo.jsp");
        }
        else
        {
            response.sendRedirect("errorpage.jsp");
        }
    }
}
```

10. 配置 RegisterServlet

同 LoginServlet，该 servlet 也必须在 web.xml 中配置后才能使用。

配置代码如下：

```xml
<servlet>
    <servlet-name>RegisterServlet</servlet-name>
    <servlet-class>com.register.RegisterServlet</servlet-class>
</servlet>
<servlet-mapping>
    <servlet-name>RegisterServlet</servlet-name>
    <url-pattern>/registerservlet</url-pattern>
```

```
</servlet-mapping>
```

11. 编制 register.jsp 文件

先将 UserRegister_2 工程中的页面文件 register.jsp 复制到本工程中,对其进行修改。在提交表单时,原先请求执行的是 doregister.jsp 页面文件,现在要执行的是一个名为 registerservlet 的 servlet 控制器。

修改代码如下:

```
...
<form name="registerform" method="post" action="registerservlet">
...
</form>
```

12. 发布并测试程序

将 UserRegister_3 应用发布至服务器,打开浏览器,在地址栏中输入 http://localhost:8080/UserRegister_3/login.jsp,运行并测试注册部分。

任务四 用 Struts 模式实现

【任务描述】

Struts 模式继承了 MVC(模型-视图-控制器)设计模式的特性,可以看作是对 JSP+JavaBean+Servlet 模式的进一步封装。用 Struts 模式实现时,程序中的数据显示与输入仍由页面负责,数据存储、处理由 JavaBean 组件和表单 bean 负责,流程控制由自定义动作 Action 类负责。因此,本任务可以在 JSP+JavaBean+Servlet 实现模式的基础上,编制自定义动作类来代替 Servlet 类负责流程控制,编制表单 bean 类获取页面表单中的数据,在配置文件中对表单 bean 类、动作类进行配置,并对相关页面文件进行修改。

【任务分析】

1. 程序运行基本流程

Struts 模式下程序运行基本流程图如图 4-13 所示。

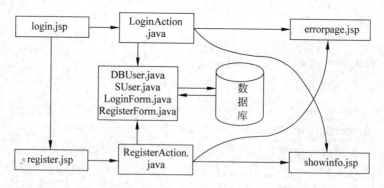

图 4-13 Struts 模式下程序运行基本流程图

2. 程序文件清单

Struts 模式下程序文件清单如表 4-5 所示。

项目四　网上用户注册程序

表 4-5　Struts 模式下程序文件清单

组件类别	文　件　名	功　能　描　述
视图层（View）	login.jsp	用户登录的页面
	register.jsp	用户输入注册信息的页面
	showinfo.jsp	显示用户注册信息的页面
	errorpage.jsp	登录和注册失败时显示错误信息的页面
控制器层（Controller）	LoginAction.java	处理用户登录的动作类
	RegisterAction.java	处理用户注册的动作类
	sturts-config.xml	Sturts 配置文件
	web.xml	应用程序通用配置文件
模型层（Model）	LoginForm.java	用户登录的表单 bean
	RegisterForm.java	用户注册的表单 bean
	DBUser.java	连接及操纵数据库的组件，一个工具类
	SUser.java	代表用户的实体类
过滤器类	SetCharacterEncodingFilter.java	处理字符编码的过滤器类

3．Action 映射表

Action 映射表见表 4-6。

表 4-6　Action 映射表

动作（Action）	入口	表单 bean（ActionForm）	出口
LoginAction	login.jsp	LoginForm	showinfo.jsp errorpage.jsp
RegisterAction	register.jsp	RegisterForm	showinfo.jsp errorpage.jsp

4．项目目录结构

Struts 模式下项目目录结构图如图 4-14 所示。

【任务实施】

1．新建工程

启动 MyEclipse，选择"New->Web Project"，新建一个名称为 UserRegister_4 的工程。

2．使工程支持 Struts

选中 UserRegister_4，右击，在弹出的菜单中选择"MyEclpse…,"选择"Add Structs Capabilities"；修改对话框中配置，如图 4-15 所示；单击"Finish"按钮，使工程具有支持 Struts 的功能。

在以上对话框完成后，会看到在工程的编译路径中引入了 Struts 的一些库文件，在 Web-inf 文件夹中增加了 Struts 的配置文件及标签库文件，并且修改了应用程序的配置文件。

3．将数据库复制到工程中

选中 Web 文件夹，右击，选择"New->Folder"，新建名称为 database 的文件夹，将 UserRegister_1 工程中的数据库文件 UserRegister.mdb 复制到文件夹中。

4．新建包结构

选中 src 文件夹，右击，选择"New->Package"，在 src 文件夹下新建一个名为 com.register 的包结构，JavaBean 类文件及动作类文件均存放在该包中。

图 4-14 Struts 模式下项目目录结构图

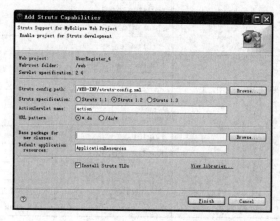

图 4-15 使工程具有支持 Struts 能力对话框

5. 编制 JavaBean 组件

与 UserRegister_3 工程中的 SUser.java、DBUser.java 文件完全一致，可以直接将其复制到本工程中。

6. 编制过滤器类并配置过滤器

为使系统支持汉字的传递，可使用过滤器。

（1）新建并编制过滤器类 SetCharacterEncodingFilter.java

实现代码如下：

```
package com.register;

import java.io.IOException;
import javax.servlet.*;
public class SetCharacterEncodingFilter implements Filter
{
    protected String encoding = null;
    protected FilterConfig filterConfig = null;
    protected boolean ignore = true;

    public void destroy()
    {
        this.encoding = null;
        this.filterConfig = null;
    }
    public void doFilter( ServletRequest request, ServletResponse response, FilterChain chain ) throws
IOException,ServletException
    {
        if (ignore || (request.getCharacterEncoding() == null))
        {
            String encoding = selectEncoding(request);
            if (encoding != null)
            {
                request.setCharacterEncoding(encoding);
            }
        }
        chain.doFilter(request,response);
```

项目四　网上用户注册程序

```java
            }
    public void init(FilterConfig filterConfig) throws ServletException
    {
        this.filterConfig = filterConfig;
        this.encoding = filterConfig.getInitParameter("encoding");
        String value = filterConfig.getInitParameter("ignore");
        if (value == null)
        {
            this.ignore = true;
        }
        else if (value.equalsIgnoreCase("true"))
        {
            this.ignore = true;
        }
        else if (value.equalsIgnoreCase("yes"))
        {
            this.ignore = true;
        }
        else
        {
            this.ignore = false;
        }
    }
    protected String selectEncoding(ServletRequest request)
    {
        return (this.encoding);
    }
}
```

(2) 配置过滤器

在 web.xml 文件中配置过滤器,配置代码放置在 web.xml 文件中的＜servlet＞元素对前。
实现代码如下:

```xml
<filter>
        <description>SetCharacterEncodingFilter</description>
        <display-name>SetCharacterEncodingFilter</display-name>
        <filter-name>SetCharacterEncodingFilter</filter-name>
        <filter-class>
            com.register.SetCharacterEncodingFilter
        </filter-class>
        <init-param>
            <param-name>encoding</param-name>
            <param-value>utf-8</param-value>
        </init-param>
</filter>
<filter-mapping>
        <filter-name>SetCharacterEncodingFilter</filter-name>
        <url-pattern>/*</url-pattern>
</filter-mapping>
```

7. 编制用于登录处理的相关文件

(1) 新建并编制登录表单 bean 类文件 LoginForm.java

在 Struts 中,表单 bean 类均是 ActionForm 的子类。LoginForm 类就是一个表单 bean,其主要作用是在提交表单时,配合 Action 对象获取表单中提交的数据,数据的获取是自动的。该类的编制必须按照 JavaBean 的约定,属性 uname、upass 分别对应登录页面 login.jsp 中的表单元素 uname、upass,类中还必须有与属性配合的获取(getter)和设置(setter)属性的方法。

实现代码如下:

```java
package com.register;
```

```java
import org.apache.struts.action.ActionForm;
public class LoginForm extends ActionForm
{
    private String uname;
    private String upass;

    public String getUname()
    {
        return uname;
    }
    public void setUname(String uname)
    {
        this.uname = uname;
    }

    public String getUpass()
    {
        return upass;
    }
    public void setUpass(String upass)
    {
        this.upass = upass;
    }
}
```

(2) 新建并编制动作类文件 LoginAction.java

这是 Struts 中的自定义动作类,它们均从 Action 类继承。该类的主要作用是处理从 login.jsp 页面提交过来的登录请求,并借助 LoginForm 表单 bean 获取到的数据对用户进行合法性检查,根据检查的结果转向不同的页面。

实现代码如下:

```java
package com.register;
import javax.servlet.http.*;
import org.apache.struts.action.*;
public class LoginAction extends Action //Action 的子类
{
    //重新定义方法 execute()
    public ActionForward execute ( ActionMapping mapping, ActionForm form, HttpServletRequest request, HttpServletResponse response) throws Exception
    {
        LoginForm loginform = (LoginForm) form;
        String uname = loginform.getUname();
        String upass = loginform.getUpass();

        HttpSession session = request.getSession();
        DBUser dbuser = new DBUser(servlet.getServletContext());
        SUser suser = dbuser.getSUser(uname,upass);
        if(suser!= null)
        {
            session.setAttribute("suser",suser);
            return   mapping.findForward("Success");
        }
        else
        {
            return  mapping.findForward("Error");
        }
    }
}
```

项目四　网上用户注册程序

（3）配置表单 bean 及动作 Action 类

表单 bean 及动作 Action 类均是 Struts 的组件，它们都需要在 Sturts 的配置文件 sturts-config.xml 中配置后才能使用。

配置 LoginForm 表单 bean 的代码应写在＜form-beans＞元素对中，代码如下：

```
<form-bean name="loginform"    type="com.register.LoginForm"/>
```

配置 LoginAction 动作类映射关系的代码应写在＜action-mappings＞元素对中，代码如下：

```
<action path="/loginaction"   type="com.register.LoginAction"   name="loginform"    scope="request">
    <forward name="Success"    path="/showinfo.jsp"/>
    <forward name="Error"      path="/errorpage.jsp"/>
</action>
```

（4）编制 3 个页面文件 login.jsp、showinfo.jsp、errorpage.jsp

先将 UserRegister_3 工程中的 3 个页面文件 login.jsp、showinfo.jsp、errorpage.jsp 复制到本工程中，只需修改 login.jsp，在提交表单时，将表单 form 的属性 action 的值由原来对 Servlet 类 LoginServlet 的调用改为对动作类文件 LoginAction 的调用。

修改代码如下：

```
…
<form name="loginform" method="post" action=" loginaction.do ">
…
</form>
```

8．发布并测试程序

启动 Tomcat 服务器，将 UserRegister_4 应用发布至服务器；打开浏览器，在地址栏中输入 http://localhost:8080/UserRegister_4/login.jsp，运行并测试登录部分。

9．编制用于注册处理的相关文件

（1）新建并编制注册表单 bean 类文件 RegisterForm.java

类似于 LoginForm 类，RegisterForm 中一组属性是与注册页面 regitser.jsp 中的表单元素 uname、upass 相对应的，类中还必须有与属性配合的获取（getter）和设置（setter）属性的方法。

实现代码如下：

```java
package com.register;

import org.apache.struts.action.ActionForm;
public class RegisterForm extends ActionForm
{
    private String uname;
    private String upass;
    private String usex;
    private int uage;
    private String ulevel;
    private String uaddress;
    private String umail;
    private String uemail;
    private String uphone;

    public String getUname()
    {
        return uname;
    }
    public void setUname(String uname)
    {
        this.uname = uname;
```

        }
    ...
}
```

(2) 新建并编制动作类文件 RegisterAction.java

该类的主要作用是处理从 register.jsp 页面提交过来的注册请求,并借助 RegisterForm 表单 bean 获取到的数据向数据库增加一个用户,根据执行的结果转向不同的页面。

实现代码如下:

```java
package com.register;

import javax.servlet.http.*;
import org.apache.struts.action.*;
public class RegisterAction extends Action
{
    public ActionForward execute(ActionMapping mapping, ActionForm form, HttpServletRequest request, HttpServletResponse response) throws Exception
    {
        RegisterForm userform = (RegisterForm)form;

        String uname = userform.getUname();
        String upass = userform.getUpass();
        String usex = userform.getUsex();
        int uage = userform.getUage();
        String ulevel = userform.getUlevel();
        String uaddress = userform.getUaddress();
        String umail = userform.getUmail();
        String uemail = userform.getUemail();
        String uphone = userform.getUphone();

        HttpSession session = request.getSession();
        DBUser dbuser = new DBUser(servlet.getServletContext());
        SUser suser = new SUser(uname,upass,usex,uage,ulevel,uaddress,umail,uemail,uphone);
        boolean flag = dbuser.addSUser(suser);
        if(flag)
        {
            session.setAttribute("suser",suser);
            return (mapping.findForward("Success"));
        }
        else
        {
            return mapping.findForward("Error");
        }
    }
}
```

(3) 配置表单 bean 及动作 Action 类

配置 RegisterForm 表单 bean 的代码应写在 struts-config.xml 文件的＜form-beans＞元素对中。

配置代码如下:

```xml
<form-bean name="registerform" type="com.register.RegisterForm" />
```

配置 RegisterAction 动作类映射关系的代码应写在 struts-config.xml 文件的＜form-beans＞元素对中。配置代码如下:

```xml
<action path="/registeraction" type="com.register.RegisterAction" name="registerform" scope="request">
```
```

```
        <forward name = "Success"      path = "/showinfo.jsp"/>
        <forward name = "Error"        path = "/errorpage.jsp"/>
</action>
```

（4）编制 register.jsp 文件

先将 UserRegister_3 工程中的 register.jsp 复制到本工程中，在提交表单时，将表单 form 的属性 action 的值由原来对 Servlet 类 RegisterServlet 的调用改为对动作类文件 RegisterAction 的调用。

修改代码如下：

```
...
<form name = "registerform" method = "post" action = "registeraction.do ">
...
</form>
```

10．发布并测试程序

将 UserRegister_4 应用发布至服务器，打开浏览器，在地址栏中输入 http://localhost:8080/UserRegister_4/login.jsp，运行并测试注册部分。

项目小结

本项目通过对于一个网上用户注册程序的实践，详细介绍了如何利用四种开发模式来开发这一项目程序的基本方法及步骤。这是一个基础项目，主要是为了让读者了解和掌握利用 JSP 技术开发 Web 应用程序的一些基础性知识和技能。

项目五 学生基本信息维护系统

项目描述

学生基本信息维护系统是一个 Java Web 应用系统,主要实现对学生基本信息的浏览及维护。

项目预览

在浏览器地址栏中输入 http://localhost:8080/Student_2,程序运行时,出现如图 5-1 所示的 index.jsp 首页面。

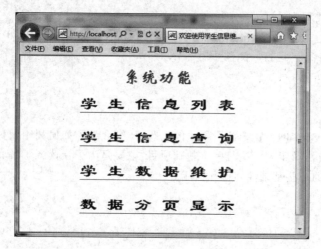

图 5-1 系统运行首页面

分别单击各项超链接后,可以实现学生基本信息的浏览、查询及增、删、改等维护操作,还可以进行分页浏览。

项目分析

面目四网上用户注册程序主要实现对已注册用户的查询(登录)及新用户的增加(注册),是一个比较简单的 Web 应用程序;而在本项目中,可以实现对批量学生的基本信息进行浏览、按姓名查询以及学生基本信息的增、删、改等维护操作,还可以进行分页浏览,是一个功能相对齐全的 Web 应用程序。我们将用 3 种开发模式来具实现这一项目程序,将更多地涉及用 JSP 技术及框架技术开发 Web 应用程序的许多基础知识及技巧。

项目设计

1. 页面总体流程的设计

页面总体流程图如图 5-2 所示。

2. 数据库的设计

数据库的设计比较简单,只要在其中设计一张学生基本信息表即可。数据库名称为 StudentDB,表名为 studenttb,表结构见表 5-1。

项目五 学生基本信息维护系统

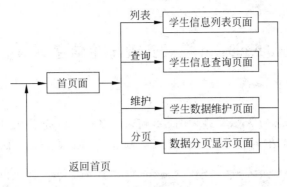

图 5-2 页面总体流程图

表 5-1 学生基本信息表 studenttb 的表结构

列名	数据类型	长度	允许为空	主键	含义
ssno	文本	6	否	是	学号
ssname	文本	10	否		姓名
sssex	文本	2	是		性别
ssage	数字		是		年龄
ssmajor	文本	50	是		所学专业
ssfav	文本	50	是		爱好
ssaddress	文本	50	是		家庭住址

环境需求

操作系统：Windows XP 或 Windows 7；

数据库平台：Microsoft Access；

JDK 版本：JDK Version1.6；

集成开发环境：MyEclipse 8.5；

Web 服务器：Tomcat 6.0；

浏览器：IE6.0 及以上版本。

项目知识点

① JSP 的页面结构，常用指令、动作的使用方法；

② JSP 常用内置对象的使用方法；

③ JavaScript 脚本语言的基本使用方法；

④ JSP＋JavaBean 模式、JSP＋JavaBean＋Servlet 模式、Struts 模式开发程序的基本方法及步骤；

⑤ JavaBean 的概念及编制方法；

⑥ Servelt 的概念、编制方法及配置方法；

⑦ Struts 中表单 Bean、动作 Action 类的概念、编制方法及配置方法；

⑧ JSP 中使用 JDBC 连接、操纵数据库的基本方法；

⑨ 页面之间传递数据的方法；

⑩ 页面数据的分页显示方法。

任务一 用 JSP＋JavaBean 模式实现学生信息列表

【任务描述】

本任务用 JSP＋JavaBean 模式实现如图 5-3 所示的学生信息列表，因此要编制连接及操纵数据库的 JavaBean 组件、代表学生的实体类及用于控制流程、显示数据的页面文件。

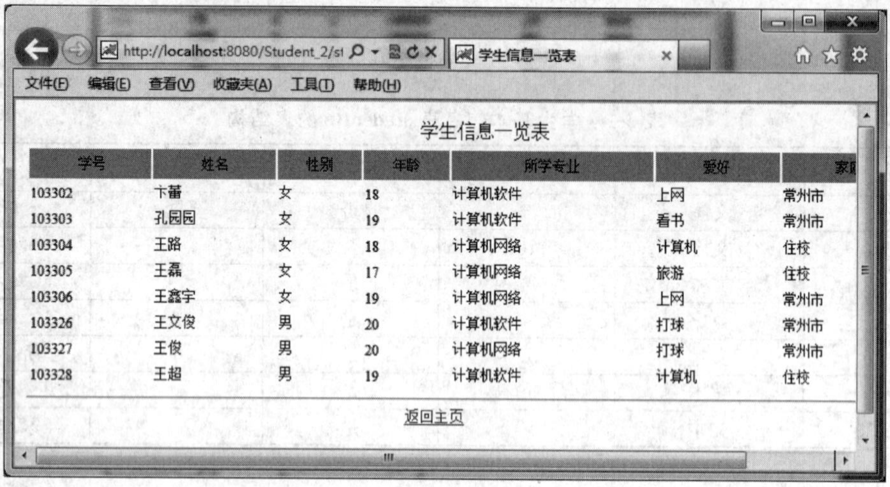

图 5-3 学生信息列表页面

【任务分析】

1. 程序运行基本流程

JSP＋JavaBean 模式下程序运行基本流程图如图 5-4 所示。

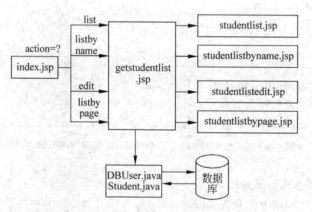

图 5-4 JSP＋JavaBean 模式下程序运行基本流程图

2. 全部程序文件清单

JSP＋JavaBean 模式下全部程序文件清单见表 5-2。

项目五 学生基本信息维护系统

表 5-2　JSP+JavaBean 模式下全部程序文件清单

组件类别	文件名	功能描述
JSP 页面	index.jsp	系统首页面
	getstudentlist.jsp	获取所有学生数据,并根据参数转至相应页面
	studentlist.jsp	显示所有学生基本信息的页面
	studentlistbyname.jsp	按姓名进行查询并显示查询结果的页面
	studentlistedit.jsp	对所有学生进行增、删、改操作的页面
	studentlistbypage.jsp	分页显示所有学生基本信息的页面
	getstudentlistbyname.jsp	获取姓名并进行查询的页面
	studentnew.jsp	新增时输入学生信息的页面
	studentmodify.jsp	修改时编辑学生信息的页面
	addstudent.jsp	获取一个学生信息并插入数据库的页面
	modifystudent.jsp	获取一个学生信息并更新至数据库的页面
	deletestudent.jsp	删除一个学生记录的页面
	error.jsp	操作失败时显示出错信息的页面
JavaBean(模型 Model)	DBUser.java	连接及操纵数据库的组件,一个工具类
	Student.java	代表学生的实体类

3．项目目录结构

JSP+JavaBean 模式下项目目录结构图如图 5-5 所示。

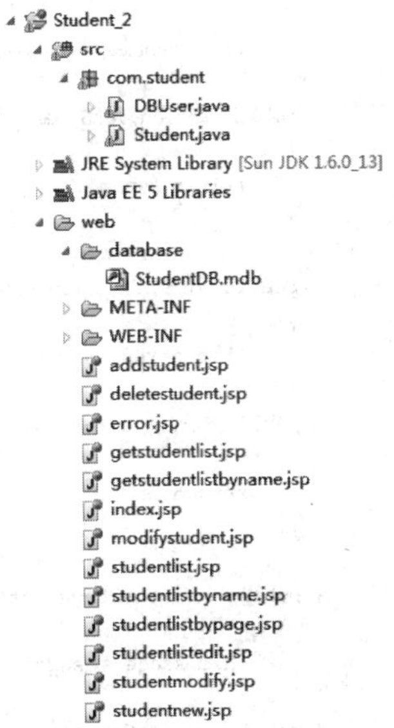

【任务实施】

1．准备数据库

在 Microsoft Access 中创建一个数据库,名称为 StudentDB.mdb,创建一张名称为 studenttb 的表,表结构见表 5-1,并输入几条记录。

2．新建工程

启动 MyEclipse,选择"New->Web Project",新建一个名称为 Student_2 的工程。

3．将数据库复制到工程中

选中 Web 文件夹,右击,选择"New->Folder",新建名称为 database 的文件夹,将步骤 1 中创建的数据库文件 StudentDB.mdb 复制到文件夹中。

4．新建包结构

选中 src 文件夹,右击,选择"New->Package",在 src 文件夹下新建一个名为 com.student 的包结构,JavaBean 类文件均存放在该包中。

图 5-5　JSP+JavaBean 模式下项目目录结构图

5．新建并编制 JavaBean 组件

（1）新建并编制 Student.java 文件

选中 com.student 包,右击,选择"New->File",新建 Student.java 类文件,这是一个实体 JavaBean 组件,一个该类的对象代表一个具体的学生,包含该学生的基本信息。

实现代码如下:

```
package com.student;
```

```java
import java.io.Serializable;
public class Student implements Serializable
{
    //私有学生属性
    private String ssno;
    private String ssname;
    private String sssex;
    private int ssage;
    private String ssmajor;
    private String ssfav;
    private String ssaddress;

    public Student()
    {

    }
    //一组 get 和 set 方法
    public void setSsno(String ssno)
    {
        this.ssno = ssno;
    }
    public String getSsno()
    {
        return ssno;
    }

    public void setSsname(String ssname)
    {
        this.ssname = ssname;
    }
    public String getSsname()
    {
        return ssname;
    }

    public void setSssex(String sssex)
    {
        this.sssex = sssex;
    }
    public String getSssex()
    {
        return sssex;
    }

    public void setSsage(int ssage)
    {
        this.ssage = ssage;
    }
    public int getSsage()
    {
        return ssage;
    }

    public void setSsmajor(String ssmajor)
    {
        this.ssmajor = ssmajor;
    }
    public String getSsmajor()
    {
        return ssmajor;
    }

    public void setSsfav(String ssfav)
```

项目五 学生基本信息维护系统

```java
        {
            this.ssfav = ssfav;
        }
        public String getSsfav()
        {
            return ssfav;
        }

        public void setSsaddress(String ssaddress)
        {
            this.ssaddress = ssaddress;
        }
        public String getSsaddress()
        {
            return ssaddress;
        }

}
```

（2）新建并编制 DBUser.java 文件

这是一个工具 JavaBean 组件，它封装了对数据库的所有操作，包括在构造方法中加载驱动程序、获取所有学生列表及增、删、改、查询操作等方法。在其中先定义一个构造方法，其他方法在后续编制具体模块时再行定义。

实现代码如下：

```java
package com.student;

import javax.servlet.ServletContext;
import java.sql.*;
import java.util.ArrayList;
import java.util.List;
public class DBUser
{
    private ServletContext application;
    //定义构造方法，加载驱动程序，接收全局应用对象
    public DBUser(ServletContext application)
    {
        try
        {
            Class.forName("sun.jdbc.odbc.JdbcOdbcDriver");
        }
        catch(ClassNotFoundException e)
        {
            e.printStackTrace();
        }
        catch(Exception e)
        {
            e.printStackTrace();
        }
        this.application = application;
    }
```

6. 新建并编制 index.jsp 文件

如图 5-1 所示，页面采用两个表格嵌套布局，表格在页面中上下、左右居中。"学生信息列表"、"学生信息查询"、"学生数据维护"、"数据分页显示"为超链接，分别通过传递不同的参数链接至 getstudentlist.jsp 页面。

实现代码如下：

```jsp
<%@ page contentType="text/html;charset=utf-8" %>
```

```html
<html>
  <head>
    <title>欢迎使用学生信息维护系统</title>
  </head>
  <body>
    <table width="100%" height="100%">
      <tr>
        <td align="center" valign="middle">
          <table>
            <tr>
              <td align="center"><font face="华文新魏" color="#ff0000" size="6"><b>系统功能</b></font></td>
            </tr>
            <tr height="60">
              <td align="center"><font face="隶书" size="6"><b><a href="getstudentlist.jsp?action=list">学 生 信 息 列 表</a></b></font></td>
            </tr>
            <tr height="60">
              <td align="center"><font face="隶书" size="6"><b><a href="getstudentlist.jsp?action=listbyname">学 生 信 息 查 询</a></b></font></td>
            </tr>
            <tr height="60">
              <td align="center"><font face="隶书" size="6"><b><a href="getstudentlist.jsp?action=edit">学 生 数 据 维 护</a></b></font></td>
            </tr>
            <tr height="60">
              <td align="center"><font face="隶书" size="6"><b><a href="getstudentlist.jsp?action=listbypage">数 据 分 页 显 示</a></b></font></td>
            </tr>
          </table>
        </td>
      </tr>
    </table>
  </body>
</html>
```

7. 发布并测试程序

启动 Tomcat 服务器,将 Student_2 应用发布至服务器;打开浏览器,在地址栏中输入 http://localhost:8080/Student_2,运行并测试系统首页面。

8. 学生信息列表功能的实现

(1) 修改 DBUser 类

在其中定义一个方法,方法为 getStudent()。该方法的作用是取出所有学生数据,每个学生的数据存放在 Student 类型的一个对象中,所有学生数据存放在一个 List 类型的对象中,并返回。

实现代码如下:

```java
public List getStudent()
{
    List sStudent = new ArrayList();
    Connection con = null;
    Student student = null;
    try
    {
        String url = "jdbc:odbc:driver={Microsoft Access Driver (*.mdb)};DBQ=" + application.getRealPath("") + "/database/StudentDB.mdb";
        con = DriverManager.getConnection(url,"","");

        Statement stmt = con.createStatement();
        ResultSet result = stmt.executeQuery("select * from studenttb order by ssno");
```

项目五 学生基本信息维护系统

```java
            while(result.next())
            {
                student = new Student();
                student.setSsno(result.getString("ssno"));
                student.setSsname(result.getString("ssname"));
                student.setSssex(result.getString("sssex"));
                student.setSsage(result.getInt("ssage"));
                student.setSsmajor(result.getString("ssmajor"));
                student.setSsfav(result.getString("ssfav"));
                student.setSsaddress(result.getString("ssaddress"));

                sStudent.add(student);
            }
        }
        catch(SQLException e)
        {
            e.printStackTrace();
        }
        catch(Exception e)
        {
            e.printStackTrace();
        }
        finally
        {
            if(con!= null)
            {
                try
                {
                    con.close();
                }
                catch(Exception e)
                {
                    e.printStackTrace();
                }
            }
        }
        return sStudent;
    }
```

（2）新建并编制 getstudentlist.jsp 文件

在页面中首先获取从 index.jsp 页面中传递过来的参数 action 的值,调用 DBUser 类中的 getStudent()方法,取出所有学生数据,并存放至会话 session 级变量 StudentList 中,根据参数 action 的值引导至 4 个不同的页面。它实际上起一个控制器的作用,并不直接形成显示页面输出。

实现代码如下：

```jsp
<%@page contentType="text/html;charset=utf-8" %>
<%@ page import="com.student.*,java.util.List" %>
<%
    request.setCharacterEncoding("utf-8");

    String action = request.getParameter("action");
    DBUser dbuser = new DBUser(this.getServletContext());
    List sStudent = dbuser.getStudent();
    session.setAttribute("StudentList",sStudent);
    if(action.equals("list"))                          //学生信息列表
    {
        response.sendRedirect("studentlist.jsp");
    }
    else if(action.equals("listbyname"))               //学生信息查询
    {
        response.sendRedirect("studentlistbyname.jsp");
```

```jsp
        }
        else if(action.equals("edit"))                    //学生信息维护
        {
            response.sendRedirect("studentlistedit.jsp");
        }
        else if(action.equals("listbypage"))              //学生数据分页
        {
            response.sendRedirect("studentlistbypage.jsp");
        }
%>
```

（3）新建并编制 studentlist.jsp 文件

如图 5-3 所示，该页面从会话级变量 StudentList 中取出所有学生数据，以图示的形式显示出来。

实现代码如下：

```jsp
<%@page contentType="text/html;charset=utf-8" %>
<%@page import="com.student.*,java.util.List" %>
<%
    List sStudent = (List)session.getAttribute("StudentList");
%>
<html>
    <head>
        <title>学生信息一览表</title>
    </head>
    <body>
        <table width="800"  align="center">
            <caption>学生信息一览表</caption>
            <tr height="25" style="font-size:12px" bgcolor="#22cccc" align="center">
                <td>学号</td>
                <td>姓名</td>
                <td>性别</td>
                <td>年龄</td>
                <td>所学专业</td>
                <td>爱好</td>
                <td>家庭住址</td>
            </tr>
            <%
                if(sStudent != null)
                {
                    for(int i = 0;i < sStudent.size();i++)
                    {
                        Student student = (Student)sStudent.get(i);
            %>
            <tr height="20" style="font-size:12px"  align="left">
                <td><%= student.getSsno() %></td>
                <td><%= student.getSsname() %></td>
                <td><%= student.getSssex() %></td>
                <td><%= student.getSsage() %></td>
                <td><%= student.getSsmajor() %></td>
                <td><%= student.getSsfav() %></td>
                <td><%= student.getSsaddress() %></td>
            </tr>
            <%
                    }
                }
            %>
        </table>
        <hr  width="800">
        <center><font size="2"><a href="index.jsp">返回主页</a></font></center>
    </body>
</html>
```

项目五 学生基本信息维护系统

由代码可以看出,在页面的开头处,将会话session级变量StudentList中的所有学生数据取出,存至一个List类型的变量中;在数据行处,对List类型的变量中的数据进行循环,逐个取出学生数据并显示出来。

(4) 发布并测试程序

将Student_2应用发布至服务器,打开浏览器,在地址栏中输入http://localhost:8080/Student_2,运行并测试学生信息列表部分。

任务二 用JSP+JavaBean模式实现学生信息查询

【任务描述】

本任务用JSP+JavaBean模式实现如图5-6所示的学生信息查询,因此要编制JavaBean组件中查询数据的方法及用于控制流程、查询并显示数据的页面文件。

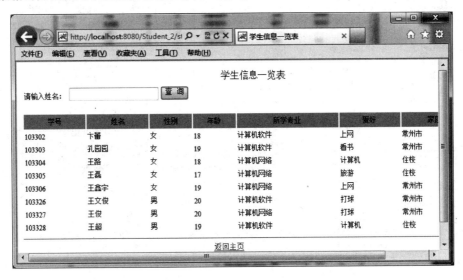

图5-6 按姓名查询学生页面

【任务分析】

JSP+JavaBean模式下按姓名查询学生信息流程图如图5-7所示。

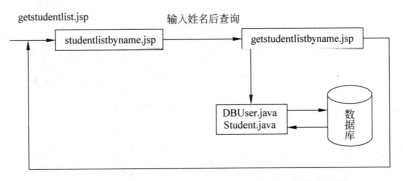

图5-7 JSP+JavaBean模式下按姓名查询学生信息流程图

· 127 ·

【任务实施】

1. 新建并编制 studentlistbyname.jsp 文件

如图 5-6 所示,该页面从会话 session 级变量 StudentList 中取出查询到的满足条件的学生数据,以图示的形式显示出来。编制要求基本同 studentlist.jsp 页面,可先将 studentlist.jsp 文件另存为 studentlistbyname.jsp 文件,再在表格的标题行的上面加一行一列,在其中放一个表单,表单中的文本域用于输入姓名。

实现代码如下:

```jsp
<%@page contentType="text/html;charset=utf-8" %>
<%@page import="com.student.*,java.util.List" %>
<%
    List sStudent = (List)session.getAttribute("StudentList");
%>
<html>
    <head>
        <title>学生信息一览表</title>
    </head>
    <body>
        <table width="800" align="center">
            <caption>学生信息一览表</caption>
            <tr height="15" style="font-size:12px" align="left">
                <td colspan="7">
                    <form name="searchform" method="post" action="getstudentlistbyname.jsp">
                        请输入姓名:
                        <input type="text" name="ssname" value=""/>
                        <input type="submit" name="btsearch" value="查 询"/>
                    </form>
                </td>
            </tr>
            <tr height="25" style="font-size:12px" bgcolor="#22cccc" align="center">
                <td>学号</td>
                <td>姓名</td>
                <td>性别</td>
                <td>年龄</td>
                <td>所学专业</td>
                <td>爱好</td>
                <td>家庭住址</td>
            </tr>
            <%
                if(sStudent != null)
                {
                    for(int i = 0;i < sStudent.size();i++)
                    {
                        Student student = (Student)sStudent.get(i);
            %>
            <tr height="20" style="font-size:12px" align="left">
                <td><%= student.getSsno() %></td>
                <td><%= student.getSsname() %></td>
                <td><%= student.getSssex() %></td>
                <td><%= student.getSsage() %></td>
                <td><%= student.getSsmajor() %></td>
                <td><%= student.getSsfav() %></td>
                <td><%= student.getSsaddress() %></td>
            </tr>
            <%
                }
```

项目五 学生基本信息维护系统

```
            }
        %>
        </table>
        <hr width = "800">
        <center><font size = '2'><a href = "index.jsp">返回主页</a></font></center>
    </body>
</html>
```

2. 修改 DBUser 类

在其中定义一个方法,方法为 getStudentByName(String ssname);该方法的作用是根据传递过来的姓名参数取出满足条件的学生数据,并返回。

实现代码如下:

```java
public List  getStudentByName(String ssname)
{
        List   sStudent = new ArrayList();
        Connection con = null;
        Student student = null;
        try
        {
            String url = "jdbc:odbc:driver = {Microsoft Access Driver ( * .mdb)};DBQ = " + application.getRealPath("") + "/database/StudentDB.mdb";
            con = DriverManager.getConnection(url,"","");
            Statement stmt = con.createStatement();
            String sql;
            if(ssname == null || ssname.equals(""))         //未输入,取全部学生
            {
                sql = "select * from studenttb order by ssno";
            }
            else                                            //有效输入,查询对应学生
            {
                sql = "select * from studenttb where ssname like '%" + ssname + "%' order by ssno";
            }
            ResultSet  result = stmt.executeQuery(sql);
            while(result.next())
            {
                student = new Student();
                student.setSsno(result.getString("ssno"));
                student.setSsname(result.getString("ssname"));
                student.setSssex(result.getString("sssex"));
                student.setSsage(result.getInt("ssage"));
                student.setSsmajor(result.getString("ssmajor"));
                student.setSsfav(result.getString("ssfav"));
                student.setSsaddress(result.getString("ssaddress"));

                sStudent.add(student);
            }
        }
        catch(SQLException e)
        {
            e.printStackTrace();
        }
        catch(Exception e)
        {
            e.printStackTrace();
        }
        finally
        {
            if(con!= null)
            {
                try
                {
```

```
                    con.close();
                }
                catch(Exception e)
                {
                    e.printStackTrace();
                }
            }
        }
        return sStudent;
    }
```

3．新建并编制 getstudentlistbyname.jsp 文件

在页面中首先获取从 studentlistbyname.jsp 页面中传递过来的姓名，调用 DBUser 类中的 getStudentByName()方法，取出满足条件的学生数据，并存放至会话 session 级变量 StudentList 中，转至 studentlistbyname.jsp 页面。它本身并不直接形成显示页面输出。

实现代码如下：

```jsp
<%@ page contentType = "text/html;charset = utf-8" %>
<%@ page import = "com.student.*,java.util.List" %>
<%
    request.setCharacterEncoding("utf-8");
    String ssname = request.getParameter("ssname");

    DBUser dbuser = new DBUser(application);
    List sStudent = dbuser.getStudentByName(ssname);
    session.setAttribute("StudentList",sStudent);
    response.sendRedirect("studentlistbyname.jsp");
%>
```

4．发布并测试程序

将 Student_2 应用发布至服务器，打开浏览器，在地址栏中输入 http://localhost:8080/Student_2，运行并测试学生信息查询部分。

任务三　用 JSP＋JavaBean 模式实现学生数据维护

【任务描述】

本任务用 JSP＋JavaBean 模式实现如图 5-8 至图 5-11 所示的学生数据维护，因此要编制 JavaBean 组件中增、删、改数据的方法及用于控制流程、编辑数据并显示数据的页面文件。

图 5-8　学生基本信息维护页面

项目五 学生基本信息维护系统

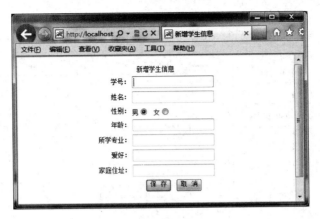

图 5-9 新增学生信息页面

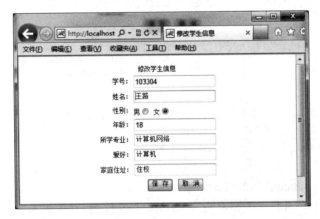

图 5-10 修改学生信息页面

图 5-11 出错信息显示页面

【任务分析】

JSP+JavaBean 模式下学生基本信息维护流程图如图 5-12 所示。

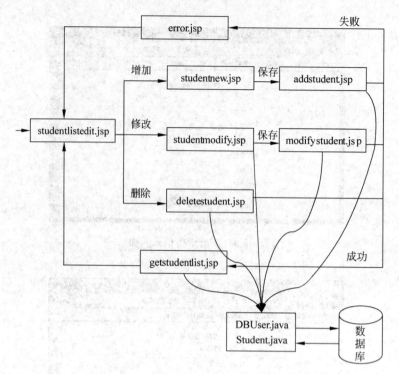

图 5-12　JSP＋JavaBean 模式下学生基本信息维护流程图

【任务实施】

1. 新建并编制 studentlistedit.jsp 文件

如图 5-8 所示，该编制的基本要求同 studentlist.jsp 页面，可先将 studentlist.jsp 文件另存为 studentlistedit.jsp 文件，将表格的列数改为 5 列，后面 3 个列分别改成"增加"、"修改"、"删除"；列中数据分别为链接至相应页面的超链接，"增加"超链接直接链接至新增页面 studentnew.jsp，后两个链接暂时为空。

实现代码如下：

```jsp
<%@page contentType="text/html;charset=utf-8" %>
<%@page import="com.student.*,java.util.List" %>
<%
    List sStudent = (List)session.getAttribute("StudentList");
%>
<html>
    <head>
        <title>学生信息一览表</title>
    </head>
    <body>
        <table width="600" align="center">
            <caption>学生信息一览表</caption>
            <tr height="25" style="font-size:12px" bgcolor="#22cccc" align="center">
                <td>学号</td>
                <td>姓名</td>
                <td>增加记录</td>
                <td>修改记录</td>
                <td>删除记录</td>
            </tr>
<%
```

```jsp
                        if(sStudent != null)
                        {
                            for(int i = 0;i<sStudent.size();i++)
                            {
                                Student student = (Student)sStudent.get(i);
%>
                        <tr height="20" style="font-size:12px" align="left">
                            <td><%=student.getSsno()%></td>
                            <td><%=student.getSsname()%></td>
                            <td><a href="studentnew.jsp">增加</a></td>
                            <td><a href="">修改</a></td>
                            <td><a href="">删除</a></td>
                        </tr>
<%
                            }
                        }
%>
        </table>
        <hr width="600">
        <center><font size='2'><a href="index.jsp">返回主页</a></font></center>
    </body>
</html>
```

2．实现增加记录功能

(1) 新建并编制 studentnew.jsp 文件

如图 5-9 所示,该页面用来输入一个学生数据,单击"保存"按钮后将数据提交至 addstudent.jsp 页面保存至数据库中;在提交前,应对学生的学号、姓名进行合法性验证。

实现代码如下:

```jsp
<%@page contentType="text/html;charset=utf-8" %>
<%@page import="java.sql.*" %>
<html>
    <head>
        <title>新增学生信息</title>
        <script type="text/javascript">
            function checkdata()
            {
                var ssno = document.studentform.ssno.value;
                if(ssno=="" || ssno.length!=6 || !isNumberic(ssno))
                {
                    alert("学号不能为空,必须输入,并且只能是6位数字!");
                    document.studentform.ssno.focus();
                    return false;
                }

                var ssname = document.studentform.ssname.value;
                if(ssname=="" )
                {
                    alert("学生姓名不能为空,必须输入!");
                    document.studentform.ssname.focus();
                    return false;
                }

                return true;
            }

            function isNumberic(str)
            {
                var len = str.length;
                for(var i=0;i<len;i++)
                    if(str.charAt(i)<'0' || str.charAt(i)>'9')
```

```
                    return false;
                return true;
            }
        </script>
    </head>
    <body onload="document.studentform.ssno.focus()">
        <form name="studentform" method="post" action="addstudent.jsp">
            <table align="center" style="font-size:12px">
                <tr>
                    <td colspan="2" align="center">新增学生信息</td>
                </tr>
                <tr>
                    <td align="right">学号：</td>
                    <td><input type="text" name="ssno"></td>
                </tr>
                <tr>
                    <td align="right">姓名：</td>
                    <td><input type="text" name="ssname"></td>
                </tr>
                <tr>
                    <td align="right">性别：</td>
                    <td>男<input type="radio" name="sssex" value="男" checked>   女<input type="radio" name="sssex" value="女"></td>
                </tr>
                <tr>
                    <td align="right">年龄：</td>
                    <td><input type="text" name="ssage"></td>
                </tr>
                <tr>
                    <td align="right">所学专业：</td>
                    <td><input type="text" name="ssmajor"></td>
                </tr>
                <tr>
                    <td align="right">爱好：</td>
                    <td><input type="text" name="ssfav"></td>
                </tr>
                <tr>
                    <td align="right">家庭住址：</td>
                    <td><input type="text" name="ssaddress"></td>
                </tr>
                <tr>
                    <td> </td>
                    <td align="center"><input type="submit" name="btok" value="保存" onclick="return checkdata()" />
                          <input type="button" name="btcancel" value="取消" onclick="javascript:history.back()"/>
                    </td>
                </tr>
            </table>
        </form>
    </body>
</html>
```

(2) 修改 DBUser 类

在其中定义一个方法，方法为 addStudent(Student student)，该方法的作用是将传递过来的学生对象参数保存至数据库中。

实现代码如下：

```
public boolean addStudent(Student student)
{
    Connection con = null;
    boolean flag = false;
```

```java
        try
        {
            String url = " jdbc: odbc: driver = {Microsoft Access Driver ( * . mdb)}; DBQ = " + application.
getRealPath("") + "/database/StudentDB.mdb";
            con = DriverManager.getConnection(url,"","");

            Statement stmt = con.createStatement();

            String sql = "insert into studenttb values('" + student.getSsno() + "','"
            + student.getSsname() + "','"
            + student.getSssex() + "',"
            + student.getSsage() + ",'"
            + student.getSsmajor() + "','"
            + student.getSsfav() + "','"
            + student.getSsaddress()
            + "')";

            int row = stmt.executeUpdate(sql);
            if(row == 1)
                flag = true;
        }
        catch(SQLException e)
        {
            e.printStackTrace();
        }
        catch(Exception e)
        {
            e.printStackTrace();
        }
        finally
        {
            if(con!= null)
            {
                try
                {
                    con.close();
                }
                catch(Exception e)
                {
                    e.printStackTrace();
                }
            }
        }
        return flag;
    }
```

(3) 新建并编制 addstudent.jsp 文件

在页面中首先使用 JSP 的动作获取 studentnew.jsp 页面中传递过来的学生数据，调用 DBUser 类中的方法 addStudent()，将数据保存至数据库中；成功则通过 getstudentlist.jsp 页面重取数据后转至 studentlistedit.jsp 页面，失败则转至 error.jsp 页面。它本身并不直接形成显示页面输出。

实现代码如下：

```jsp
<%@page contentType = "text/html;charset = utf - 8" %>
<%@page import = "java.sql.*,com.student.*" %>

<%   request.setCharacterEncoding("utf - 8");   %>
<jsp:useBean id = "student"   class = "com.student.Student" scope = "page"   />
<jsp:setProperty name = "student"   property = " * " />
<html>
    <head>
```

```
            <title>新增学生信息</title>
        </head>
        <body>
            <%
                DBUser dbuser = new DBUser(application);
                boolean flag = dbuser.addStudent(student);
                if(flag)
                {
                    response.sendRedirect("getstudentlist.jsp?action=edit");
                }
                else
                {
                    response.sendRedirect("error.jsp");
                }
            %>
        </body>
</html>
```

(4) 新建并编制 error.jsp 文件

如图 5-11 所示，该页面用来显示新增、修改、删改操作失败时的出错信息。

实现代码如下：

```
<%@page contentType="text/html;charset=utf-8" %>
<html>
    <head>
        <title>结果页面</title>
    </head>
    <body>
        <center>
            <h3>新增、修改、删改操作失败!</h3><br>
            <a href="studentlistedit.jsp">返回</a>
        </center>
    </body>
</html>
```

(5) 发布并测试程序

将 Student_2 应用发布至服务器，打开浏览器，在地址栏中输入 http://localhost:8080/Student_2，运行并测试学生记录增加部分。

3. 实现修改记录功能

(1) 编制 studentlistedit.jsp 页面文件中"修改"超链接

将超链接改为：

```
<a href="studentmodify.jsp?ssno=<%=student.getSsno()%>">修改</a>
```

可以看出，通过超链接传递了待修改学生记录的学生学号。

(2) 修改 DBUser 类

在其中定义一个方法，方法为 getStudentBySsno(String ssno)，该方法的作用是根据传递过来的学生学号从数据库取出学生数据。

实现代码如下：

```
public Student getStudentBySsno(String ssno)
{
    Connection con = null;
    Student student = null;
    try
    {
        String url = "jdbc:odbc:driver={Microsoft Access Driver (*.mdb)};DBQ=" + application.getRealPath("") + "/database/StudentDB.mdb";
        con = DriverManager.getConnection(url,"","");
```

项目五　学生基本信息维护系统

```java
        Statement stmt = con.createStatement();
        ResultSet result = stmt.executeQuery("select * from studenttb where ssno = '" + ssno + "'");
        while(result.next())
        {
            student = new Student();

            student.setSsno(result.getString("ssno"));
            student.setSsname(result.getString("ssname"));
            student.setSssex(result.getString("sssex"));
            student.setSsage(result.getInt("ssage"));
            student.setSsmajor(result.getString("ssmajor"));
            student.setSsfav(result.getString("ssfav"));
            student.setSsaddress(result.getString("ssaddress"));
        }
    }
    catch(SQLException e)
    {
        e.printStackTrace();
    }
    catch(Exception e)
    {
        e.printStackTrace();
    }
    finally
    {
        if(con!= null)
        {
            try
            {
                con.close();
            }
            catch(Exception e)
            {
                e.printStackTrace();
            }
        }
    }
    return student;
}
```

(3) 新建并编制 studentmodify.jsp 文件

如图 5-10 所示,该页面用来修改学生数据。在页面中先获取传递过来学生学号,调用 DBUser 类中的方法 getStudentBySsno()取出学生数据,将取出的学生数据设置至页面中供修改,单击"保存"按钮后将数据提交至 modifystudent.jsp 页面更新至数据库中。在提交前,应对学生的姓名进行合法性验证,学号不能修改。具体编制时可在新增页面的基础上进行修改。

实现代码如下:

```jsp
<%@page contentType = "text/html;charset = utf - 8" %>
<%@page import = "java.sql.*,com.student.*" %>
<%
    request.setCharacterEncoding("utf - 8");
        String ssno = request.getParameter("ssno");
    DBUser dbuser = new DBUser(application);
    Student student = dbuser.getStudentBySsno(ssno);
%>
<html>
    <head>
        <title>修改学生信息</title>
        <script type = "text/javascript">
            function checkdata()
            {
```

```jsp
            var ssname = document.studentform.ssname.value;
            if(ssname == "" )
            {
                alert("学生姓名不能为空,必须输入!");
                document.studentform.ssname.focus();
                return false;
            }

            return true;
        }
    </script>
</head>
<body onload = "document.studentform.ssname.focus()" >
    <form name = "studentform" method = "post" action = "modifystudent.jsp">
        <table align = "center" style = "font-size:12px">
            <tr>
                <td colspan = "2" align = "center">修改学生信息</td>
            </tr>
            <tr>
                <td align = "right">学号:</td>
                <td><input type = "text" name = "ssno" value = "<% = student.getSsno() %>" readonly = "true"/></td>
            </tr>
            <tr>
                <td align = "right">姓名:</td>
                <td><input type = "text" name = "ssname" value = "<% = student.getSsname() %>" /></td>
            </tr>
            <tr>
                <td align = "right">性别:</td>
                <td>
                    男<input type = "radio" name = "sssex" value = "男"   <% = (student.getSssex().equals("男")?"checked":"") %> />
                    女<input type = "radio" name = "sssex" value = "女"   <% = (student.getSssex().equals("女")?"checked":"") %> />
                </td>
            </tr>
            <tr>
                <td align = "right">年龄:</td>
                <td><input type = "text" name = "ssage" value = "<% = student.getSsage() %>" /></td>
            </tr>
            <tr>
                <td align = "right">所学专业:</td>
                <td><input type = "text" name = "ssmajor" value = "<% = student.getSsmajor() %>" /></td>
            </tr>
            <tr>
                <td align = "right">爱好:</td>
                <td><input type = "text" name = "ssfav" value = "<% = student.getSsfav() %>" /></td>
            </tr>
            <tr>
                <td align = "right">家庭住址:</td>
                <td><input type = "text" name = "ssaddress" value = "<% = student.getSsaddress() %>" /></td>
            </tr>
            <tr>
                <td> </td>
                <td  align = "center"><input type = "submit" name = "btok" value = "保 存"  onclick = "return checkdata()" />
                      <input type = "button" name = "btcancel" value = "取 消" onclick = "javascript:history.back()"/>
                </td>
            </tr>
```

```
            </table>
        </form>
    </body>
</html>
```

(4) 修改 DBUser 类

在其中定义一个方法,方法为 modifyStudent(Student student),该方法的作用是将传递过来的学生对象参数更新至数据库中。

实现代码如下:

```java
public boolean modifyStudent(Student student)
{
        Connection con = null;
        boolean flag = false;

        try
        {
                String url = "jdbc:odbc:driver={Microsoft Access Driver (*.mdb)};DBQ=" + application.getRealPath("") + "/database/StudentDB.mdb";
                con = DriverManager.getConnection(url,"","");
                Statement stmt = con.createStatement();
                String sql = "update studenttb  set ssname = '" + student.getSsname()
                    + "',sssex = '" + student.getSssex()
                    + "',ssage = " + student.getSsage()
                    + ",ssmajor = '" + student.getSsmajor()
                    + "',ssfav = '" + student.getSsfav()
                    + "',ssaddress = '" + student.getSsaddress()
                    + "' where ssno = '" + student.getSsno() + "'";

                int row = stmt.executeUpdate(sql);
                if(row == 1)
                    flag = true;
        }
        catch(SQLException e)
        {
            e.printStackTrace();
        }
        catch(Exception e)
        {
            e.printStackTrace();
        }
        finally
        {
            if(con!= null)
            {
                try
                {
                    con.close();
                }
                catch(Exception e)
                {
                    e.printStackTrace();
                }
            }
        }
        return flag;
}
```

(5) 新建并编制 modifystudent.jsp 文件

在页面中首先使用 JSP 的动作获取 studentmodify.jsp 页面中传递过来的学生数据,调用 DBUser 类中的方法 modifyStudent(),将数据更新至数据库中;成功则通过 getstudentlist.jsp 页面重取数据后转至 studentlistedit.jsp 页面,失败则转至 error.jsp 页面。它本身并不直接形成显示页面输出。

实现代码如下:

```jsp
<%@page contentType="text/html;charset=utf-8" %>
<%@page import="java.sql.*,com.student.*" %>

<%  request.setCharacterEncoding("utf-8"); %>
<jsp:useBean id="student"   class="com.student.Student" scope="page"  />
<jsp:setProperty name="student"   property="*" />

<html>
    <head>
        <title>修改学生信息</title>
    </head>
    <body>
        <%
            DBUser dbuser = new DBUser(application);
            boolean flag = dbuser.modifyStudent(student);
            if(flag)
            {
                response.sendRedirect("getstudentlist.jsp?action=edit");
            }
            else
            {
                response.sendRedirect("error.jsp");
            }
        %>
    </body>
</html>
```

(6)发布并测试程序

将 Student_2 应用发布至服务器,打开浏览器,在地址栏中输入 http://localhost:8080/Student_2,运行并测试学生记录修改部分。

4. 实现删除记录功能

(1)编制 studentlistedit.jsp 页面文件中"删除"超链接

在删除记录数据前,一般应先进行删除确认,可以将"删除"超链接引导至一个 javascript 函数中进行删除前的确认。先在页面的<head>标签对中定义一个 javascript 函数,实现代码如下:

```html
<script type="text/javascript">
    function Delete(ssno)
    {
        if(confirm("你是否确定要删除当前记录数据?")==true)
        location.href="deletestudent.jsp?ssno="+ssno;
    }
</script>
```

再将超链接改为:

```html
<a href="javascript:Delete(<%=student.getSsno()%>)">删除</a>
```

(2)修改 DBUser 类

在其中定义一个方法,方法名为 deleteStudent(String ssno),该方法的作用是根据传递过来的学号将该学生从数据库中删除。

实现代码如下:

```java
public boolean deleteStudent(String ssno)
{
    Connection con = null;
```

```
        boolean flag = false;
        try
        {
            String url = "jdbc:odbc:driver={Microsoft Access Driver (*.mdb)};DBQ=" + application.
getRealPath("") + "/database/StudentDB.mdb";
            con = DriverManager.getConnection(url,"","");

            Statement stmt = con.createStatement();
            String sql = "delete from studenttb where ssno ='" + ssno + "'";
            int row = stmt.executeUpdate(sql);
            if(row == 1)
                flag = true;
        }
        catch(SQLException e)
        {
            e.printStackTrace();
        }
        catch(Exception e)
        {
            e.printStackTrace();
        }
        finally
        {
            if(con!= null)
            {
                try
                {
                    con.close();
                }
                catch(Exception e)
                {
                    e.printStackTrace();
                }
            }
        }
        return flag;
    }
```

（3）新建并编制 deletestudent.jsp 文件

在页面中首先获取传递过来学生学号，调用 DBUser 类中的方法 deleteStudent()，将该学生从数据库中删除；成功则通过 getstudentlist.jsp 页面重取数据后转至 studentlistedit.jsp 页面，失败则转至 error.jsp 页面。它本身并不直接形成显示页面输出。

实现代码如下：

```
<%@page contentType="text/html;charset=utf-8" %>
<%@page import="java.sql.*,com.student.*" %>
<html>
    <head>
        <title>学生信息一览表</title>
    </head>
    <body>
        <%
            request.setCharacterEncoding("utf-8");
            String ssno = request.getParameter("ssno");

            DBUser dbuser = new DBUser(application);
            boolean flag = dbuser.deleteStudent(ssno);
            if(flag)
            {
                response.sendRedirect("getstudentlist.jsp?action=edit");
            }
            else
            {
```

```
            response.sendRedirect("error.jsp");
        }
    %>
    </body>
</html>
```

（4）发布并测试程序

将 Student_2 应用发布至服务器，打开浏览器，在地址栏中输入 http://localhost:8080/Student_2，运行并测试学生记录删除部分。

任务四 用 JSP+JavaBean 模式实现学生数据分页显示

【任务描述】

本任务用 JSP+JavaBean 模式实现如图 5-13 所示的学生数据分页显示。

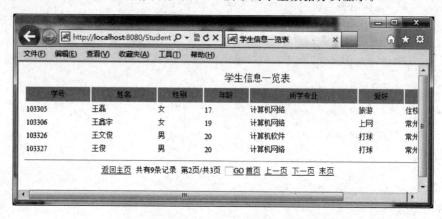

图 5-13 学生基本信息分页显示页面

【任务分析】

数据的分页显示在网页展现数据时非常重要。要实现分页显示，主要是要明确页面数据的来源、每页中数据的数量及如何实现跳转。在本任务中，页面数据存放在 List 类型的对象中，对象中的每一个元素是一个 Student 类型的对象，代表一个学生，具体的分页处理和跳转均是在页面中完成的。

【任务实施】

1. 新建并编制 studentlistbypage.jsp 文件

如图 5-13 所示，该页面从会话级变量 StudentList 中取出所有学生数据，按照数据分页处理的要求，以图示的形式显示出来。

实现代码如下：

```
<%@page contentType="text/html;charset=utf-8" %>
<%@page import="java.sql.*,com.student.*,java.util.List" %>
<%
    List sStudent = (List)session.getAttribute("StudentList");
%>
<html>
    <head>
```

```jsp
            <title>学生信息一览表</title>
            <script>
                function Goto()
                {
location.href = "studentlistbypage.jsp?currentPage = " + document.getElementById("currentPage").value;
                }
            </script>
    </head>
    <body>
        <table width = "800" align = "center">
            <caption>学生信息一览表</caption>
            <tr height = "25" style = "font-size:12px" bgcolor = "#22cccc" align = "center">
                <td>学号</td>
                <td>姓名</td>
                <td>性别</td>
                <td>年龄</td>
                <td>所学专业</td>
                <td>爱好</td>
                <td>家庭住址</td>
            </tr>
            <%
             int currentPage = 1;                            //当前默认页为第一页

                String strCurrentPage = request.getParameter("currentPage");
                if(strCurrentPage == null)
                    strCurrentPage = "1";
                try
                {
                    currentPage = Integer.parseInt(strCurrentPage);
                }
                catch(Exception e)
                {
                    currentPage = 1;
                }

                int recordCount = 0;                         //记录总数
                int pageSize = 4;                            //每页记录数
                int pageCount = 0;                           //总页数
                if(sStudent != null )
                {
                    recordCount = sStudent.size();
                    //获取总页数
                    if( recordCount % pageSize == 0)
                        pageCount = recordCount / pageSize;
                    else
                        pageCount = recordCount/pageSize + 1;

                    int start;
                    //把记录指针移指当前页第一条记录之前
                    if( (currentPage - 1) * pageSize == 0 )
                        start = 0;
                    else
                        start = (currentPage - 1) * pageSize;

                    int n = 0;
                    for(int i = start;i < sStudent.size();i++)
                    {
                        Student student = (Student)sStudent.get(i);

            %>
                        <tr height = "20" style = "font-size:12px"  align = "left">
                            <td><% = student.getSsno() %></td>
                            <td><% = student.getSsname() %></td>
                            <td><% = student.getSssex() %></td>
                            <td><% = student.getSsage() %></td>
                            <td><% = student.getSsmajor() %></td>
```

```
                <td><% = student.getSsfav()%></td>
                <td><% = student.getSsaddress()%></td>
            </tr>
        <%
            n++;
            if(n >= pageSize)          //当前页显示完后退出循环
                break;
        }
    }
    %>
    </table>
    <hr width = "800">
    <%
    out.print("<center><font size = '2'>");
        out.print("<a href = 'index.jsp'>返回主页</a>   ");
        out.print("共有" + recordCount + "条记录   第" + currentPage + "页/共" + pageCount + "页   ");
        out.print("<input type = 'text' name = 'currentPage' style = 'width:15px;height:15px'/><a href = 'javascript:Goto()'>GO</a>   ");

        if(pageCount == 1)
            ;
        else
        {
            if( currentPage == 1)
            {
                out.print("<a href = 'studentlistbypage.jsp?currentPage = " + (currentPage + 1) + "'>下一页</a>   ");
                out.print("<a href = 'studentlistbypage.jsp?currentPage = " + pageCount + "'>末页</a>");
            }
            else if( currentPage == pageCount)
            {
                out.print("<a href = 'studentlistbypage.jsp?currentPage = 1'>首页</a>   ");
                out.print("<a href = 'studentlistbypage.jsp?currentPage = " + (currentPage - 1) + "'>上一页</a>   ");
            }
            else
            {
                out.print("<a href = 'studentlistbypage.jsp?currentPage = 1'>首页</a>   ");
                out.print("<a href = 'studentlistbypage.jsp?currentPage = " + (currentPage - 1) + "'>上一页</a>   ");
                out.print("<a href = 'studentlistbypage.jsp?currentPage = " + (currentPage + 1) + "'>下一页</a>   ");
                out.print("<a href = 'studentlistbypage.jsp?currentPage = " + pageCount + "'>末页</a>");
            }
        }
        out.print("</center></font>");
    %>
    </body>
</html>
```

2. 发布并测试程序

将 Student_2 应用发布至服务器，打开浏览器，在地址栏中输入 http://localhost:8080/Student_2，运行并测试数据分页显示部分。

任务五　用 JSP＋JavaBean＋Servlet 模式实现学生信息列表

【任务描述】

本任务用 JSP＋JavaBean＋Servlet 模式实现如图 5-3 所示的学生信息列表，可以在 JSP＋JavaBean 模式的基础上，编制 Servlet 类来负责流程控制，并对页面文件进行相应修改。

项目五 学生基本信息维护系统

【任务分析】

1．程序运行基本流程

JSP＋JavaBean＋Servlet 模式下程序运行基本流程图如图 5-14 所示。

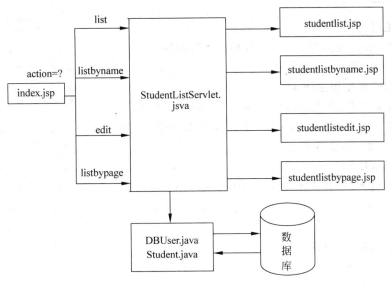

图 5-14　JSP＋JavaBean＋Servlet 模式下程序运行基本流程图

2．全部程序文件清单

JSP＋JavaBean＋Servlet 模式下全部程序文件清单见表 5-3。

表 5-3　JSP＋JavaBean＋Servlet 模式下全部程序文件清单

组件类别	文 件 名	功 能 描 述
视图层 （View）	index.jsp	系统首页面
	studentlist.jsp	显示所有学生基本信息的页面
	studentlistbyname.jsp	按姓名进行查询并显示查询结果的页面
	studentlistedit.jsp	对所有学生进行增、删、改操作的页面
	studentlistbypage.jsp	分页显示所有学生基本信息的页面
	studentnew.jsp	新增时输入学生信息的页面
	studentmodify.jsp	修改时编辑学生信息的页面
	error.jsp	操作失败时显示出错信息的页面
控制器层 （Controller）	StudentListServlet.java	获取所有学生数据，并根据参数转至相应页面的 Servlet 类
	StudentListByNameServlet.java	获取姓名并进行查询的 Servlet 类
	StudentEditServlet.java	处理数据编辑请求的 Servlet 类
	StudentSaveServlet.java	处理数据保存请求的 Servlet 类
	StudentDeleteServlet.java	处理数据删除请求的 Servlet 类
	PageServlet.java	分页处理的 Servlet 类
	web.xml	应用程序的通用配置文件
模型层 （Model）	DBUser.java	连接及操纵数据库的组件，一个工具类
	Student.java	代表学生的实体类
常量类	Constants.java	存放分页处理一些数据的常量类

3. 项目目录结构

JSP+JavaBean+Servlet 模式下项目目录结构图如图 5-15 所示。

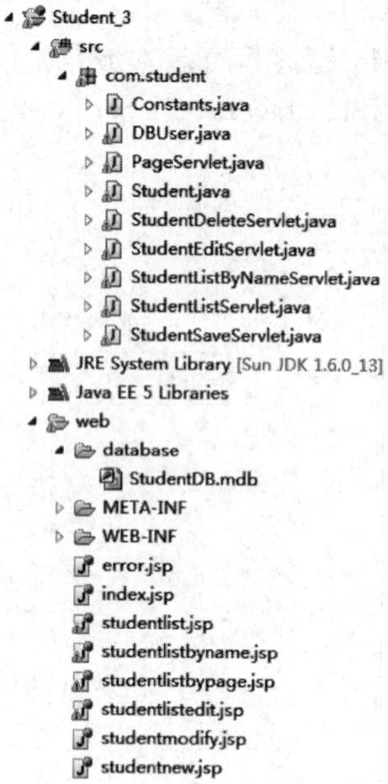

图 5-15 JSP+JavaBean+Servlet 模式下项目目录结构图

【任务实施】

1. 新建工程

启动 MyEclipse，选择“New->Web Project”，新建一个名称为 Student_3 的工程。

2. 将数据库复制到工程中

选中 Web 文件夹，右击，选择“New->Folder”，新建名称为 database 的文件夹，将 Student_2 工程中的数据库文件 StudentDB. mdb 复制到文件夹中。

3. 新建包结构

选中 src 文件夹，右击，选择“New->Package”，在 src 文件夹下新建一个名为 com. student 的包结构，JavaBean 类文件及 Servlet 类文件均存放在该包中。

4. 编制 JavaBean 组件

DBUser. java 与 Student_2 工程中的 DBUser. java 文件完全一致，可以直接将其复制到本工程中。

Student. Java 与 Student_2 工程中的 Student. Java 略有差别，可以先将其直接复制到本工程中，再在类中增加一个如下代码所示的构造方法：

```
public Student (String ssno, String ssname, String sssex, int ssage, String ssmajor, String ssfav, String
ssaddress)
{
        this. ssno = ssno;
        this. ssname = ssname;
        this. sssex = sssex;
        this. ssage = ssage;
        this. ssmajor = ssmajor;
        this. ssfav = ssfav;
        this. ssaddress = ssaddress;
}
```

5. 编制 index. jsp 文件

先将 Student_2 工程中的 index. jsp 文件复制过来，对其进行修改。将 4 个超链接中原来对 getstudentlist. jsp 页面的请求改成对 Servlet 类 StudentListServlet 的请求，传递的参数不变。

主要实现代码如下：

```
…
<tr height = "60">
            <td align = "center"  ><font  face = "隶书"  size = "6"><b><a href = "studentlistservlet?
action = list">学  生  信  息  列  表</a></b></font></td>
</tr>
<tr height = "60">
            <td align = "center"><font  face = "隶书"  size = "6"><b><a href = "studentlistservlet?
action = listbyname">学  生  信  息  查  询</a></b></font></td>
</tr>
<tr height = "60">
            <td align = "center"><font  face = "隶书"  size = "6"><b><a href = "studentlistservlet?
action = edit">学  生  数  据  维  护</a></b></font></td>
```

```
</tr>
<tr height = "60">
        <td align = "center"><font face = "隶书" size = "6"><b><a href = "studentlistservlet?action = listbypage">数 据 分 页 显 示</a></b></font></td>
</tr>
…
```

6．发布并测试程序

启动 Tomcat 服务器，将 Student_3 应用发布至服务器；打开浏览器，在地址栏中输入 http://localhost:8080/Student_3，运行并测试系统首页面。

7．学生信息列表功能的实现

（1）新建并编制 Servlet 类文件 StudentListServlet.java

StudentListServlet 类的功能与原先 getstudentlist.jsp 页面功能一致。在该类中，首先获取从 index.jsp 页面中传递过来的参数 action 的值，调用 DBUser 类中的 getStudent() 方法，取出所有学生数据，并存放至会话 session 级变量 StudentList 中，根据参数 action 的值引导至 4 个不同的页面。

实现代码如下：

```java
package com.student;

import java.util.List;
import java.io.*;
import javax.servlet.*;
import javax.servlet.http.*;
public class StudentListServlet extends HttpServlet
{
    public void doGet(HttpServletRequest request, HttpServletResponse response) throws ServletException, IOException
    {
        doPost(request,response);
    }
    public void doPost(HttpServletRequest request, HttpServletResponse response) throws ServletException, IOException
    {
        request.setCharacterEncoding("utf-8");
        String action = request.getParameter("action");

        DBUser dbuser = new DBUser(this.getServletContext());
        List sStudent = dbuser.getStudent();
        HttpSession session = request.getSession();
        session.setAttribute("StudentList",sStudent);
        if(action.equals("list"))
        {
            response.sendRedirect("studentlist.jsp");
        }
        else if(action.equals("listbyname"))
        {
            response.sendRedirect("studentlistbyname.jsp");
        }
        else if(action.equals("edit"))
        {
            response.sendRedirect("studentlistedit.jsp");
        }
        else if(action.equals("listbypage"))
        {
            session.setAttribute(Constants.Search_AllList,sStudent);
            response.sendRedirect("pageservlet?pageId = 1&forwardPage = studentlistbypage.jsp");
        }
    }
}
```

（2）配置 StudentListServlet

一个 Servlet 类就是一个组件，必须在 web.xml 中配置后才能使用。对于每一个 servlet 必须创建一个 servlet 声明和一个对应的 servlet 映射。

配置代码如下：

```
<servlet>
    <servlet-name>StudentListServlet</servlet-name>
    <servlet-class>com.student.StudentListServlet</servlet-class>
</servlet>
<servlet-mapping>
    <servlet-name>StudentListServlet</servlet-name>
    <url-pattern>/studentlistservlet</url-pattern>
</servlet-mapping>
```

（3）编制 studentlist.jsp 文件

studentlist.jsp 文件与 Student_2 中 studentlist.jsp 文件完全一致，直接将 Student_2 工程中的 studentlist.jsp 文件复制到本工程中。

（4）发布并测试程序

将 Student_3 应用发布至服务器，打开浏览器，在地址栏中输入 http://localhost:8080/Student_3，运行并测试学生信息列表部分。

任务六　用 JSP＋JavaBean＋Servlet 模式实现学生信息查询

【任务描述】

本任务用 JSP＋JavaBean＋Servlet 模式实现如图 5-6 所示的学生信息查询，可以在 JSP＋JavaBean 模式的基础上，编制 Servlet 类来负责流程控制，并对页面文件进行相应修改。

【任务分析】

JSP＋JavaBean＋Servlet 模式下按姓名查询流程图如图 5-16 所示。

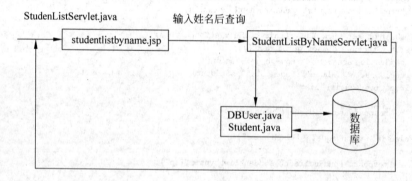

图 5-16　JSP＋JavaBean＋Servlet 模式下按姓名查询流程图

【任务实施】

1. 新建并编制 Servlet 类文件 StudentListByNameServlet.java

StudentListByNameServlet 类的功能与原先 getstudentlistbyname.jsp 页面功能一致。在该类中，首先获取从 studentlistbyname.jsp 页面中传递过来的姓名，调用 DBUser 类中的 getStudentByName()

方法,取出满足条件的学生数据,并存放至会话 session 级变量 StudentList 中,转至 studentlistbyname.jsp 页面。

实现代码如下:

```java
package com.student;

import java.util.List;
import java.io.*;
import javax.servlet.*;
import javax.servlet.http.*;
public class StudentListByNameServlet extends HttpServlet
{
    public void doGet(HttpServletRequest request, HttpServletResponse response) throws ServletException, IOException
    {
        doPost(request,response);
    }
    public void doPost(HttpServletRequest request, HttpServletResponse response) throws ServletException, IOException
    {
        request.setCharacterEncoding("utf-8");
        String ssname = request.getParameter("ssname");

        HttpSession session = request.getSession();
        DBUser dbuser = new DBUser(this.getServletContext());
        List sStudent = dbuser.getStudentByName(ssname);
        session.setAttribute("StudentList",sStudent);
        response.sendRedirect("studentlistbyname.jsp");
    }
}
```

2. 配置 StudentListByNameServlet

同 StudentListServlet 一样,该 servlet 也必须在 web.xml 中配置后才能使用。

配置代码如下:

```xml
<servlet>
        <servlet-name>StudentListByNameServlet</servlet-name>
<servlet-class>com.student.StudentListByNameServlet</servlet-class>
</servlet>
<servlet-mapping>
        <servlet-name>StudentListByNameServlet</servlet-name>
        <url-pattern>/studentlistbynameservlet</url-pattern>
</servlet-mapping>
```

3. 编制 getstudentlistbyname.jsp 文件

先将 Student_2 工程中的页面文件 studentlistbyname.jsp 复制到本工程中,对其进行修改。在提交表单时,原先请求执行的是 getstudentlistbyname.jsp 页面文件,现在要执行的是一个名为 studentlistbynameservlet 的 servlet 控制器。

修改代码如下:

```
...
<form name="searchform" method="post" action="studentlistbynameservlet">
...
</form>
```

4. 发布并测试程序

将 Student_3 应用发布至服务器,打开浏览器,在地址栏中输入 http://localhost:8080/Student_3,运行并测试学生信息查询部分。

任务七　用 JSP+JavaBean+Servlet 模式实现学生数据维护

【任务描述】

本任务用 JSP+JavaBean+Servlet 模式实现图 5-8 至图 5-11 所示的学生数据维护，可以在 JSP+JavaBean 模式的基础上，编制 Servlet 类来负责增、删、改数据的流程控制，并对页面文件作相应修改。

【任务分析】

JSP+JavaBean+Servlet 模式下学生基本信息维护流程图如图 5-17 所示。

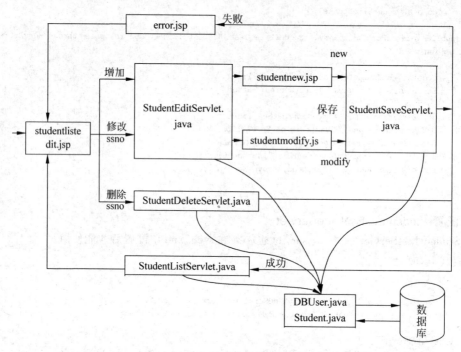

图 5-17　JSP+JavaBean+Servlet 模式下学生基本信息维护流程图

【任务实施】

1. 编制 studentlistedit.jsp 文件

先将 Student_2 工程中的 studentlistedit.jsp 文件复制过来，对其进行修改。将页面中"增加"、"修改"超链接由原来对页面的调用改为对相应的 Servlet 类的调用。

将"增加"超链接改为：

<a href="studenteditservlet">增加</a>

将"修改"超链接改为：

<a href="studenteditservlet?ssno=<%=student.getSsno()%>">修改</a>

其中，studenteditservlet 为用于编辑的 Servlet 类的逻辑名。

项目五　学生基本信息维护系统

2．实现增加、修改记录功能

（1）新建并编制 Servlet 类文件 StudentEditServlet.java

StudenEditServlet.Java 类的主要作用是处理数据编辑请求。在该类中，首先获取从 studentlistedit.jsp 页面中传递过来的学号 ssno，判断学号是否为空。如果为空，说明是新增，直接转至新增页面 studentnew.jsp；否则，说明是修改，调用 DBUser 类中的方法 getStudentBySsno()取出学生数据，并存放至会话 session 级变量中，转至修改页面 studentmodify.jsp。

实现代码如下：

```java
package com.student;

import java.io.*;
import javax.servlet.*;
import javax.servlet.http.*;
public class StudentEditServlet extends HttpServlet
{
    //doGet
    public void doGet(HttpServletRequest request, HttpServletResponse response) throws ServletException, IOException
    {
        doPost(request,response);
    }
    //doPost
    public void doPost(HttpServletRequest request, HttpServletResponse response) throws ServletException, IOException
    {
        request.setCharacterEncoding("utf-8");
        String ssno = request.getParameter("ssno");

        HttpSession session = request.getSession();
        if(ssno == null || ssno.equals("") )              //新增
            response.sendRedirect("studentnew.jsp");
        else                                               //修改
        {
            DBUser dbuser = new DBUser(this.getServletContext());
            Student student = dbuser.getStudentBySsno(ssno);
            session.setAttribute("student",student);
            response.sendRedirect("studentmodify.jsp");
        }
    }
}
```

（2）编制 studentnew.jsp 文件

先将 Student_2 工程中的页面文件 studentnew.jsp 复制到本工程中，对其进行修改。在提交表单时，将表单 form 的属性 action 的值由原来对页面 addstudent.jsp 的调用改为对 Servlet 类 studentsaveservlet 的调用，并将参数 action 的值 new 传递过去。

修改代码如下：

```
…
<form name="studentform" method="post" action="studentsaveservlet?action=new">
…
</form>
```

（3）编制 studentmodify.jsp 文件

先将 Student_2 工程中的页面文件 studentmodify.jsp 复制到本工程中，对其进行修改。

一是将页面开始处的 Java 代码段改为：

```
<%
```

```
    Student student = (Student)session.getAttribute("student");
%>
```

这是因为在 StudentEditServlet 类中已经将待修改的学生数据取出并存放在会话 session 级变量 student 中了，这里只要直接取出即可。

二是在提交表单时，将表单 form 的属性 action 的值由原来对页面 modifystudent.jsp 的调用改为对 Servlet 类 studentsaveservlet 的调用，并将参数 action 的值 modify 传递过去。

修改代码如下：

```
...
<form name = "studentform" method = "post" action = "studentsaveservlet?action = modify">
...
</form>
```

(4) 新建并编制 Servlet 类文件 StudentSaveServlet.java

StudentSaveServlet 类的主要作用是处理数据保存请求。在该类中，先获取从 studentnew.jsp 或 studentmodify.jsp 页面中传递过来参数 action 的值及表单提交的所有数据，根据 action 的值判断是新增还是修改。如果是新增，调用 DBUser 类中的方法 addStudent()，将数据保存至数据库中；如果是修改，调用 DBUser 类中的方法 modifyStudent()，将数据更新至数据库中；保存成功则调用 Servlet 类 StudentListServlet 重取数据后转至 studentlistedit.jsp 页面，失败则转至 error.jsp 页面。

实现代码如下：

```java
package com.student;

import java.io.*;
import javax.servlet.*;
import javax.servlet.http.*;
public class StudentSaveServlet extends HttpServlet
{
    //doGet
    public void doGet(HttpServletRequest request, HttpServletResponse response) throws ServletException, IOException
    {
        doPost(request, response);
    }
    //doPost
    public void doPost(HttpServletRequest request, HttpServletResponse response) throws ServletException, IOException
    {
        request.setCharacterEncoding("utf-8");
        String action = request.getParameter("action");

        String ssno = request.getParameter("ssno");
        String ssname = request.getParameter("ssname");
        String sssex = request.getParameter("sssex");
        String ssage = request.getParameter("ssage");
        String ssmajor = request.getParameter("ssmajor");
        String ssfav = request.getParameter("ssfav");
        String ssaddress = request.getParameter("ssaddress");

        Student student = new Student(ssno,ssname,sssex,Integer.parseInt(ssage),ssmajor,ssfav,ssaddress);
        HttpSession session = request.getSession();
        DBUser dbuser = new DBUser(this.getServletContext());
        if(action.equals("new"))                     //新增
        {
            boolean flag = dbuser.addStudent(student);
            if(flag)
                response.sendRedirect("studentlistservlet?action = edit");
```

```
        else
            response.sendRedirect("error.jsp");
    }
    else if( action.equals("modify") )                    //修改
    {
        boolean flag = dbuser.modifyStudent(student);
        if(flag)
            response.sendRedirect("studentlistservlet?action = edit");
        else
            response.sendRedirect("error.jsp");
    }
}
```

(5) 配置 StudentEditServlet 及 StudentSaveServlet

同 StudentListServlet,这两个 servlet 也必须在 web.xml 中配置后才能使用。

配置代码如下:

```
<servlet>
    <servlet-name>StudentEditServlet</servlet-name>
    <servlet-class>com.student.StudentEditServlet</servlet-class>
</servlet>
<servlet-mapping>
    <servlet-name>StudentEditServlet</servlet-name>
    <url-pattern>/studenteditservlet</url-pattern>
</servlet-mapping>

<servlet>
    <servlet-name>StudentSaveServlet</servlet-name>
    <servlet-class>com.student.StudentSaveServlet</servlet-class>
</servlet>
<servlet-mapping>
    <servlet-name>StudentSaveServlet</servlet-name>
    <url-pattern>/studentsaveservlet</url-pattern>
</servlet-mapping>
```

(6) 编制 error.jsp 文件

error.jsp 文件与 Student_2 中 error.jsp 文件完全一致,直接将 Student_2 工程中的 error.jsp 文件复制到本工程中。

(7) 发布并测试程序

将 Student_3 应用发布至服务器,打开浏览器,在地址栏中输入 http://localhost:8080/Student_3,运行并测试学生记录增加及修改部分。

3. 实现删除记录功能

(1) 编制 studentlistedit.jsp 文件

将页面中"删除"超链接由原来对页面的调用改为对相应的 Servlet 类的调用。

"删除"超链接只需要将 Delete()函数中 location 锚的 href 属性的值改为:

```
location.href = "studentdeleteservlet?ssno = " + ssno
```

其中,studentdeleteservlet 为用于删除操作的 Servlet 类的逻辑名。

(2) 新建并编制 Servlet 类文件 StudentDeleteServlet.java

StudentDeleteServlet.java 类的主要作用是处理记录删除请求。它的功能与原先 deletestudent.jsp 页面功能一致,首先获取传递过来学生学号,调用 DBUser 类中的方法 deleteStudent(),将该学生从数据库中删除,成功则调用 Servlet 类 StudentListServlet 重取数据后转至 studentlistedit.jsp 页面,失败则转至 error.jsp 页面。

实现代码如下:

```java
package com.student;

import java.io.*;
import javax.servlet.*;
import javax.servlet.http.*;
public class StudentDeleteServlet extends HttpServlet
{
    public void doGet(HttpServletRequest request,HttpServletResponse response) throws ServletException,IOException
    {
        doPost(request,response);
    }
    public void doPost(HttpServletRequest request,HttpServletResponse response) throws ServletException,IOException
    {
        request.setCharacterEncoding("utf-8");
        String ssno = request.getParameter("ssno");

        HttpSession session = request.getSession();
        DBUser dbuser = new DBUser(this.getServletContext());
        boolean flag = dbuser.deleteStudent(ssno);
        if(flag)
            response.sendRedirect("studentlistservlet?action=edit");
        else
            response.sendRedirect("error.jsp");
    }
}
```

（3）配置 StudentDeleteServlet

同 StudentListServlet，该 servlet 也必须在 web.xml 中配置后才能使用。

配置代码如下：

```xml
<servlet>
    <servlet-name>StudentDeleteServlet</servlet-name>
    <servlet-class>com.student.StudentDeleteServlet</servlet-class>
</servlet>
<servlet-mapping>
    <servlet-name>StudentDeleteServlet</servlet-name>
    <url-pattern>/studentdeleteservlet</url-pattern>
</servlet-mapping>
```

（4）发布并测试程序

将 Student_3 应用发布至服务器，打开浏览器，在地址栏中输入 http://localhost:8080/Student_3，运行并测试学生记录删除部分。

任务八　用 Struts 模式实现学生信息列表

【任务描述】

本任务用 Struts 模式实现图 5-3 所示的学生信息列表。Struts 模式继承了 MVC（模型-视图-控制器）设计模式的特性，可以看作是对 JSP+JavaBean+Servlet 模式的进一步封装。用 Struts 模式实现时，程序中数据显示与输入仍由页面负责，数据存储和处理 JavaBean 组件和表单 bean 负责，流程控制由自定义动作类负责。因此可以在 JSP+JavaBean+Servlet 模式实现学生信息列表的基础上，编制自定义动作类来代替 Servlet 类负责流程控制，在配置文件中对动作类进行配置，并对页面文件作相应修改。

项目五 学生基本信息维护系统

【任务分析】

1. 程序运行基本流程

Struts 模式下程序运行基本流程图如图 5-18 所示。

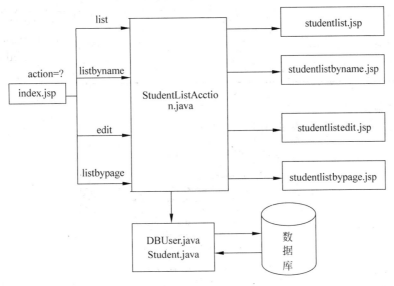

图 5-18 Struts 模式下程序运行基本流程图

2. 全部程序文件清单

Struts 模式下全部程序文件清单见表 5-4。

表 5-4 Struts 模式下全部程序文件清单

组件类别	文件名	功能描述
视图层 （View）	index.jsp	系统首页面
	studentlist.jsp	显示所有学生基本信息的页面
	studentlistbyname.jsp	按姓名进行查询并显示查询结果的页面
	studentlistedit.jsp	对所有学生进行增、删、改操作的页面
	studentnew.jsp	新增时输入学生信息的页面
	studentmodify.jsp	修改时编辑学生信息的页面
	error.jsp	操作失败时显示出错信息的页面
控制器层 （Controller）	StudentListAction.java	获取所有学生数据，并根据参数转至相应页面的动作类
	StudentListByNameAction.java	获取姓名并进行查询的动作类
	StudentEditAction.java	处理数据编辑请求的动作类
	StudentSaveAction.java	处理数据保存请求的动作类
	StudentDeleteAction.java	处理数据删除请求的动作类
	sturts-config.xml	Sturts 配置文件
	web.xml	应用程序的通用配置文件
模型层 （Model）	StudentForm.java	新增、修改学生的表单 bean
	DBUser.java	连接及操纵数据库的组件，一个工具类
	Student.java	代表学生的实体类
过滤器类	SetCharacterEncodingFilter.java	处理字符编码的过滤器类

3. Action 映射表

Action 映射表见表 5-5。

表 5-5　Action 映射表

动作(Action)	入口	表单 bean(ActionForm)	出口
StudentListAction	index.jsp		studentlist.jsp studentlistbyname.jsp studentlistedit.jsp studentlistbypage.jsp

4. 项目目录结构

Struts 模式下项目目录结构图如图 5-19 所示。

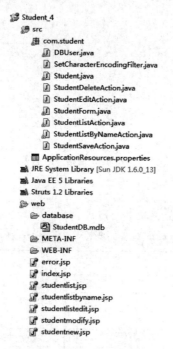

图 5-19　Struts 模式下项目目录结构图

【任务实施】

1. 新建工程

启动 MyEclipse，选择"New->Web Project"，新建一个名称为 Student_4 的工程。

2. 使工程支持 Struts

选中 Student_4，右击，在弹出菜单中选择"MyEclpse…"，选 Add Structs Capabilities，修改对话框中配置，如图 5-20 所示，单击"Finish"，使工程具有支持 Struts 的能力。

3. 将数据库复制到工程中

选中 Web 文件夹，右击，选择"New->Folder"，新建名称为 database 的文件夹，将 Student_3 工程中的数据库文件 StudentDB.mdb 复制到文件夹中。

4. 新建包结构

选中 src 文件夹，右击，选择"New->Package"，在 src 文件夹下新建一个名为 com.student 的

项目五　学生基本信息维护系统

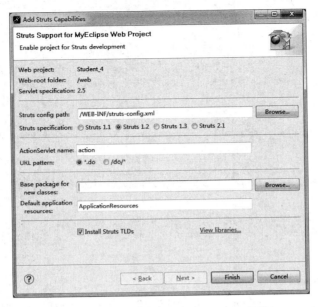

图 5-20　使工程具有支持 Struts 能力对话框

包结构,JavaBean 类文件及动作类文件均存放在该包中。

5．编制 JavaBean 组件

与 Student_3 工程中的 Student.java、DBUser.java 文件完全一致,可以直接将其复制至本工程中。

6．编制过滤器类并配置过滤器

为使系统支持汉字的传递,可使用过滤器。

（1）编制过滤器类 SetCharacterEncodingFilter.java

与 UserRegister_4 工程中的 SetCharacterEncodingFilter.java 文件完全一致,可以直接将其复制到本工程中。

（2）配置过滤器

过滤器在 web.xml 文件中配置过滤器,配置代码在 web.xml 文件中＜servlet＞元素对前。

实现代码如下:

```xml
<filter>
    <description>SetCharacterEncodingFilter</description>
    <display-name>SetCharacterEncodingFilter</display-name>
    <filter-name>SetCharacterEncodingFilter</filter-name>
    <filter-class>
        com.student.SetCharacterEncodingFilter
    </filter-class>
    <init-param>
        <param-name>encoding</param-name>
        <param-value>utf-8</param-value>
    </init-param>
</filter>
<filter-mapping>
    <filter-name>SetCharacterEncodingFilter</filter-name>
    <url-pattern>/*</url-pattern>
</filter-mapping>
```

7．编制 index.jsp 文件

先将 Student_3 工程中的 index.jsp 文件复制过来,对其进行修改。将四个超链接中原来对 Servle 类 StudentListServlet 的请求改成对 Action 类 StudentListAction 的请求,传递的参数不变。

Java & JSP 应用程序实例开发

主要实现代码如下：

```
...
<tr height = "60">
              <td align = "center" ><font face = "隶书" size = "6"><b><a href =
"studentlistaction.do?action = list">学  生  信  息  列  表</a></b></font></td>
</tr>
<tr height = "60">
              <td align = "center"><font face = "隶书" size = "6"><b><a href =
"studentlistaction.do?action = listbyname">学  生  信  息  查  询</a></b></font></td>
</tr>
<tr height = "60">
              <td align = "center"><font face = "隶书" size = "6"><b><a href =
"studentlistaction.do?action = edit">学  生  数  据  维  护</a></b></font></td>
</tr>
<tr height = "60">
              <td align = "center"><font face = "隶书" size = "6"><b><a href =
"studentlistaction.do?action = listbypage">数  据  分  页  显  示</a></b></font></td>
</tr>
...
```

8．发布并测试程序

启动 Tomcat 服务器，将 Student_4 应用发布至服务器，打开浏览器，在地址栏中输入 http://localhost:8080/Student_4，运行并测试系统首页面。

9．学生信息列表功能的实现

（1）新建并编制动作类文件 StudentListAction.java

StudentListActio 类的功能与原先 Servlet 类 StudentListServlet 功能一致。首先获取从 index.jsp 页面中传递过来的参数 action 的值，调用 DBUser 类中的 getStudent() 方法，取出所有学生数据，并存放至会话 session 级变量 StudentList 中，根据参数 action 的值引导至四个不同的页面。

实现代码如下：

```
package com.student;

import java.util.List;
import javax.servlet.http.*;
import org.apache.struts.action.*;
public class StudentListAction extends Action
{
    public ActionForward execute(ActionMapping mapping, ActionForm form, HttpServletRequest request,
HttpServletResponse response) throws Exception
    {
        String action = request.getParameter("action");

        DBUser dbuser = new DBUser(servlet.getServletContext());
        List sStudent = dbuser.getStudent();
        HttpSession session = request.getSession();
        session.setAttribute("StudentList",sStudent);

        String forward = "";
        if(action.equals("list"))
        {
            forward = "List";
        }
        else if(action.equals("listbyname"))
        {
            forward = "ListByName";
        }
        else if(action.equals("edit"))
        {
            forward = "Edit";
        }
```

```
        else if(action.equals("listbypage"))
        {
            forward = "ListByPage";
        }
        return mapping.findForward(forward);
    }
}
```

(2) 配置动作类 StudentListAction

动作 Action 类是 Struts 的组件，需要在 Sturts 的配置文件 sturts-config.xml 中配置后才能使用。

配置 StudentListAction 动作类映射关系的代码应写在＜action-mappings＞元素对中。

配置代码如下：

```
<action path = "/studentlistaction" type = "com.student.StudentListAction">
    <forward name = "List"       path = "/studentlist.jsp"/>
    <forward name = "ListByName" path = "/studentlistbyname.jsp"/>
    <forward name = "Edit"       path = "/studentlistedit.jsp"/>
    <forward name = "ListByPage" path = "/studentlistbypage.jsp"/>
</action>
```

(3) 编制 studentlist.jsp 文件

studentlist.jsp 文件与 Student_3 中 studentlist.jsp 文件完全一致，直接将 Student_3 工程中的 studentlist.jsp 文件复制到本工程中。

(4) 发布并测试程序

将 Student_4 应用发布至服务器，打开浏览器，在地址栏中输入 http://localhost:8080/Student_4，运行并测试学生信息列表部分。

任务九　用 Struts 模式实现学生信息查询

【任务描述】

本任务用 Struts 模式实现图 5-6 所示的学生信息查询。可以在 JSP＋JavaBean＋Servlet 模式的基础上，编制自定义动作类来代替 Servlet 类负责流程控制，在配置文件中对动作类进行配置，并对页面文件作相应修改。

【任务分析】

1. 程序运行基本流程

Struts 模式下按姓名查询基本流程图如图 5-21 所示。

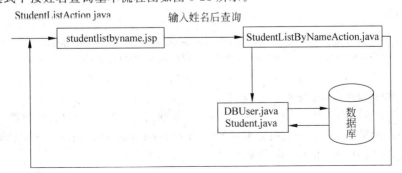

图 5-21　Struts 模式下按姓名查询基本流程图

2. Action 映射表

Action 映射表见表 5-6。

表 5-6 Action 映射表

动作(Action)	入口	表单 bean(ActionForm)	出口
StudentListByNameAction	studentlistbyname.jsp		studentlistbyname.jsp

【任务实施】

1. 新建并编制动作类文件 StudentListByNameAction.java

StudentListByNameAction 类的功能与原先 Servlet 类 StudentListByNameServlet 功能一致，首先获取从 studentlistbyname.jsp 页面中传递过来的姓名，调用 DBUser 类中的 getStudentByName()方法，取出满足条件的学生数据，并存放至会话 session 级变量 StudentList 中，转至 studentlistbyname.jsp 页面。

实现代码如下：

```java
package com.student;

import java.util.List;
import javax.servlet.http.*;
import org.apache.struts.action.*;
public class StudentListByNameAction extends Action
{
    public ActionForward execute(ActionMapping mapping, ActionForm form, HttpServletRequest request, HttpServletResponse response) throws Exception
    {
        String ssname = request.getParameter("ssname");

        DBUser dbuser = new DBUser(servlet.getServletContext());
        List sStudent = dbuser.getStudentByName(ssname);
        HttpSession session = request.getSession();
        session.setAttribute("StudentList",sStudent);

        return  mapping.findForward("ListByName");
    }
}
```

2. 配置动作类 StudentListByNameAction

配置 StudentListByNameAction 动作类映射关系的代码应写在 struts-config.xml 文件的＜action-mappings＞元素对中。

配置代码如下：

```xml
<action path="/studentlistbynameaction" type="com.student.StudentListByNameAction">
        <forward name="ListByName" path="/studentlistbyname.jsp"/>
</action>
```

3. 编制 getstudentlistbyname.jsp 文件

先将 Student_3 工程中的 studentlistbyname.jsp 复制到本工程中，对其进行修改。在提交表单时，将表单 form 的属性 action 的值由原来对 Servlet 类 StudentListByNameServlet 的调用改为对动作类文件 StudentListByNameAction 的调用。

项目五 学生基本信息维护系统

修改代码如下：

…
< form name = "searchform" method = "post" action = "studentlistbynameaction.do" >
…
</form >

4．发布并测试程序

将 Student_3 应用发布至服务器，打开浏览器，在地址栏中输入 http://localhost:8080/Student_3，运行并测试学生信息查询部分。

任务十 用 Struts 模式实现学生数据维护

【任务描述】

本任务用 Struts 模式实现图 5-8 至图 5-11 所示的学生数据维护。可以在 JSP＋JavaBean＋Servlet 模式的基础上，编制自定义动作类来代替 Servlet 类负责增、删、改数据的流程控制，编制表单 bean 类获取页面表单中的数据，在配置文件中对表单 bean 类、动作类进行配置，并对页面文件作相应修改。

【任务分析】

1．程序运行基本流程

图 5-22 所示为 Struts 模式下学生数据维护流程图。

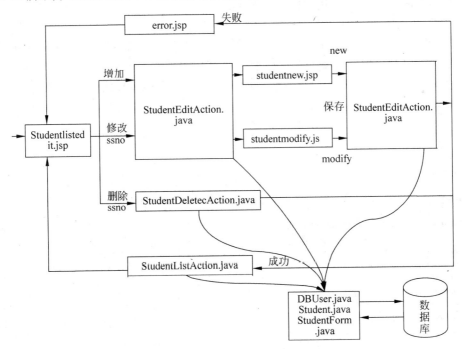

图 5-22 Struts 模式下学生数据维护流程图

2．Action 映射表

Action 映射表见表 5-7。

表 5-7 Action 映射表

动作（Action）	入口	表单 bean（ActionForm）	出口
StudentEditAction	studentlistedit.jsp	StudentForm	studentnew.jsp studentmodify.jsp
StudentSaveAction	studentnew.jsp studentmodify.jsp	StudentForm	studentlistedit.jsp error.jsp
StudentDeleteAction	studentlistedit.jsp		studentlistedit.jsp error.jsp

【任务实施】

1. 编制 studentlistedit.jsp 文件

先将 Student_3 工程中的页面文件 studentlistedit.jsp 复制到本工程中，对其进行修改。将页面中"增加"、"修改"超链接由原来对 Servlet 类的调用改为对相应的动作类的调用。

将"增加"超链接改为：

```
< a href = "studenteditaction.do">增加</a>
```

将"修改"超链接改为：

```
< a href = "studenteditaction.do?ssno = <% = student.getSsno() %>">修改</a>
```

其中 studenteditaction 为用于编辑的动作类的逻辑名。

2. 新建并编制表单 bean 类文件 StudentForm.java

在 Struts 中，表单 bean 类均是 ActionForm 的子类，StudentForm 类就是一个表单 bean。在这里，有两个作用，一是在新增、修改页面中输入数据后提交表单时，配合 Action 对象获取表单中提交的数据，数据的获取是自动的；二是在修改时，如果事先将待修改的数据存入表单 bean，那么在转至修改页面时，这些数据将自动显示在页面中。该类的编制必须按照 JavaBean 的约定，各个属性的名称分别对应于页面中表单元素，类中还必须有与属性配合的获取（getter）和设置（setter）属性的方法。

实现代码如下：

```java
package com.student;

import javax.servlet.http.HttpServletRequest;
import org.apache.struts.action.*;
public class StudentForm extends ActionForm
{
    private String ssno;
    private String ssname;
    private String sssex;
    private int ssage;
    private String ssmajor;
    private String ssfav;
    private String ssaddress;

    public void setSsno(String ssno)
    {
        this.ssno = ssno;
    }
    public String getSsno()
    {
```

```java
        return ssno;
    }
    public void setSsname(String ssname)
    {
        this.ssname = ssname;
    }
    public String getSsname()
    {
        return ssname;
    }

    public void setSssex(String sssex)
    {
        this.sssex = sssex;
    }
    public String getSssex()
    {
        return sssex;
    }

    public void setSsage(int ssage)
    {
        this.ssage = ssage;
    }
    public int getSsage()
    {
        return ssage;
    }

    public void setSsmajor(String ssmajor)
    {
        this.ssmajor = ssmajor;
    }
    public String getSsmajor()
    {
        return ssmajor;
    }

    public void setSsfav(String ssfav)
    {
        this.ssfav = ssfav;
    }
    public String getSsfav()
    {
        return ssfav;
    }

    public void setSsaddress(String ssaddress)
    {
        this.ssaddress = ssaddress;
    }
    public String getSsaddress()
    {
        return ssaddress;
    }
}
```

3．实现增加、修改记录功能

（1）新建并编制动作类文件 StudentEditAction.java

该类的主要作用是处理数据编辑请求。在该类中，先获取从 studentlistedit.jsp 页面中传递过来的学号 ssno，判断学号是否为空。如果为空，说明是新增，直接转至 studentnew.jsp；否则，说明是修改,调用 DBUser 类中的方法 getStudentBySsno() 取出学生数据,将数据设置至关联的表单 bean 中,转至 studentmodify.jsp 页面。

实现代码如下:

```java
package com.student;

import javax.servlet.http.*;
import org.apache.struts.action.*;
public class StudentEditAction extends Action
{
    public ActionForward execute ( ActionMapping mapping, ActionForm form, HttpServletRequest request, HttpServletResponse response) throws Exception
    {
        String ssno = request.getParameter("ssno");
        StudentForm studentform = (StudentForm)form;

        if(ssno == null || ssno.equals("")  )
        {
            return  mapping.findForward("StudentNew");
        }
        else
        {
            DBUser dbuser = new DBUser(servlet.getServletContext());
            Student student = dbuser.getStudentBySsno(ssno);
            if(student!= null)
            {
                studentform.setSsno(student.getSsno());
                studentform.setSsname(student.getSsname());
                studentform.setSssex(student.getSssex());
                studentform.setSsage(student.getSsage());
                studentform.setSsmajor(student.getSsmajor());
                studentform.setSsfav(student.getSsfav());
                studentform.setSsaddress(student.getSsaddress());
            }
            return  mapping.findForward("StudentModify");
        }
    }
}
```

(2) 编制 studentnew.jsp 文件

先将 Student_3 工程中的 studentnew.jsp 文件复制过来,对其进行修改。

一是用 Struts 的 HTML 标签改写页面,这需要在页面开始处用 taglib 命令导入标签库;二是在提交表单时,将表单 form 的属性 action 的值由原来对 Servlet 类 studentsaveservlet 的调用改为对动作类 studentsaveaction 的调用,并将参数 action 的值 new 传递给 Servlet。

实现代码如下:

```jsp
<%@page contentType="text/html;charset=utf-8" %>
<%@ taglib uri="http://struts.apache.org/tags-html" prefix="html" %>
<html>
    <head>
        <title>新增学生信息</title>
        <script type="text/javascript">
            function checkdata()
            {
                var ssno = document.studentform.ssno.value;
                if (ssno=="" || ssno.length!=6  ||!isNumberic(ssno) )
                {
                    alert("学号不能为空,必须输入,并且只能是6位数字!");
                    document.studentform.ssno.focus();
```

项目五 学生基本信息维护系统

```
            return false;
        }
    var ssname=document.studentform.ssname.value;
    if(ssname=="")
    {
        alert("学生姓名不能为空,必须输入!");
        document.studentform.ssname.focus();
        return false;
    }
    return true;
}
function isNumberic(str)
{
    var len=str.length;
    for(var i=0;i<len;i++)
        if(str.charAt(i)<'0'||str.charAt(i)>'9')
            return false;
    return true;
}
</script>
</head>
<body onload="document.studentform.ssno.focus();document.all('sssex')[0].checked=true;">
    <html:form method="post" action="/studentsaveaction.do?action=new">
        <table align="center" style="font-size:12px">
            <tr>
                <td colspan="2" align="center">新增学生信息</td>
            </tr>
            <tr>
                <td align="right">学号:</td>
                <td><html:text property="ssno" /></td>
            </tr>
            <tr>
                <td align="right">姓名:</td>
                <td><html:text property="ssname" /></td>
            </tr>
            <tr>
                <td>性别:</td>
                <td><html:radio property="sssex" value="男" />男<html:radio property="sssex" value="女" />女</td>
            </tr>
            <tr>
                <td align="right">年龄:</td>
                <td><html:text property="ssage" /></td>
            </tr>
            <tr>
                <td align="right">所学专业:</td>
                <td><html:text property="ssmajor" /></td>
            </tr>
            <tr>
                <td align="right">爱好:</td>
                <td><html:text property="ssfav" /></td>
            </tr>
            <tr>
                <td align="right">家庭住址:</td>
                <td><html:text property="ssaddress" /></td>
            </tr>
            <tr>
                <td> </td>
                <td align="center"><html:submit property="btok" value="保 存" onclick="return checkdata()" />
                      <html:button property="btcancel" value="取 消" onclick="javascript:history.back()"/>
                </td>
```

Java & JSP应用程序实例开发

```
        </tr>
    </table>
</html:form>
</body>
</html>
```

(3) 编制 studentmodify.jsp 文件

先将 Student_3 工程中的 studentmodify.jsp 文件复制过来,对其进行修改。

一是用 Struts 的 HTML 标签改写页面,这需要在页面开始处用 taglib 命令导入标签库;二是在提交表单时,将表单 form 的属性 action 的值由原来对 Servlet 类 studentsaveservlet 的调用改为对动作类 studentsaveaction 的调用,并将参数 action 的值 modify 传递给动作类。

实现代码如下:

```jsp
<%@page contentType="text/html;charset=utf-8" %>
<%@ taglib uri="http://struts.apache.org/tags-html"  prefix="html" %>

<html>
    <head>
        <title>修改学生信息</title>
        <script type="text/javascript">
            function checkdata()
            {
                var ssname=document.studentform.ssname.value;
                if (ssname=="" )
                {
                    alert("学生姓名不能为空,必须输入!");
                    document.studentform.ssname.focus();
                    return false;
                }
                return true;
            }
        </script>
    </head>
    <body onload="document.studentform.ssname.focus()">
        <html:form  method="post" action="/studentsaveaction.do?action=modify">
            <table align="center" style="font-size:12px">
                <tr>
                    <td colspan="2" align="center">修改学生信息</td>
                </tr>
                <tr>
                    <td align="right">学号:</td>
                    <td><html:text property="ssno"  readonly="true"/></td>
                </tr>
                <tr>
                    <td align="right">姓名:</td>
                    <td><html:text property="ssname" /></td>
                </tr>
                <tr>
                    <td>性别:</td>
                    <td><html:radio property="sssex" value="男" />男<html:radio property="sssex" value="女" />女</td>
                </tr>
                <tr>
                    <td align="right">年龄:</td>
                    <td><html:text property="ssage" /></td>
                </tr>
                <tr>
                    <td align="right">所学专业:</td>
                    <td><html:text property="ssmajor" /></td>
```

项目五 学生基本信息维护系统

```
            </tr>
            <tr>
                <td align = "right">爱好：</td>
                <td><html:text property = "ssfav" /></td>
            </tr>
            <tr>
                <td align = "right">家庭住址：</td>
                <td><html:text property = "ssaddress" /></td>
            </tr>
            <tr>
                <td> </td>
                <td align = "center"><html:submit property = "btok" value = "保 存" onclick = "return checkdata()"/>
                              <html:button property = "btcancel" value = "取 消" onclick = "javascript:history.back()"/>
                </td>
            </tr>
        </table>
    </html:form>
</body>
</html>
```

（4）配置表单 bean 及动作 Action 类

配置 StudentForm 表单 bean 的代码应写在 struts-config.xml 文件的＜form-beans＞元素对中。

配置代码如下：

```
<form-bean name = "studentform"    type = "com.student.StudentForm" />
```

配置 StudentEditAction 动作类映射关系的代码应写在 struts-config.xml 文件的＜action-mappings＞元素对中。

配置代码如下：

```
<action path = "/studenteditaction"      type = "com.student.StudentEditAction"   name = "studentform"   scope = "request"    validate = "false">
        <forward name = "StudentNew"    path = "/studentnew.jsp"/>
        <forward name = "StudentModify"   path = "/studentmodify.jsp"/>
</action>
```

（5）发布并测试程序

请注意，为了在测试时能正确地显示新增、修改页面，可以先将 studentnew.jsp 与 studentmodify.jsp 页面表单属性 action 值中的 studentsaveaction 改为 studenteditaction，这是因为 studentsaveaction 尚未编制和配置。将 Student_4 应用发布至服务器，打开浏览器，在地址栏中输入 http://localhost:8080/Student_4，运行并测试学生记录增加及修改页面是否能正确地显示出来。

（6）新建并编制动作类文件 StudentSaveAction.java

该类的主要作用是处理数据保存请求。在该类中，先获取从 studentnew.jsp 或 studentmodify.jsp 页面中传递过来参数 action 的值，再通过关联的表单 bean 获取提交的所有数据。根据 action 的值判断是新增还是修改，如果是新增，调用 DBUser 类中的方法 addStudent()，将数据插入至数据库中；如果是修改，调用 DBUser 类中的方法 modifyStudent()，将数据更新至数据库中；保存成功则调用 Action 类 StudentListAction 重取数据后转至 studentlistedit.jsp 页面，失败则转至 error.jsp 页面。

实现代码如下：

```
package com.student;

import javax.servlet.http.*;
```

Java & JSP应用程序实例开发

```java
import org.apache.struts.action.*;
public class StudentSaveAction extends Action
{
    public ActionForward execute(ActionMapping mapping, ActionForm form, HttpServletRequest request,
HttpServletResponse response) throws Exception
    {
        String action = request.getParameter("action");

        StudentForm studentform = (StudentForm)form;
        String ssno = studentform.getSsno();
          String ssname = studentform.getSsname();
          String sssex = studentform.getSssex();
          int    ssage = studentform.getSsage();
          String ssmajor = studentform.getSsmajor();
          String ssfav = studentform.getSsfav();
          String ssaddress = studentform.getSsaddress();

        Student student = new
        Student(ssno,ssname,sssex,ssage,ssmajor,ssfav,ssaddress);
        DBUser dbuser = new DBUser(servlet.getServletContext());
        if(action.equals("new")  )
        {
            boolean flag = dbuser.addStudent(student);
            if(flag)
                return   mapping.findForward("StudentList");
        }
        else if( action.equals("modify")  )
        {
            boolean flag = dbuser.modifyStudent(student);
            if(flag)
                return   mapping.findForward("StudentList");
        }
        return   mapping.findForward("Error");
    }
}
```

（7）配置动作类 StudentSaveAction

配置 StudentSaveAction 动作类映射关系的代码应写在 struts-config.xml 文件的＜action-mappings＞元素对中。

配置代码如下：

```xml
<action path = "/studentsaveaction"     type = "com.student.StudentSaveAction"    name = "studentform"
scope = "request"  validate = "true" input = "/studentnew.jsp">
<forward name = "StudentList"     path = "/studentlistaction.do?action = edit"/>
     <forward name = "Error"      path = "/error.jsp"/>
</action>
```

（8）编制 error.jsp 文件

error.jsp 文件与 Student_3 中 error.jsp 文件完全一致，直接将 Student_3 工程中的 error.jsp 文件复制到本工程中。

（9）发布并测试程序

将 Student_4 应用发布至服务器，打开浏览器，在地址栏中输入 http://localhost:8080/Student_4，运行并测试学生记录增加及修改部分。

4．实现删除记录功能

（1）编制 studentlistedit.jsp 文件

将页面中"删除"超链接由原来对 Servlet 类的调用改为对相应的动作类的调用。

"删除"超链接只需要将 Delete() 函数中的 location 锚的 href 属性的值改为：

项目五 学生基本信息维护系统

```
location.href = "studentdeleteaction?ssno = " + ssno
```

其中,studentdeleteaction 为用于删除操作的动作类的逻辑名。

(2) 新建并编制动作类文件 StudentDeleteAction.java

该类的主要作用是处理记录删除请求。在该类中,先获取传递过来学生学号,调用 DBUser 类中的方法 deleteStudent(),将该学生从数据库中删除,成功则调用动作类 StudentListAction 重取数据后转至 studentlistedit.jsp 页面,失败则转至 error.jsp 页面。

实现代码如下:

```java
package com.student;

import javax.servlet.http.*;
import org.apache.struts.action.*;
public class StudentDeleteAction extends Action
{
    public ActionForward execute(ActionMapping mapping, ActionForm form, HttpServletRequest request, HttpServletResponse response) throws Exception
    {
        String ssno = request.getParameter("ssno");
        DBUser dbuser = new DBUser(servlet.getServletContext());
        boolean flag = dbuser.deleteStudent(ssno);
        if(flag)
            return  mapping.findForward("StudentList");
        else
            return  mapping.findForward("Error");
    }
}
```

(3) 配置动作类 StudentDeleteAction

配置 StudentDeleteAction 动作类映射关系的代码应写在 struts-config.xml 文件的＜action-mappings＞元素对中。

配置代码如下:

```xml
<action path = "/studentdeleteaction" type = "com.student.StudentDeleteAction">
    <forward name = "StudentList"  path = "/studentlistaction.do?action = edit"/>
    <forward name = "Error"  path = "/error.jsp"/>
</action>
```

(4) 发布并测试程序

将 Student_4 应用发布至服务器,打开浏览器,在地址栏中输入 http://localhost:8080/Student_4,运行并测试学生记录删除部分。

项目小结

本项目通过一个学生基本信息维护系统的实践,进一步介绍了如何利用常用的三种开发模式来开发项目程序的基本方法及步骤,其中更多地涉及到了 Web 应用程序开发过程中的数据库中数据的查询和数据的增、删、改及数据的分页显示处理等,有利于读者进一步熟悉和掌握基本的开发技术。

项目六　教师测评系统(Struts 模式)

项目描述

教师教育教学情况测评系统(以下简称教师测评系统)是一个 B/S 结构的 Java Web 系统,主要有三个子系统组成,一是学生对班主任及任课教师进行测评的学生测评子系统,二是教师对测评情况进行查询的数据查询子系统,三是在测评前对所需基本数据进行维护的系统管理子系统。

项目预览

在浏览器地址栏中输入 http://localhost:8080/TeacherTest,程序运行时出现如图 6-1 所示的 index.jsp 首页面。

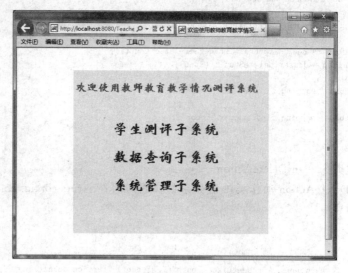

图 6-1　教师测评系统运行首页面

项目分析

为了对班主任的德育工作和任课教师的教学工作情况有一个客观的了解、掌握和评价,许多学校都开展了对教师的测评工作,其中学生对教师的评价是一个很重要的方面。原先的操作方法是,每一位同学对照例如表 6-1 所示的评分规则,在纸上给班主任和任课教师打分,学校通过汇总得出每一位教师的最终得分,这都是人工操作完成的,工作量大且繁杂费时。后来也使用过答题卡打分的形式,虽然省去了一部分汇总计算的时间,但组织准备工作量也大。随着信息技术的发展和计算机的普及,通过网络对教师进行测评并不是遥不可及的,教师测评系统就是在这样的背景下开发出来的。在系统管理子系统中准备好基本数据之后,学生通过登录测评子系统就可以对班主任和任课教师进行测评了,学校领导和教师随时可以通过数据查询子系统查询数据,做到了实时、高效、方便、快捷。系统主要使用 Struts 框架技术实现,其中还使用了 Ajax 技术,是一个相对比较复杂的 Web 应用程序。通过本系统的学习和开发实践,将使广大读者进一步掌握利用 JSP 技术开发 Web 应用程序的知识与技巧。

项目六　教师测评系统(Struts模式)

表 6-1　教师测评系统的评分细则

类别	评分项目	序号	评 分 细 则	得分 A	B	C	D
班主任德育情况测评	班主任工作评价	1	班主任师德表现	10	8	5	2
		2	早晨、中午、自习课或教育活动课组织有序	10	8	5	2
		3	定期组织班会及指导开展班级活动	15	10	5	2
		4	与班级学生交流谈心活动	5	3	2	1
		5	关心班级学生的学习、生活、心理健康	10	8	5	2
		6	经常与学生家庭联系并及时进行家访	10	8	5	2
		7	班级班风、学风的总体情况	10	8	5	2
		8	班主任对班级的管理情况及工作方法	10	8	5	2
		9	体罚和变相体罚现象(A.无　B.有)	5	0		
		10	对班主任总体满意程度(A.满意　B.较满意　C.基本满意　D.不满意)	15	10	5	2
教师教学情况测评	教学态度	1	老师总是按时上、下课,不随意调课、停课或缺课	10	8	5	2
		2	老师讲课(实训辅导)很认真	5	3	2	1
		3	老师既严格要求又关心我们	5	3	2	1
		4	老师能认真批改我们的作业,耐心解答我们提出的问题	5	3	2	1
	教学内容方法及教师素质	5	老师熟悉本专业知识和技能(实训教师专业水平高),教学水平高	10	8	5	2
		6	老师讲课突出重点和难点,难易适度,紧扣教学主题,理论与实践结合	5	3	2	1
		7	老师能经常鼓励我们积极思维、组织课堂讨论和有效的自主学习,而不是满堂灌	5	3	2	1
		8	老师能运用现代教育技术教学,板书工整、布局有序	5	3	2	1
	教学效果	9	我们对学习本课程有兴趣	5	3	2	1
		10	从老师身上我们学到了不少做人的道理	5	3	2	1
		11	教学效果好	10	8	5	2
		12	你对学科老师的总体满意程度(A.满意　B.较满意　C.基本满意　D.不满意)	30	20	10	0

对任课教师、班主任、学校的意见、建议和简要评价(不能空白):

项目设计

1. 系统功能模块的设计

教师测评系统功能模块图如图 6-2 所示。

2. 数据库的设计

本系统使用的数据库平台为 SQL Server 2005。为了实现系统所要求的功能,数据库中设计了 9 张表,其中教师、课程、班级、用户四个基本实体分别对应一张表,班级所上的课程和对班主任及任课教师的测评数据分别组织在对应的表中,为便于查询统计,对班主任及任课教师测评后的汇总数据存放在两张汇总表中,存放测评数据的 4 张表中存在数据冗余,有列名相同的两张表之间均存在关联关系。

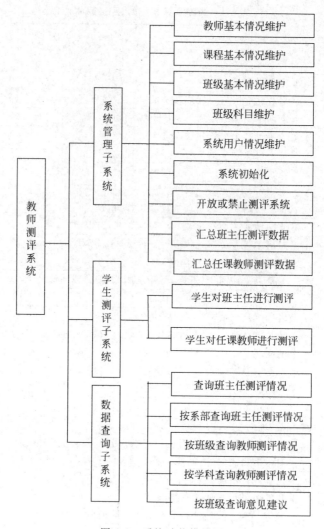

图 6-2　系统功能模块图

课程情况表,表名 scourse,表结构见表 6-2。

表 6-2　课程情况表 scourse 的表结构

列名	数据类型	长度	允许为空	主键	含义
sono	char	3	否	是	课程代码
soname	varchar	50	否		课程名称

教师情况表,表名 steacher,表结构见表 6-3。

表 6-3　教师情况表 steacher 的表结构

列名	数据类型	长度	允许为空	主键	含义
stno	char	3	否	是	教师代码
stname	varchar	8	否		教师姓名
sdname	varchar	20	是		所在部门

班级情况表,表名 sclass,表结构见表 6-4。

项目六 教师测评系统(Struts模式)

表 6-4 班级情况表 sclass 的表结构

列名	数据类型	长度	允许为空	主键	含义
scno	char	4	否	是	班级号
snno	char	2	否		年级代码
scname	varchar	50	否		专业名称
stno	char	3	否		班主任代码
scpnumb	tinyint		是		班级人数
scnote	varchar	50	是		备注
scout	char	1	否		标志
sdname	varchar	20	否		所在系部

其中：scout 为是否允许测评的标志，"1"为允许，"0"表示不允许，默认为"0"。

班级选课表，表名 sccourse，表结构见表 6-5。

表 6-5 班级选课表 sccourse 的表结构

列名	数据类型	长度	允许为空	主键	含义
scno	char	4	否	是	班级号
sono	char	3	否	是	所选课程代码
stno	char	3	否		任课教师代码

班主任教育情况测评表，表名 scyn，表结构见表 6-6。

表 6-6 班主任教育情况测评表 scyn 的表结构

列名	数据类型	长度	允许为空	主键	含义
scid	int		否	是	内码,自动增长
ssno	char	6	否		学生学号
scno	char	4	否		班级号
snno	char	2	否		年级代码
stno	char	3	否		班主任代码
stname	varchar	8	否		班主任姓名
item1	decimal(5,2)		是		分数1
item2	decimal(5,2)		是		分数2
item3	decimal(5,2)		是		分数3
item4	decimal(5,2)		是		分数4
item5	decimal(5,2)		是		分数5
item6	decimal(5,2)		是		分数6
item7	decimal(5,2)		是		分数7
item8	decimal(5,2)		是		分数8
item9	decimal(5,2)		是		分数9
item10	decimal(5,2)		是		分数10
sctext	varchar	250	是		意见及建议

教师教学情况测评表，表名 styn，表结构见表 6-7。

表 6-7 教师教学情况测评表 styn 的表结构

列名	数据类型	长度	允许为空	主键	含义
stid	int		否	是	内码,自动增长
ssno	char	6	否		学生学号
scno	char	4	否		班级号
snno	char	2	否		年级代码
sono	char	3	否		课程代码
soname	varchar	50	否		课程名称
stno	char	3	否		任课教师代码
stname	varchar	8	否		任课教师姓名
item1	decimal(5,2)		是		分数1
item2	decimal(5,2)		是		分数2
item3	decimal(5,2)		是		分数3
item4	decimal(5,2)		是		分数4
item5	decimal(5,2)		是		分数5
item6	decimal(5,2)		是		分数6
item7	decimal(5,2)		是		分数7
item8	decimal(5,2)		是		分数8
item9	decimal(5,2)		是		分数9
item10	decimal(5,2)		是		分数10
item11	decimal(5,2)		是		分数11
item12	decimal(5,2)		是		分数12

班主任教育情况测评汇总表,表名 scynsum,表结构见表 6-8。

表 6-8 班主任教育情况测评汇总表 scynsum 的表结构

列名	数据类型	长度	允许为空	主键	含义
scno	char	4	否	是	班级号
snno	char	2	否		年级代码
stno	char	3	否		班主任代码
stname	varchar	8	否		班主任姓名
num	tinyint		是		测评人数
itsum	decimal(5,2)		是		总分
itorder	smallint		是		名次
item1	decimal(5,2)		是		分数1
item2	decimal(5,2)		是		分数2
item3	decimal(5,2)		是		分数3
item4	decimal(5,2)		是		分数4
item5	decimal(5,2)		是		分数5
item6	decimal(5,2)		是		分数6
item7	decimal(5,2)		是		分数7
item8	decimal(5,2)		是		分数8
item9	decimal(5,2)		是		分数9
item10	decimal(5,2)		是		分数10

教师教学情况测评汇总表,表名 stynsum,表结构见表 6-9。

项目六 教师测评系统(Struts模式)

表 6-9 教师教学情况测评汇总表 stynsum 的表结构

列名	数据类型	长度	允许为空	主键	含义
scno	char	4	否	是	班级号
sono	char	3	否	是	课程代码
stno	char	3	否		任课教师代码
snno	char	2	否		年级代码
soname	varchar	50	否		课程名称
stname	varchar	8	否		任课教师姓名
num	tinyint		是		测评人数
itorder	smallint		是		名次
itsum	decimal(5,2)		是		总分
item1	decimal(5,2)		是		分数1
item2	decimal(5,2)		是		分数2
item3	decimal(5,2)		是		分数3
item4	decimal(5,2)		是		分数4
item14	decimal(5,2)		是		大类分1
item5	decimal(5,2)		是		分数5
item6	decimal(5,2)		是		分数6
item7	decimal(5,2)		是		分数7
item8	decimal(5,2)		是		分数8
item58	decimal(5,2)		是		大类分2
item9	decimal(5,2)		是		分数9
item10	decimal(5,2)		是		分数10
item11	decimal(5,2)		是		分数11
item12	decimal(5,2)		是		分数12
Item912	decimal(5,2)		是		大类分3

系统用户表,表名 suser,表结构见表 6-10。

表 6-10 系统用户表 suser 的表结构

列名	数据类型	长度	允许为空	主键	含义
uid	tinyint		否	是	内码,自动增长
username	varchar	16	否		用户名
userpwd	varchar	16	否		密码
isadmin	tinyint		否		标志

其中:isadmin 为是否系统管理员的标志,数值为 1 代表是系统管理员,0 代表不是系统管理员。

3. CSS 的设计

在网页中使用 CSS,可以使所有页面中统一使用某种显示格式,从而简化显示风格设置,增强网页的可维护性。本系统中设计了两种 CSS 格式,一种是在三个子系统的主页面 main.jsp 中统一使用的 main.css,还有一种主要是统一表格样式的 style_css.css。

(1) main.css 的实现代码

```
/*主页面 css 文档 */
body
{
    font:12px;
```

```css
    margin:0px;
    text-align:center;
    background-color:#FFFFFF;
}
#main
{
    width:1000px;
    background-color:#FFFFFF;

}
#top
{
    width:1000px;
    height:60px;
    background-color:#F5EFE7;
    border:1px solid #E7E7E7;
}
#mid
{
    width:1000px;
    height:532px;
    margin-top:0px;
    margin-right:0px;
    margin-bottom:0px;
    margin-left:0px;
    background-color:#FFFFFF;
}
#left
{
    width:200px;
    height:532px;
    text-align:left;
    float:left;
    clear:none;
    overflow:hidden;
    background-color:#F5EFE7;
    border:1px solid #E7E7E7;
}
#right
{
    width:797px;
    height:532px;
    text-align:left;
    clear:none;
    overflow:hidden;
    background-color:#F5EFE7;
    border:1px solid #E7E7E7;
}
#end
{
    width:1000px;
    height:30px;
    margin-top:0px;
    margin-right:0px;
    margin-bottom:0px;
    margin-left:0px;
    background-color:#F5EFE7;
    border:1px solid #E7E7E7;
}
```

（2）style_css.css 的实现代码

```css
table {BORDER-RIGHT:0px; BORDER-TOP:0px; BORDER-LEFT:0px; BORDER-BOTTOM:0px}
```

项目六 教师测评系统(Struts模式)

```
caption{font-family:"楷体";font-style:bold;font-size:16px; COLOR:#150300}
td {font-family:"宋体";font-size:12px}

a:link {color:#0000cc; text-decoration:none;}
a:visited{color:#0000cc; text-decoration:none;}
    a:hover {color:#a71c8b; text-decoration:underline;}
```

4．Struts介绍及工作流程

　　Struts是由Apache开发的基于Java技术的JSP Web应用开发框架,它符合MVC设计模式。由JavaBean组件构成模型部分,由ActionServlet和Action实现控制器部分,由一组JSP文件、ActionForm组件和Struts标签库组成视图部分,有时ActionForm组件也作为模型部分。其工作流程如图6-3所示。

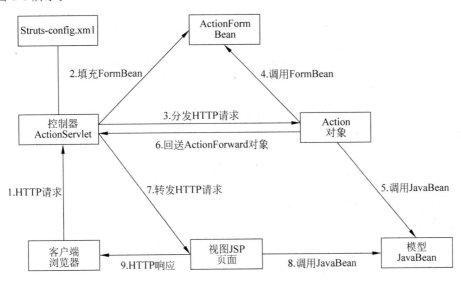

图6-3　Struts框架工作流程图

5．项目目录结构及功能

　　教师测评系统项目目录结构图如图6-4所示。
　　各个包及文件夹说明如下:
　　actions包中存放所有的动作类文件;
　　comms包中存放所有的普通JavaBean类文件、DBUser类文件、常量类Constants文件及用于分页处理的两个类文件;
　　filter包中存放过滤器类文件;
　　forms包中存放所有的表单bean类文件;
　　bas文件夹中存放系统管理子系统中所有页面文件;
　　test文件夹中存放学生测评子系统中所有页面文件;
　　search文件夹中存放数据查询子系统中所有页面文件;
　　images文件夹中存放系统中使用的一个图片文件;
　　css文件夹中存放系统中使用的两个样式表文件;
　　js文件夹中存放系统中使用的一个javascript源程序文件。

6．程序文件清单

　　(1)视图层组件(JSP页面)
　　视图层组文件清单见表6-11。

图6-4　教师测评系统项目目录结构图

表 6-11 视图层组件文件清单

文 件 名	功 能 描 述
\bas\classlist.jsp	分页显示班级列表的页面
\bas\classmody.jsp	修改班级信息的页面
\bas\classnew.jsp	新增班级信息的页面
\bas\courseedit.jsp	编辑课程信息的页面
\bas\courselist.jsp	分页显示课程列表的页面
\bas\first.jsp	显示欢迎信息的页面
\bas\info.jsp	系统中操作成功后显示的页面
\bas\login.jsp	用户登录的页面
\bas\loginfail.jsp	系统中用户登录失败后显示的页面
\bas\main.jsp	主框架页面
\bas\sccourselist.jsp	显示班级所选课程列表的页面
\bas\sccoursemody.jsp	修改班级选课信息的页面
\bas\sccoursenew.jsp	新增班级选课信息的页面
\bas\scynsum.jsp	汇总班主任测评数据的页面
\bas\selectcoursebyname.jsp	按名称查询并选择课程的页面
\bas\selectteacherbyname.jsp	按姓名查询并选择教师的页面
\bas\stynsum.jsp	汇总任课教师测评数据的页面
\bas\sysinit.jsp	系统初始化页面
\bas\sysyesno.jsp	开放或禁止测评系统的页面
\bas\teacheredit.jsp	编辑教师信息的页面
\bas\teacherlist.jsp	分页显示教师列表的页面
\bas\useredit.jsp	编辑系统用户信息的页面
\bas\userlist.jsp	显示系统用户列表的页面
\bas\wrong.jsp	系统中操作失败后显示的页面
\search\first.jsp	显示欢迎信息的页面
\search\login.jsp	用户登录的页面
\search\main.jsp	主框架页面
\search\managerall.jsp	显示班主任测评数据的页面
\search\managerbysdname.jsp	按部门查询班主任测评数据的页面
\search\sctextbyscno.jsp	按班级显示意见、建议的页面
\search\teacherbyscno.jsp	按班级查询任课教师测评数据的页面
\search\teacherbysono.jsp	按课程查询任课教师测评数据的页面
\test\first.jsp	显示欢迎信息的页面
\test\login.jsp	用户登录的页面
\test\main.jsp	主框架页面
\test\managertest.jsp	对班主任进行测评的页面
\test\teachertest.jsp	对任课教师进行测评的页面

（2）控制器层组件（动作类）

控制器层组件文件清单见表 6-12。

项目六 教师测评系统(Struts模式)

表 6-12 控制器层组件文件清单

文 件 名	功 能 描 述
AdminLoginAction.java	处理系统管理用户登录的动作类
ClassDeleteAction.java	处理班级数据删除请求的动作类
ClassEditAction.java	处理班级数据编辑请求的动作类
ClassListAction.java	处理班级数据列表显示请求的动作类
ClassSaveAction.java	处理班级数据保存请求的动作类
CourseDeleteAction.java	处理课程数据删除请求的动作类
CourseEditAction.java	处理课程数据编辑请求的动作类
CourseListAction.java	处理课程数据列表显示请求的动作类
CourseSaveAction.java	处理课程数据保存请求的动作类
ManagerAllAction.java	查询全部班主任测评数据的动作类
ManagerBySdnameAction.java	按部门查询班主任测评数据的动作类
ManagerPreAction.java	测评班主任之前进行预处理的动作类
ManagerTestAction.java	处理班主任测评数据保存请求的动作类
ScCourseDeleteAction.java	处理班级科目数据删除请求的动作类
ScCourseEditAction.java	处理班级科目数据编辑请求的动作类
ScCourseListAction.java	处理班级科目列表显示请求的动作类
ScCourseSaveAction.java	处理班级科目数据保存请求的动作类
SctextByScnoAction.java	按班级号查询意见、建议的动作类
ScynsumAction.java	汇总班主任测评数据的动作类
SearchLoginAction.java	处理查询用户登录的动作类
StudentLoginAction.java	处理学生测评用户登录的动作类
StynsumAction.java	汇总任课教师测评数据的动作类
SysInitAction.java	处理系统初始化请求的动作类
SysYesnoAction.java	处理开放或禁止测评系统请求的动作类
TeacherByScnoAction.java	按班级查询教师测评数据的动作类
TeacherBySonoAction.java	按课程查教师测评数据的动作类
TeacherDeleteAction.java	处理教师数据删除请求的动作类
TeacherEditAction.java	处理教师数据编辑请求的动作类
TeacherListAction.java	处理教师数据列表显示请求的动作类
TeacherPreAction.java	测评教师之前进行预处理的动作类
TeacherSaveAction.java	处理教师数据保存请求的动作类
TeacherTestAction.java	处理教师测评数据保存请求的动作类
UserDeleteAction.java	处理系统用户数据删除请求的动作类
UserEditAction.java	处理系统用户数据编辑请求的动作类
UserListAction.java	处理系统用户列表显示请求的动作类
UserSaveAction.java	处理系统用户数据保存请求的动作类
PageAction.java	对数据分页处理的动作类
SearchAction.java	按名称查询教师或课程的动作类

(3) 模型层组件(Javabean 类及表单 bean 类)

模型层组件文件清单见表 6-13。

表 6-13 模型层组件文件清单

文 件 名	功 能 描 述
Course.java	代表课程的实体类
DBUser.java	连接及操纵数据库的组件,一个工具类
ManagerAll.java	代表班主任测评数据的实体类
PageControl.java	分页处理的一个工具类
ScCourse.java	代表所选课程的实体类
SClass.java	代表班级的实体类
Suser.java	代表用户的实体类
Teacher.java	代表教师的实体类
TeacherAll.java	代表任课教师测评数据的实体类
ClassForm.java	编辑班级信息时用的表单 bean 类
CourseForm.java	编辑课程信息时用的表单 bean 类
LoginForm.java	用户登录的表单 bean 类
ManagerTestForm.java	测评班主任的表单 bean 类
ScCourseForm.java	编辑班级科目时用的表单 bean 类
TeacherForm.java	编辑教师信息时用的表单 bean 类
TeacherTestForm.java	测评任课教师的表单 bean 类
UserForm.java	编辑用户信息时用的表单 bean 类

（4）其他文件

教师测评系统其他文件清单见表 6-14。

表 6-14 教师测评系统其他文件清单

文 件 名	功 能 描 述
sturts-config.xml	Sturts 配置文件
web.xml	应用程序通用配置文件
Constants.java	定义系统中所用会话变量的常量类
SetCharacterEncodingFilter.java	处理字符编码的过滤器类

环境需求

操作系统：Windows XP 或 Windows 7；

数据库平台：Microsoft SQL Server 2005；

JDK 版本：JDK Version1.6；

集成开发环境：MyEclipse 8.5；

Web 服务器：Tomcat 6.0；

浏览器：IE6.0 及以上版本。

项目知识点

① CSS＋DIV 基本概念、设计及使用方法；

② JavaScript 脚本语言的使用方法；

③ Ajax 的基本概念及使用方法；

④ SQL 语言中较复杂查询、插入、更新、删除语句的使用方法；

⑤ Struts 模式的特点、流程及开发的方法、步骤；

⑥ Struts 中较复杂的表单 Bean、动作 Action 类的编制方法及配置方法；

⑦ Struts 中通过配置数据源连接、操纵数据库的基本方法；

⑧ 页面之间传递数据的方法；

⑨ 页面数据的分页显示方法。

任务一　创建工程、准备数据库、编制首页面

【任务描述】

创建一个新的工程，准备相关文件并配置环境，编制并测试系统首页面。

【任务分析】

教师测评系统首页面运行流程图如图 6-5 所示。

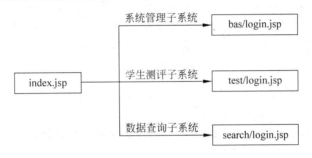

图 6-5　教师测评系统首页面运行流程图

【任务实施】

1. 准备数据库

① 启动 SQL Server 2005 Management Studio，创建名称为 TeacherTest 的数据库。

② 在数据库中创建表 6-2 至表 6-10 所示表结构的表。

③ 分别在教师情况表 steacher、课程情况表 scourse、班级情况表 sclass、系统用户表 suser 表中输入几条记录。

④ 备份数据库。打开 SQL Server 2005 Management Studio，选中 TeacherTest 数据库，右键菜单选所有任务，备份数据库，文件名为 TeacherTest.bak。

2. 新建工程

启动 MyEclipse，选择"New > Web Project"，新建一个名称为 TeacherTest 的工程。

3. 使工程支持 Struts

选中 TeacherTest，右击，在弹出菜单中选择"MyEclpse…"，选择"Add Structs Capabilities"，修改对话框中配置如图 6-6 所示，单击"Finish"按钮，使工程具有支持 Struts 的能力。

在以上对话框的操作完成后，会看到在工程的编译路径中引入了 Struts 的一些库文件，在 WEB-INF 文件夹增加了 Struts 的配置文件及标签库文件，并且修改了应用程序的配置文件。

4. 新建包结构

选中 src 文件夹，右击，选择"New->Package"，在 src 文件夹下新建如图 6-4 所示的包结构，JavaBean 类文件及动作类文件均存放在该包中。

图 6-6　使工程具有支持 Struts 能力对话框

5. 配置数据源

① 将用于连接数据库的 2 个文件 sqljdbc.jar、naming-factory-dbcp.jar 复制到 WEB-INF 文件夹下的 lib 文件夹中；

② 在 Struts 配置文件 struts-config.xml 的＜data-sources＞元素对中进行配置。

配置代码如下：

```xml
<data-sources>
    <data-source key="dbkey" type="org.apache.tomcat.dbcp.dbcp.BasicDataSource">
        <set-property property="driverClassName" value="com.microsoft.sqlserver.jdbc.SQLServerDriver" />
        <set-property property="url"    value="jdbc:sqlserver://127.0.0.1:1433;DatabaseName=TeacherTest" />
        <set-property property="maxActive"  value="50"/>
        <set-property property="username"  value="sa"/>
        <set-property property="password"  value="123"/>
        <set-property property="autoCommit"  value="true"/>
    </data-source>
</data-sources>
```

从以上配置代码可以看出，为简单起见，连接数据库所用的是"sa"用户，密码被设成了"123"。

6. 新建并编制 JavaBean 组件 DBUser.java 文件

这是一个工具 JavaBean 组件，它封装了对数据库的所有操作。在其中先定义一个构造方法，其他方法在后续编制具体模块时再行定义。

实现代码如下：

```java
package com.teach.comms;

import java.util.*;
import javax.sql.DataSource;
import java.sql.*;
//对数据库操作的类,封装了对数据库操作的所有方法
public class DBUser
{
```

项目六 教师测评系统(Struts模式)

```
    //数据源
    private DataSource dataSource;

    //构造方法
    public DBUser(DataSource dataSource)
    {
        this.dataSource = dataSource;
    }
}
```

7. 新建 web 文件夹下的各个子文件夹

选中 web 文件夹,右击,选择"New->Folder",在 web 文件夹下新建如图 6-4 所示的各个子文件夹。将创建的两个样式表文件 main.css、style_css.css 复制到文件夹中。

8. 配置过滤器

先将项目四或项目五中所用的过滤器类 SetCharacterEncodingFilter.java 文件复制到 com.teach.com 包中,再将项目四或项目五中 web.xml 文件的配置代码复制过来并修改。

9. 新建并编制 index.jsp 文件

如图 6-1 所示,该页面为系统首页面。用嵌套表格布局,表格在页面中居中。"学生测评子系统"、"数据查询子系统"、"系统管理子系统"分别为一超链接,分别链接至对应子系统下的 login.jsp 页面。

实现代码如下:

```
<%@page contentType="text/html;charset=utf-8" %>
<html>
    <head>
        <title>欢迎使用教师教育教学情况测评系统</title>
        <link href="css/style_css.css" rel="stylesheet" type="text/css">
    </head>
    <body>
        <table width="100%" height="100%">
            <tr>
                <td align="center" valign="middle">
                    <table bgcolor="#F5EFE7">
                        <tr height="20">
                            <td width="100">  </td><td> </td><td> </td>
                        </tr>
                        <tr>
                            <td>
                                <font face="华文新魏" color="#ff0000" size="5">
                                    <b>欢迎使用教师教育教学情况测评系统</b>
                                </font>
                            </td>
                        </tr>
                        <tr height="50">
                            <td> </td>
                        </tr>
                        <tr>
                            <td align="center">
                                <font face="华文新魏" size="6">
                                    <b><a href="test/login.jsp">学生测评子系统</a></b>
                                </font>
                            </td>
                        </tr>
                        <tr height="20">
                            <td> </td>
                        </tr>
```

```
            <tr height = "20">
                <td align = "center">
                    <font face = "华文新魏" size = "6">
                        <b><a href = "search/login.jsp">数据查询子系统</a></b>
                    </font>
                </td>
            </tr>
            <tr height = "20">
                <td> </td>
            </tr>
            <tr height = "20">
                <td align = "center">
                    <font face = "华文新魏"  size = "6">
                        <b><a href = "bas/login.jsp">系统管理子系统</a></b>
                    </font>
                </td>
            </tr>
            <tr height = "80">
                <td> </td>
            </tr>
        </table>
      </td>
    </tr>
  </table>
 </body>
</html>
```

10．发布并测试程序

启动 Tomcat 服务器，将 TeacherTest 应用发布至服务器，打开浏览器，在地址栏中输入 http://localhost:8080/TeacherTest，运行并测试系统首页面。

任务二　编制系统管理子系统登录模块

【任务描述】

本任务主要编制系统管理子系统登录模块中的相关文件并测试运行。单击首页面中"系统管理子系统"超链接，出现如图 6-7 所示的登录页面。

输入用户名及密码，登录成功后进入如图 6-8 所示的系统管理子系统主页面。

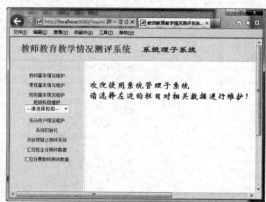

图 6-7　系统管理员登录页面　　　　　　图 6-8　系统管理子系统主页面

【任务分析】

1. 程序运行基本流程

系统管理子系统登录模块流程图如图 6-9 所示。

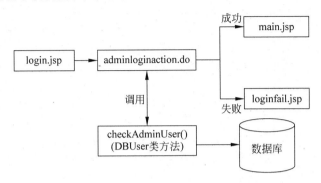

图 6-9　系统管理子系统登录模块流程图

2. Action 映射表

系统管理子系统 Action 映射表见表 6-15。

表 6-15　系统管理子系统 Action 映射表

动作（Action）	入口	表单 bean（ActionForm）	出口
AdminLoginAction	login.jsp	LoginForm	main.jsp loginfail.jsp

【任务实施】

1. 新建并编制 login.jsp 文件

如图 6-7 所示，该页面为系统管理员登录的页面。用嵌套表格布局，表格在页面中居中，表单元素用标准 html 语言编制。提交前用 javascript 代码进行客户端验证，用户名及密码不能为空。

实现代码如下：

```
<%@page contentType="text/html;charset=utf-8"%>
<html>
    <head>
        <title>教师教育教学情况测评系统---系统维护子系统</title>
        <link href="../css/style_css.css" rel=stylesheet type="text/css">
        <script type="text/javascript">
            function checkdata()
            {
                var username=document.loginform.username.value;
                if (username=="" )
                {
                  alert("用户名不能为空!");
                  document.loginform.username.focus();
                  return false;
                }

                var userpwd=document.loginform.userpwd.value;
                if (userpwd=="")
                {
                  alert("密码不能为空!");
                  document.loginform.userpwd.focus();
```

```
                    return false;
                }
                return true;
            }
        </script>
    </head>
    <body onload = "document.loginform.username.focus()">
        <table width = "100%" height = "100%">
            <tr>
                <td align = "center" valign = "middle">
                    <form name = "loginform" method = "post" action = "adminloginaction.do">
                        <table bgcolor = "#f1fcf7">
                            <tr>
                                <td colspan = "2" align = "center">系统管理员登录</td>
                            </tr>
                            <tr>
                                <td align = "center">用户名  </td>
                                <td><input type = "text" name = "username" style = "width:125px;" autocomplete = "off"></td>
                            </tr>
                            <tr>
                                <td align = "center">密   码  </td>
                                <td><input type = "password" name = "userpwd" style = "width:125px;" autocomplete = "off"></td>
                            </tr>
                            <tr>
                                <td> </td>
                                <td align = "center"><input type = "submit" value = "登 录" onclick = "return checkdata()"></td>
                            </tr>
                        </table>
                    </form>
                </td>
            </tr>
        </table>
    </body>
</html>
```

2. 新建并编制表单 bean 类文件 LoginForm.java

在 Struts 中，表单 bean 类均是 ActionForm 的子类，LoginForm 类就是一个表单 bean，其主要作用是在提交表单时，配合 Action 对象获取表单中提交的数据，数据的获取是自动的。该类的编制必须遵守 JavaBean 的约定，属性 username、userpwd 分别对应了登录页面 login.jsp 中表单元素 username、userpwd，类中还必须有与属性配合的获取（getter）和设置（setter）属性的方法。

实现代码如下：

```java
package com.teach.forms;
import org.apache.struts.action.ActionForm;
public class LoginForm extends ActionForm
{
    int uid;
    private String username;
    private String userpwd;
    private int isadmin;

    public int getUid()
    {
        return uid;
    }
    public void setUid(int uid)
    {
```

项目六 教师测评系统(Struts模式)

```
        this.uid = uid;
    }
    …//省略与其属性对应的 get 和 set 方法
```

3．修改 DBUser 类

在其中定义一个方法，方法为 checkAdminUser(username,userpwd)，参数为用户名及密码，该方法的作用是根据传入的用户名及密码，判断该系统管理员用户是否存在，返回一布尔值。

实现代码如下：

```
public boolean checkAdminUser(String username,String userpwd)
{
        Connection con = null;
        boolean flag = false;
        try
        {
            con = dataSource.getConnection();       //获取连接
            Statement stat = con.createStatement();
            ResultSet result = stat.executeQuery("select * from suser where username = '" + username +
"' and userpwd = '" + userpwd + "' and isadmin = 1");
            if( result.next())
            {
                flag = true;
            }
        }
        catch(SQLException e)
        …//省略部分同 CheckAdminUser(username,userpwd)方法中的类同部分
return flag;
}
```

4．新建并编制整个系统中所使用的常量类文件 Constants.java

在 com.teach.com 包中新建 Constants.java，其中定义了整个系统中所用的会话级变量。

实现代码如下：

```
package com.teach.comms;

//Session keys
public final class Constants
{
    //存放操作成功与出错的消息
    public static final String Show_Message = "Show_Message";
    public static final String Error_Message = "Error_Message";

    //存放查询结果的记录集
    public static final String Search_AllList = "Search_AllList";

    //分页处理
    public static final int    Rows_PerPapge = 20;                  //常量,每页行数
    public static final String Search_PageList = "Search_PageList"; //存放当前页的记录集
    public static final String PageNo_Cur = "PageNo_Cur";           //当前页号
    public static final String Pages_All = "Pages_All";             //总页数
    public static final String Records_All = "Records_All";         //记录数

    //当前用户名
    public static final String Login_Curuser = "Login_Curuser";

    //系统管理子系统
    //全部班级代号
    public static final String Bas_ScnoAll   = "Bas_ScnoAll";
    //全部班主任姓名
```

```java
    public static final String Bas_StnameAll = "Bas_StnameAll";
    //班级选课时的当前班级号
    public static final String Bas_ScCourse_CurScno = "Bas_ScCourse_CurScno";
    //全部教师
    public static final String Bas_SteacherAll  = "Bas_SteacherAll";

    //学生测评子系统
    //参与测评的当前学生学号
    public static final String Test_CurSsno = "Test_CurSsno";
    //所在班级
    public static final String Test_CurScno = "Test_CurScno";
    //班级所属年级
    public static final String Test_CurSnno = "Test_CurSnno";
    //被测班主任代号
    public static final String Test_CurManagerNo = "Test_CurManagerNo";
    //被测班主任姓名
    public static final String Test_CurManagerName = "Test_CurManagerName";
    //该班主任是否已被测评过,是为班主任代号,否为 * * *
    public static final String Test_ManagerNo = "Test_ManagerNo";
    //班级须测评的课程代码
    public static final String Test_ScCourseAll = "Test_ScCourseAll";
    //班级须测评的课程名称
    public static final String Test_ScCourseNameAll = "Test_ScCourseNameAll";
    //班级须测评的教师代码
    public static final String Test_TeacherAll = "Test_TeacherAll";
    //班级须测评的教师姓名
    public static final String Test_TeacherNameAll = "Test_TeacherNameAll";
    //班级已测评过的课程代码
    public static final String Test_ScCourseYes = "Test_ScCourseYes";
    //班级已测评过的教师代码
    public static final String Test_TeacherYes = "Test_TeacherYes";
    //当前正在测评的课程代码
    public static final String Test_ScourseCur = "Test_ScourseCur";
    //当前正在测评的课程名称
    public static final String Test_ScourseNameCur = "Test_ScourseNameCur";

    //数据查询子系统
    //从班级表中取出的所有系部名称,不是教师表中的系部
    public static final String Search_SdnameAll = "Search_SdnameAll";
    //课程代号,从表 styn 取出
    public static final String Search_SonoAll = "Search_SonoAll";
    //课程名称,从表 styn 取出
    public static final String Search_SonameAll = "Search_SonameAll";
    //根据部门查询班主任测评情况时的当前部门名称
    public static final String Search_ManagerByCurSdname = "Search_ManagerByCurSdname";
    //根据班级查询任课教师时的当前班级号
    public static final String Search_TeacherByCurScno = "Search_TeacherByCurScno";
    //根据课程查询任课教师时的当前班级号
    public static final String Search_TeacherByCurSono = "Search_TeacherByCurSono";
    //根据班级查询对任课教师、班主任、学校的意见、建议和简要评价
    public static final String Search_ScTextByCurScno = "Search_ScTextByCurScno";
}
```

5. 新建并编制动作类文件 AdminLoginAction.java

AdminLoginAction 类的主要作用是处理系统管理员的登录请求。在该动作类中,先取出表单 bean 中的用户名及密码,调用 DBUser 类中的方法 checkAdminUser()验证该用户的合法性。若合法则将用户名保存至会话级变量 Constants.Bas_Curuser 中,并转至主页面 main.jsp;不合法则转至 loginfail.jsp 页面,并显示错误信息。

实现代码如下：

```java
package com.teach.actions;

import java.util.Vector;
import com.teach.comms.*;
import com.teach.forms.LoginForm;
import javax.servlet.http.*;
import javax.servlet.ServletContext;
import javax.sql.DataSource;
import org.apache.struts.action.*;
public class AdminLoginAction extends Action
{
    public ActionForward execute(ActionMapping mapping,ActionForm form,
            HttpServletRequest request,HttpServletResponse response) throws Exception
    {
        LoginForm loginform = (LoginForm)form;
        String username = loginform.getUsername();
        String userpwd  = loginform.getUserpwd();

        ServletContext context = servlet.getServletContext();
        DataSource dataSource = (DataSource)context.getAttribute("dbkey");

        DBUser dbuser = new DBUser(dataSource);
        ActionMessages errors = new ActionMessages();
        if(!dbuser.checkAdminUser(username,userpwd))             //不存在该用户
        {
            errors.add(ActionMessages.GLOBAL_MESSAGE,new ActionMessage("errors.loginFail"));
            if(!errors.isEmpty())
            {
                saveErrors(request,errors);
            }
            return mapping.findForward("LoginFail");
        }
        else
        {
            HttpSession session = request.getSession();
            session.setAttribute(Constants.Login_Curuser,username);

            return mapping.findForward("Success");               //登录成功
        }
    }
}
```

6．编辑属性文件 ApplicationResources.properties

属性文件 ApplicationResources.properties 是一个资源文件，其中一般定义有系统运行过程中出现的错误信息。本系统中定义的相关错误信息如图 6-10 所示。

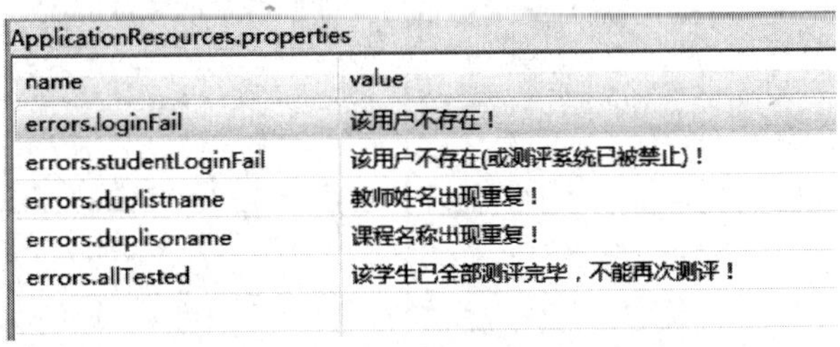

ApplicationResources.properties	
name	value
errors.loginFail	该用户不存在！
errors.studentLoginFail	该用户不存在(或测评系统已被禁止)！
errors.duplistname	教师姓名出现重复！
errors.duplisoname	课程名称出现重复！
errors.allTested	该学生已全部测评完毕，不能再次测评！

图 6-10　属性文件中定义的错误信息

7. 新建并编制 loginfail.jsp 文件

如图 6-11 所示，该页面用<html:errors/>标签显示登录过程中出现的错误信息，被三个子系统所共用。页面用嵌套表格布局页面，表格在页面中左右居中，上下方向居上。

图 6-11 显示登录失败信息的页面

实现代码如下：

```
<%@page contentType="text/html;charset=utf-8"%>
<%@taglib uri="/WEB-INF/struts-html.tld" prefix="html"%>

<html>
  <head>
    <title>教师教育教学情况测评系统</title>
    <link href="../css/style_css.css" rel=stylesheet type="text/css">
  </head>
  <body>
    <table width="100%" height="100%">
      <tr>
        <td align="center" valign="top">
          <div align=center>
            <table bgcolor="#f1fcf7">
              <tr height="30">
                <td width="400" align="center" bgcolor="#dbc2b0"><b>信息显示页面</b></td>
              </tr>
              <tr>
                <td align="center" bgcolor="#f5efe7">
                  <font size="5" color="#ff0000"><html:errors/></font>
                </td>
              </tr>
              <tr height="50">
                <td align="center" bgcolor="#f5efe7">
                  <a href="javascript:history.back()">返回</a>
                </td>
              </tr>
            </table>
            <hr width="400" color="#dbc2b0" size="2">
          </div>
        </td>
      </tr>
    </table>
  </body>
</html>
```

8. 新建并编制 first.jsp 文件

如图 6-12 所示，该页面仅用来在登录成功后在框架右侧栏中显示欢迎信息。

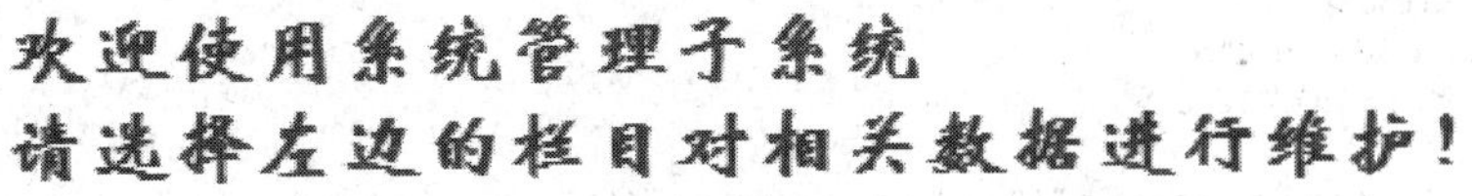

项目六 教师测评系统(Struts模式)

欢迎使用系统管理子系统
请选择左边的栏目对相关数据进行维护!

图 6-12　登录成功后显示欢迎信息的页面

实现代码如下:

```
<%@page contentType="text/html;charset=utf-8"%>
<html>
    <body>
        <br>
        <br>
        <p>
        <font face="华文新魏" color="#996633" size="5">
            <b>
                            欢迎使用系统管理子系统
            <br>
                            请选择左边的栏目对相关数据进行维护!
            </b></font>
            <br>
    </body>
</html>
```

9. 新建并编制 main.jsp 文件

如图 6-8 所示,该页面为系统主框架页面,采用 DIV+CSS 设计,分上、下、左、右四部分,左侧栏中为各模块对应的超链接,各超链接暂时为空,右侧栏用于显示具体的操作页面。

实现代码如下:

```
<%@page contentType="text/html;charset=utf-8"%>
<%@page import="com.teach.comms.Constants,java.util.Vector"%>

<%
    String username = (String)session.getAttribute(Constants.Login_Curuser);
    if(username == null || username.equals(""))
    {
        response.sendRedirect("login.jsp");
    }
%>
<html>
    <head>
        <title>教师教育教学情况测评系统---系统管理子系统</title>
        <link href="../css/main.css" rel="stylesheet" type="text/css" />
        <link href="../css/style_css.css" rel="stylesheet" type="text/css" />
    </head>

    <body>
        <div id="main">
            <div id="top">
                <table  width="100%" height="100%">
                    <tr align="center" valign="middle">
                        <td align="left">
                               
                            <font face="宋体" color="#FF0000" size="5"><b>教师教育教学情况测评系统</b></font>
                               
                            <font face="隶书" color="#FF0000" size="5"><b>系统管理子系统</b></font>
                        </td>
                    </tr>
                </table>
```

```html
        </div>
        <div id="mid">
            <div id="left">
                <table width="100%" cellspacing="1">
                    <tr height="25">
                        <td align="center"> </td>
                    </tr>
                    <tr height="25">
                        <td align="center"><a href="" target="inMain">教师数据维护</a></td>
                    </tr>
                    <tr height="25">
                        <td align="center"><a href="" target="inMain">课程数据维护</a></td>
                    </tr>
                    <tr height="25">
                        <td align="center"><a href="" target="inMain">班级数据维护</a></td>
                    </tr>
                    <tr height="25">
                        <td align="center">
                            <fieldset style="width:150px;">
                                <legend align="center">班级科目数据维护</legend>
                                <select name="select1" style="width:120px;" onchange="">
                                    <option value="">---请选择班级---</option>
                                </select>
                            </fieldset>
                        </td>
                    </tr>
                    <tr height="25">
                        <td align="center"><a href="" target="inMain">系统用户数据维护</a></td>
                    </tr>
                    <tr height="25">
                        <td align="center"><a href="" target="inMain">系统初始化</a></td>
                    </tr>
                    <tr height="25">
                        <td align="center"><a href="" target="inMain">开放或禁止测评系统</a></td>
                    </tr>
                    <tr height="25">
                        <td align="center"><a href="" target="inMain">汇总班级测评数据</a></td>
                    </tr>
                    <tr height="25">
                        <td align="center"><a href="" target="inMain">汇总教师测评数据</a></td>
                    </tr>
                </table>
            </div>
            <div id="right">
                <iframe src="first.jsp" name="inMain" id="inMain" width="100%" height="100%" frameborder=0 scrolling="yes"></iframe>
            </div>
        </div>
        <div id="end">
            <table width="100%" height="100%">
                <tr>
                    <td align="center" valign="middle">
                        <table>
                            <tr>
                                <td>
                                    版权所有 2009 XXX 高等职业技术学校
                                </td>
                            </tr>
                        </table>
                    </td>
```

```
                </tr>
            </table>
        </div>
      </div>
    </body>
</html>
```

10. 配置表单 bean 及动作 Action 类

配置 LoginForm 表单 bean 的代码应写在 struts-config. xml 文件的<form-beans >元素对中。

配置代码如下：

```
<form - bean name = "loginform"          type = "com.teach.forms.LoginForm" />
```

配置 AdminLoginAction 动作类映射关系的代码应写在 struts-config. xml 文件的<action-mappings>元素对中。

配置代码如下：

```
< action path = "/bas/adminloginaction"   type = "com.teach.actions.AdminLoginAction" name = "loginform" scope =
"request" validate = "false">
            < forward name = "Success"     path = "/bas/main.jsp"/>
            < forward name = "LoginFail"     path = "/bas/loginfail.jsp"/>
</action>
```

11. 发布并测试程序

启动 Tomcat 服务器，将 TeacherTest 应用发布至服务器，打开浏览器，在地址栏中输入 http://localhost:8080/ TeacherTest，运行并测试系统管理子系统的登录模块。

任务三　编制教师基本情况维护模块

【任务描述】

本任务主要编制教师基本情况维护模块中的相关文件并测试运行。教师基本情况维护模块主要是对教师情况表 steacher 中数据进行分页显示、添加、修改及删除操作。其运行示意图如图 6-13、图 6-14、图 6-15 所示。

教师情况表

序号	教师代码	教师姓名	所在系部
281	069	周文斌	信息工程系
282	070	吴海萍	信息工程系
283	071	是新宇	信息工程系
284	072	袁琦兰	信息工程系
285	073	高文卓	信息工程系
286	074	刘玲	信息工程系
287	075	周烨	信息工程系
288	076	林志利	信息工程系
289	077	陈士杰	信息工程系
290	078	周凯	信息工程系
291	079	王丽霞	信息工程系
292	081	钱琳媛	信息工程系
293	082	李新南	信息工程系
294	083	王影	信息工程系
295	084	徐丽平	信息工程系
296	085	黄秀娟	信息工程系
297	086	支元	信息工程系
298	087	王登科	信息工程系
299	088	李忠	信息工程系
300	089	房艳萍	信息工程系

共439条记录　第15页/共22页　GO　首页　上一页　下一页　末页　添 加　修 改　删 除

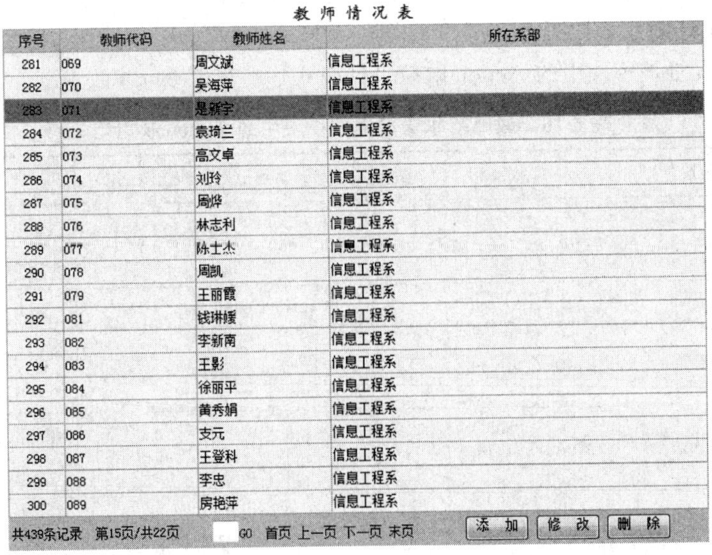

图 6-13　教师基本情况列表页面

图 6-14 新增教师信息页面

图 6-15 修改教师信息页面

【任务分析】

1. 程序运行基本流程

教师基本情况维护模块运行流程图如图 6-16 所示。

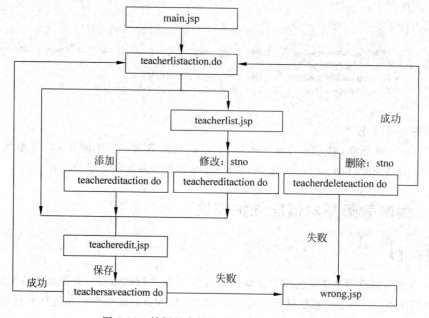

图 6-16 教师基本情况维护模块运行流程图

2. Action 映射表

教师基本情况维护模块中的 Action 映射表见表 6-16。

表 6-16 教师基本情况维护模块中的 Action 映射表

动作(Action)	入口	表单 bean（ActionForm）	出口
TeacherListAction	main.jsp		teacherlist.jsp teacheredit.jsp
TeacherEditAction	teacherlist.jsp	TeacherForm	teacheredit.jsp
TeacherSaveAction	teacheredit.jsp	TeacherForm	teacherlist.jsp wrong.jsp
TeacherDeleteAction	teacherlist.jsp		teacherlist.jsp wrong.jsp
PageAction	teacherlist.jsp	DynaActionForm	teacherlist.jsp

项目六 教师测评系统(Struts模式)

【任务实施】

1. 实现数据的列表显示

(1) 新建并编制 JavaBean 类文件 Teacher.java

这是一个代表教师的实体 JavaBean 组件。其中的私有属性与教师情况表 steacher 中的列一致,类中还要编写与属性配套的 get/set 方法。

实现代码如下:

```java
package com.teach.comms;

import java.io.Serializable;
public class Teacher implements Serializable
{
    private String stno;
    private String stname;
    private String sdname;
    //与属性对应的 get 和 set 方法
    public String getStno()
    {
        return stno;
    }
    public void setStno(String stno)
    {
        this.stno = stno;
    }
    …//省略与其余属性对应的 get 和 set 方法
}
```

(2) 修改 DBUser 类

在其中定义一个方法,方法为 getSTeacher(),该方法的作用是取出所有教师数据,每个教师的数据存放在 Teacher 类型的一个对象中,所有教师数据存放在一个 List 类型的对象中,并返回。

实现代码如下:

```java
public List getSTeacher()
{
        Connection con = null;
        List   sTeacher = new ArrayList();
        Teacher teacher = null;

        try
        {
            con = dataSource.getConnection();
            Statement stat = con.createStatement();
            ResultSet result = stat.executeQuery("select * from steacher order by sdname");
            while(result.next())
            {
                teacher = new Teacher();
                teacher.setStno(result.getString("stno"));
                teacher.setStname(result.getString("stname"));
                teacher.setSdname(result.getString("sdname"));

                sTeacher.add(teacher);
            }
        }
        catch(SQLException e)
            …//省略部分同 CheckAdminUser(username,userpwd)方法中类同部分
        return STeacher;
}
```

（3）新建并编制动作类文件 TeacherListAction.java

该类的主要作用是处理数据列表显示请求。在该类中调用 DBUser 类中的 getSTeacher()方法,取出所有教师数据,并存放至会话级变量 Search_AllList 中；如果没有教师数据,则引导至编辑页面 teacheredit.jsp 进行新增,否则引导至 teacherlist.jsp 页面进行列表显示。

实现代码如下：

```java
package com.teach.actions;

import java.util.List;
…//省略的包和类可参考 AdminLonginAction
public class TeacherListAction extends Action
{
    public ActionForward execute(ActionMapping mapping, ActionForm form, HttpServletRequest request,
    HttpServletResponse response) throws Exception
    {
        ServletContext context = servlet.getServletContext();
        DataSource dataSource = (DataSource)context.getAttribute("dbkey");
        DBUser dbuser = new DBUser(dataSource);

        HttpSession session = request.getSession();
        List sTeacher = dbuser.getSTeacher();
        session.setAttribute(Constants.Search_AllList,sTeacher);

        if (sTeacher.size() == 0)
        {
            return  mapping.findForward("TeacherEdit");
        }
        else
        {
            return (mapping.findForward("TeacherList"));      //列出所有课程
        }
    }
}
```

（4）新建并编制 teacherlist.jsp 文件

如图 6-13 所示,本页面从会话级变量 Search_AllList 中的取出所有教师数据,存至一 List 类型的变量中,在数据行处,对 List 类型的变量中的教师数据进行循环,逐个取出教师数据并以图示的形式显示出来。行选择及分页的功能暂未实现。

实现代码如下：

```jsp
<%@page contentType="text/html;charset=utf-8"%>
<%@page import="com.teach.comms.Constants,com.teach.comms.Teacher,java.util.List"%>
<%
    List sTeacher = (List)session.getAttribute(Constants.Search_AllList);
%>
<html>
    <head>
        <link href="../css/style_css.css" rel=stylesheet type="text/css">
    </head>
    <body>
        <table  width="100%" height="100%">
            <tr>
                <td  align="center" valign="top">
                    <table width="650"   bgcolor="#dec3b5" cellspacing="1"  >
                        <caption>教 师 情 况 表</caption>

                        <tr height="20" align="center">
                            <td width="50"   bgcolor="#f3d5d5">序号</td>
                            <td width="120"   bgcolor="#f3d5d5">教师代码</td>
```

项目六 教师测评系统(Struts模式)

```
            <td width="120"   bgcolor="#f3d5d5">教师姓名</td>
            <td bgcolor="#f3d5d5">所在系部</td>
        </tr>

        <%
            if(sTeacher != null )
            {
                for(int i = 0;i<sTeacher.size();i++)
                {
                    Teacher teacher = (Teacher)sTeacher.get(i);
        %>
        <tr  height="20"  bgcolor="#F1FCF7">
          <td  align="center"><%=i+1%></td>
          <td><%=teacher.getStno()%></td>
          <td><%=teacher.getStname()%></td>
          <td><%=teacher.getSdname()%></td>
        </tr>

        <%
                }
            }
        %>
      </table>
    </td>
  </tr>
</table>
</body>
</html>
```

(5) 配置动作类 TeacherListAction

配置 TeacherListAction 动作类映射关系的代码应写在＜action-mappings＞元素对中。

配置代码如下：

```
<action path="/bas/teacherlistaction"    type="com.teach.actions.TeacherListAction"   scope="request">
        <forward name="TeacherEdit"   path="/bas/teacheredit.jsp"/>
        <forward name="TeacherList"   path="/bas/teacherlist.jsp"/>
</action>
```

(6) 修改主页面文件 main.jsp

将"教师数据维护"的超链接改为：

```
<a href="teacherlistaction.do"   target="inMain">教师数据维护</a>
```

其中,"inMain"为主框架右侧栏子窗体的 id 值。

(7) 发布并测试程序

启动 Tomcat 服务器，将 TeacherTest 应用发布至服务器，打开浏览器，在地址栏中输入 http://localhost:8080/ TeacherTest,运行并测试是否能正确地列出所有教师数据。

2. 实现数据列表的行选择

(1) 新建并编制 JavaScript 程序文件 ClientCom.js

在 web 文件夹下的 js 文件夹中新建该文件，在其中定义一个函数 SetFocusIt。该函数的作用有两方面，一是将指定的行选中，即该行背景色为指定颜色不同于其他行的背景色；二是将选中行对应记录的关键字值保存下来。

实现代码如下：

```
var bakRow;
var bakColor;
```

```
var selectRowColor = "#ff99ff";
function SetFocusIt(curRow,keyNo,isFirst,keyValue)
{
    if (isFirst!== 0)
    {
    bakRow.bgColor = bakColor;
    }

    bakRow = curRow;
    bakColor = curRow.bgColor;
    curRow.bgColor = selectRowColor;
    keyValue.value = keyNo;
}
```

（2）修改 teacherlist.jsp 文件

先在页面的<head>标签对中，用以下<script>标签对将 ClientCom.js 文件包含进来：

```
<script type="text/javascript"  src="../js/ClientCom.js"></script>
```

再在页面中添加一隐藏域，该隐藏域用于存放所选中行对应记录的关键字值，可放在页面的<body>标签对中适当位置（第二个<table>标签前），名为"keyvalue"，初值为空。为了能区别各行，要在显示数据的行标签<tr>中，定义属性 id="Row<%=i+1%>"，并在数据行的单击行事件 onclick 中编写调用 ClientCom.js 中的 SetFocusIt 方法的代码来选择行，为了在第一次打开页面时选中第一行，在显示数据的行标签<.tr>结束处，也要编写调用 SetFocusIt 方法的代码，并注意代码中大括号与数据行开始处的大括号的匹配。

修改后的 teacherlist.jsp 文件实现代码如下：

```
<%@page contentType="text/html;charset=utf-8"%>
<%@page import="com.teach.comms.Constants,com.teach.comms.Teacher,java.util.List"%>
<%
List sTeacher = (List)session.getAttribute(Constants.Search_AllList);
%>
<html>
    <head>
      <link href="../css/style_css.css" rel=stylesheet type="text/css">
      <script type="text/javascript"  src="../js/ClientCom.js"></script>
    </head>
    <body>
        <table  width="100%" height="100%">
            <tr>
                <td  align="center" valign="top">
                    <input type="hidden" name="keyvalue" value="" />
                    <table width="650"  bgcolor="#dec3b5" cellspacing="1" >
                        <caption>教 师 情 况 表</caption>

                        <tr height="20" align="center" >
                        <td width="50"  bgcolor="#f3d5d5">序号</td>
                        <td width="120"  bgcolor="#f3d5d5">教师代码</td>
                        <td width="120"  bgcolor="#f3d5d5">教师姓名</td>
                        <td bgcolor="#f3d5d5"> 所在系部</td>
                        </tr>

                        <%
                            if(sTeacher != null )
                            {
                                for(int i = 0;i<sTeacher.size();i++)
                                {
                                    Teacher teacher = (Teacher)sTeacher.get(i);
```

项目六　教师测评系统(Struts模式)

```
                    %>
                         <tr height="20" style="cursor:hand" bgcolor="#F1FCF7" id="Row<%=i+
1%>" title="鼠标双击可以编辑该信息"onclick="SetFocusIt(this,'<%=teacher.getStno()%>',1,keyvalue)"
ondblclick="modify()">
                               <td align="center"><%=i+1+Constants.Rows_PerPapge*iPageNo_Cur%></td>
                               <td><%=teacher.getStno()%></td>
                               <td><%=teacher.getStname()%></td>
                               <td><%=teacher.getSdname()%></td>
                         </tr>
                         <%
                             if(i==0)
                             {
                         %>
                                 <script language="javascript">
                                     SetFocusIt(Row1,'<%=teacher.getStno()%>',0,keyvalue);
                                 </script>
                         <%
                             }
                         }
                     %>
                 </table>
             </td>
         </tr>
     </table>
 </body>
</html>
```

(3) 发布并测试程序

启动 Tomcat 服务器,将 TeacherTest 应用发布至服务器,打开浏览器,在地址栏中输入 http://localhost:8080/TeacherTest,运行并测试是否能正确地以指定颜色选择教师数据行。

3. 实现数据列表的分页显示

(1) 新建并编制分页处理类文件 PageControl.java

这是一个 Javabean 工具类,其中定义了一个成员方法 setPageData(),主要用于分页处理,即根据传入的待分页数据,分离出目标页的数据,计算出记录数、总页数等信息。

实现代码如下:

```java
package com.teach.comms;
import java.util.*;
import javax.servlet.http.HttpSession;
public class PageControl
{
    public void setPageData(List allList,HttpSession session,Integer pageId) //pageId:目标页号-1
    {
        /*  总记录数         当前记录号       总页数         */
        int iRecords_All=0,iPageNo_Cur=0,iPages_All=0;

        if(pageId!=null)
            iPageNo_Cur = pageId.intValue();

        if(iPageNo_Cur<0)
            iPageNo_Cur = 0;

        /*记录总数*/
        iRecords_All = allList.size();

        /*无记录,在当前页置为0*/
        if(iRecords_All==0)
            iPageNo_Cur = -1;
```

```java
        /*计算总页数*/
        if (allList.size() % Constants.Rows_PerPapge == 0)
        {
            iPages_All = allList.size() / Constants.Rows_PerPapge;
        }
        else
        {
            iPages_All = allList.size() / Constants.Rows_PerPapge + 1;
        }

        /*获取分页显示的记录集 */
        if ((allList.size()> iPageNo_Cur * Constants.Rows_PerPapge )&&(iPageNo_Cur >= 0))
        {
            List pageList = new ArrayList();
            for (int i = iPageNo_Cur * Constants.Rows_PerPapge; i < (iPageNo_Cur + 1) * Constants.Rows_PerPapge; i++)
            {
                if (i < allList.size())
                {
                    pageList.add(allList.get(i));
                }
            }
            /*存储分页数据至session中*/
            session.setAttribute(Constants.Search_PageList, pageList);
        }
        else
        {
            /*存储分页数据至session中*/
            session.setAttribute(Constants.Search_PageList, allList);
        }

        /*存储分页信息至session中*/
        session.setAttribute(Constants.Records_All, iRecords_All);
        session.setAttribute(Constants.PageNo_Cur, iPageNo_Cur);
        session.setAttribute(Constants.Pages_All   , iPages_All);
    }
}
```

(2) 新建并编制动作类文件PageAction.java

这是一个自定义动作类,主要用于分页控制、跳转处理等。

实现代码如下:

```java
package com.teach.comms;
import java.util.List;
…//省略的包和类可参考AdminLoginAction
public class PageAction extends Action
{
    public ActionForward execute (ActionMapping mapping, ActionForm form, HttpServletRequest request, HttpServletResponse response) throws Exception
    {
        DynaActionForm pageForm = (DynaActionForm)form;
        Integer pageId = (Integer)pageForm.get("pageId");           //pageId: 目标页号 - 1
        String forwardPage = (String)pageForm.get("forwardPage");   //mapping的目标页面

        pageId = (Integer)(pageId.intValue() - 1 );

        HttpSession session = request.getSession();
        int iPageNo_Cur = ((Integer)session.getAttribute(Constants.PageNo_Cur)).intValue();
        int iPages_All = ((Integer)session.getAttribute(Constants.Pages_All)).intValue();

        //目标页号在范围内,并且与当前页号不一致,进行分页处理
        if(pageId >= 0 && pageId < iPages_All && pageId!= iPageNo_Cur)
```

项目六 教师测评系统(Struts模式)

```
        {
            List allList = (List)session.getAttribute(Constants.Search_AllList);
            PageControl pageControl = new PageControl();
            pageControl.setPageData(allList,session,pageId);
        }
        return mapping.findForward(forwardPage);
    }
}
```

(3) 配置动态表单 bean 及动作类

配置动态表单 bean 的代码应写在 struts-config.xml 文件的<form-beans>元素对中。

配置代码如下:

```xml
<form-bean name="pageform" type="org.apache.struts.action.DynaActionForm">
        <form-property name="pageId"       type="java.lang.Integer"/>
        <form-property name="forwardPage" type="java.lang.String"/>
</form-bean>
```

配置 PageAction 动作类映射关系的代码应写在 struts-config.xml 文件的<action-mappings>元素对中。

配置代码如下:

```xml
<action path="/bas/pageaction"        type="com.teach.comms.PageAction"  name="pageform" scope="request" >
        <forward name="TeacherList"    path="/bas/teacherlist.jsp"/>
</action>
```

(4) 修改动作类文件 TeacherListAction.java

在动作类的 if…else…结构的 else 分支中,使用前面步骤定义的分页处理类 PageControl 对教师列表数据进行分页处理。

实现代码如下:

```java
…
else
{
        PageControl pageControl = new PageControl();
        pageControl.setPageData(sTeacher,session,0);

        return (mapping.findForward("TeacherList"));           //列出所有课程
}
…
```

(5) 修改 teacherlist.jsp 文件

在 teacherlist.jsp 页面开始处,修改嵌入的 Java 代码,取出分页处理后相关分页数据。

实现代码如下:

```jsp
…
<%
    List sTeacher = (List)session.getAttribute(Constants.Search_PageList);

    Integer Records_All = (Integer)session.getAttribute(Constants.Records_All);
    Integer PageNo_Cur = (Integer)session.getAttribute(Constants.PageNo_Cur);
    Integer Pages_All = (Integer)session.getAttribute(Constants.Pages_All);

    int iRecords_All = Records_All.intValue();
    int iPageNo_Cur = PageNo_Cur.intValue();
    int iPages_All = Pages_All.intValue();
%>
…
```

在页面中数据行的下面，放置一表格，表格有一行，其上有分页跳转的超链接。
实现代码如下：

```
…
<table width="650" bgcolor="#f3d5d5">
    <tr height="20">
        <td>共<%=iRecords_All%>条记录  第<%=iPageNo_Cur+1%>页/共<%=iPages_All%>页</td>
        <td>
            <input type="text" name="pageId" style="width:25px;height:20px;"><a href="javascript:Goto()">GO</a> 
            <a href="pageaction.do?pageId=1&forwardPage=TeacherList">首页</a>
            <a href="pageaction.do?pageId=<%=iPageNo_Cur%>&forwardPage=TeacherList">上一页</a>
            <a href="pageaction.do?pageId=<%=iPageNo_Cur+2%>&forwardPage=TeacherList">下一页</a>
            <a href="pageaction.do?pageId=<%=iPages_All%>&forwardPage=TeacherList">末页</a>
        </td>
    </tr>
</table>
…
```

在页面的<script>标签对中，定义一用于跳转的函数。
实现代码如下：

```
…
function Goto()
{
    location.href="pageaction.do?pageId=" + document.getElementById("pageId").value + "&forwardPage=TeacherList";
}
…
```

（6）发布并测试程序

启动Tomcat服务器，将TeacherTest应用发布至服务器，打开浏览器，在地址栏中输入http://localhost:8080/TeacherTest，运行并测试是否能正确地将教师数据进行分页显示。

4．实现数据的编辑

（1）修改teacherlist.jsp文件

在teacherlist.jsp页面中加入"添加"、"修改"、"删除"三个按钮，按钮可与分页跳转超链接放在同一行上。
实现代码如下：

```
…
<td> </td>
<td>
<input type="button" name="addbutton" value="添 加" onclick="add()">
<input type="button" name="editbutton" value="修 改" onclick="modify()">
<input type="button" name="delbutton" value="删 除" onclick="Delete()">
</td>
…
```

（2）定义"添加"、"修改"、"删除"三个按钮单击事件执行的函数

在页面的<script>标签对中，再分别定义函数add()、modify()、Delete()，"添加"时，引导执行编辑教师数据的动作类TeacherEditAction；"修改"时，引导执行编辑教师数据的动作类TeacherEditAction，并将当前选中行的教师代码作为参数传递过去；"删除"时，先让用户进行删除

确认,确认要删除后再引导执行删除教师数据的动作类 TeacherDeleteAction,并将当前选中行的教师代码作为参数传递过去。

实现代码如下:

```javascript
…
function add()
{
        location.href = "teachereditaction.do";
}
function modify()
{
        if (document.all.keyvalue.value == "")
            return false;
        location.href = "teachereditaction.do?stno = " + document.all.keyvalue.value;
}
function Delete()
{
        if (document.all.keyvalue.value == "")
            return;

        if (confirm("你是否要删除选中的记录数据?") == true)
            location.href = "teacherdeleteaction.do?stno = " + document.all.keyvalue.value;
}
…
```

(3) 新建并编制表单 bean 类文件 TeacherForm.java

表单 bean 类 TeacherForm 中的属性与教师数据编辑页面中表单元素相对应,类中还有与属性配合的获取(getter)和设置(setter)属性的方法。

实现代码如下:

```java
package com.teach.forms;

import org.apache.struts.action.*;
public class TeacherForm extends ActionForm
{
    private String stno;
    private String stname;
    private String sdname;

    public String getStno()
    {
        return stno;
    }
    public void setStno(String stno)
    {
        this.stno = stno;
    …//省略与其属性对应的 get 和 set 方法
    }
}
```

(4) 修改 DBUser 类

在其中定义一个方法,方法为 getSTeacherByStno(String stno),该方法的作用是根据教师代码取出该教师数据,并存放在 Teacher 类型的一个对象中返回。

实现代码如下:

```java
public Teacher getSTeacherByStno(String stno)
{
        Connection con = null;
        Teacher   teacher = null;
```

```
            try
            {
                con = dataSource.getConnection();
                Statement stat = con.createStatement();
                ResultSet result = stat.executeQuery("select * from steacher where stno = '" + stno + "'");
                if(result.next())
                {
                    teacher = new Teacher();
                    teacher.setStno(result.getString("stno"));
                    teacher.setStname(result.getString("stname"));
                    teacher.setSdname(result.getString("sdname"));
                }
            }
            catch(SQLException e)
            …//省略与 CheckAdminUser(username,userpwd)方法中类同部分
            return teacher;
        }
```

(5) 新建并编制动作类文件 TeacherEditAction.java

该类的主要作用是处理数据编辑请求。在该类中,先取出传递过来的教师代码,如果为空,则是添加;如果不为空,则是修改;添加时直接引导至 teacheredit.jsp 页面,修改时调用 DBUser 类中的 getSTeacherByStno()方法,取出该教师数据,存放在 TeacherForm 类的对象中,并引导至 teacheredit.jsp 页面。

实现代码如下:

```
package com.teach.actions;

import com.teach.comms.*;
…//省略的包和类可参考 AdminLoginAction
public class TeacherEditAction extends Action
{
    public ActionForward execute(ActionMapping mapping, ActionForm form, HttpServletRequest request,
    HttpServletResponse response) throws Exception
    {
        TeacherForm teacherform = (TeacherForm) form;
        String stno = request.getParameter("stno");

        ServletContext context = servlet.getServletContext();
        DataSource dataSource = (DataSource)context.getAttribute("dbkey");
        DBUser dbuser = new DBUser(dataSource);
        Teacher teacher = dbuser.getSTeacherByStno(stno);
        if(teacher!= null)
        {
            teacherform.setStno(teacher.getStno());
            teacherform.setStname(teacher.getStname());
            teacherform.setSdname(teacher.getSdname());
        }

        return mapping.findForward("TeacherEdit");
    }
}
```

(6) 新建并编制 teacheredit.jsp 文件

如图 6-14 和图 6-15 所示,用嵌套表格布局页面,表格居中。表单及表单中的域必须用 Struts 标签库中的 html 标签表示,教师代码域 stno 作为隐藏域。

实现代码如下:

```
<%@page contentType="text/html;charset=utf-8"%>
<%@taglib uri="http://struts.apache.org/tags-html" prefix="html"%>
```

项目六 教师测评系统(Struts模式)

```html
<html>
  <head>
    <link href="../css/style_css.css" rel=stylesheet type="text/css">
    <title>教师信息编辑修改</title>
    <script type="text/javascript">
      function checkdata()
      {
        var stname = document.teacherform.stname.value;
        if (stname == "" )
        {
          alert("姓名不能为空,必须输入!");
          document.teacherform.stname.focus();
          return false;
        }

        return true;
      }
      function Exit()
      {
        history.go(-1);
      }
    </script>
  </head>

  <body onload="document.teacherform.stname.focus()">
    <table width="100%" height="100%" >
      <tr align="center" valign="middle">
        <td>
        <html:form action="/bas/teachersaveaction.do">
        <table bgcolor="#f1fcf7">
          <tr>
            <td colspan="2" align="center">教师信息编辑</td>
          </tr>
          <tr>
            <td align="center">教师姓名</td><td><html:text property="stname"/></td><td><html:errors property="stname"/></td>
          </tr>
          <tr>
            <td>所在部门</td><td><html:text property="sdname"/></td>
            <html:hidden property="stno"/>
          </tr>
          <tr>
            <td></td><td align="center"><html:submit property="submit" value="保 存" onclick="return checkdata()"/>    <html:button property="exit" value="取 消" onclick="Exit()"/></td>
          </tr>
        </table>
        </html:form>
        </td>
      </tr>
    </table>
  </body>
</html>
```

(7) 配置表单 bean 及动作 Action 类

配置 TeacherForm 表单 bean 的代码应写在 struts-config.xml 文件的<form-beans>元素对中。

配置代码如下：

```
<form-bean name="teacherform" type="com.teach.forms.TeacherForm" />
```

配置 TeacherEditAction 动作类映射关系的代码应写在 struts-config.xml 文件的<action-mappings>元素对中。

配置代码如下：

```
< action path = "/bas/teachereditaction"     type = " com. teach. actions. TeacherEditAction"     name = "teacherform"   scope = "request" validate = "false">
        < forward name = "TeacherEdit"    path = "/bas/teacheredit.jsp"/>
</action>
```

（8）发布并测试程序

请注意，为了在测试时能正确地显示编辑页面，可以先将teacheredit.jsp页面表单属性action值中的teachersaveaction改为teachereditaction，这是因为teachersaveaction尚未编制和配置。启动Tomcat服务器，将TeacherTest应用发布至服务器，打开浏览器，在地址栏中输入http://localhost:8080/TeacherTest，运行并测试教师记录增加及修改页面是否能正确地显示出来。

5．实现数据的保存、删除

（1）修改DBUser类

在其中定义一个方法，方法为insertTeacher(String stname, String sdname)，该方法的作用是插入教师数据至数据库，成功返回true，失败返回false，其中教师代码stno列的值（001、002、003…）是在程序中生成的。

在其中再定义一个方法updateTeacher(String stno, String stname, String sdname)，该方法的作用是根据教师代码更新教师数据至数据库，成功返回true，失败返回false。

实现代码如下：

```java
//插入一个新教师
public boolean insertTeacher(String stname, String sdname)
{
        Connection con = null;
        boolean flag = false;
        String stno = "";
        try
        {
            con = dataSource.getConnection();
            Statement stat = con.createStatement();
            ResultSet result = stat.executeQuery("select max(stno) + 1 as stno from steacher");
            if ( result.next())
            {
                stno = result.getString("stno");
                if(stno == null || stno.equals(""))
                    stno = "001";
                else if(stno.length() == 1)
                    stno = "00" + stno;
                else if(stno.length() == 2)
                    stno = "0" + stno;
            }

            PreparedStatement pstat = con.prepareStatement("insert into steacher values(?,?,?)");
            pstat.setString(1, stno);
            pstat.setString(2, stname);
            pstat.setString(3, sdname);

            int row = pstat.executeUpdate();
            if(row == 1)
                flag = true;
            …//省略与CheckAdminUser(username, userpwd)方法中类同部分
        return   flag;
}
```

（2）新建并编制wrong.jsp文件

在系统运行过程中，很多处理中会出现操作成功或失败的提示信息，操作失败的提示信息事

先是保存在会话级变量 Constants.Error_Message 中的，然后在 wrong.jsp 页面中取出并显示内容（除了登录模块，登录失败的提示信息是在 loginfail.jsp 页面中显示的）。

实现代码如下：

```jsp
<%@page contentType="text/html;charset=utf-8" %>
<%@page import="com.teach.comms.Constants" %>
<%
    String Error_Message = (String)session.getAttribute(Constants.Error_Message);
%>
<html>
    <head>
        <title>教师教育教学情况测评系统</title>
        <link href="../css/style_css.css" rel=stylesheet type="text/css">
    </head>
    <body>
        <table width="100%" height="100%">
            <tr>
                <td align="center" valign="top">
                    <div align=center>
                        <table bgcolor="#f1fcf7">
                            <tr height="30">
                                <td width="400" align="center" bgcolor="#dbc2b0"><b>信息显示页面</b></td>
                            </tr>
                            <tr>
                                <td align="center" bgcolor="#f5efe7">
                                    <font size="5" color="#ff0000"><%=Error_Message%></font>
                                </td>
                            </tr>
                            <tr height="50">
                                <td align="center" bgcolor="#f5efe7">
                                    <a href="javascript:history.back()">返回</a>
                                </td>
                            </tr>
                        </table>
                        <hr width="400" color="#dbc2b0" size="2">
                    </div>
                </td>
            </tr>
        </table>
    </body>
</html>
```

（3）配置全局转发

系统中有多个处理在完成后均要转至 wrong.jsp 页面中显示错误信息，在 Struts 中，这可以通过在 struts-config.xml 配置全局转发来实现。

配置代码如下：

```xml
<global-forwards>
    <forward name="Wrong"         path="/bas/wrong.jsp"/>
</global-forwards>
```

（4）新建并编制动作类文件 TeacherSaveAction.java

该类的主要作用是处理数据保存请求。在该类中，先取出表单中的教师代码，如果为空，则是添加，调用 insertTeacher 方法插入数据；如果不为空，则是修改，调用 updateTeacher 方法更新数据。插入或更新数据成功后引导至 teacherlist.jsp 页面，插入或更新数据失败时在 wrong.jsp 页面中显示错误信息。

实现代码如下：

```java
package com.teach.actions;
```

```java
import com.teach.comms.*;
…//省略的包和类可参考 AdminLoginAction
public class TeacherSaveAction extends Action
{
    public ActionForward execute(ActionMapping mapping,ActionForm form,
          HttpServletRequest request,HttpServletResponse response) throws Exception
    {
        TeacherForm teacherform = (TeacherForm) form;
        String stno = teacherform.getStno();

        ServletContext context = servlet.getServletContext();
        DataSource dataSource = (DataSource)context.getAttribute("dbkey");
        DBUser dbuser = new DBUser(dataSource);

        HttpSession session = request.getSession();
        if (stno == null || stno.equals(""))                        //新增教师
        {
            boolean flag = dbuser.insertTeacher(teacherform.getStname(),teacherform.getSdname());
            if(!flag)
            {
                session.setAttribute(Constants.Error_Message,"添加教师信息失败!");
                return mapping.findForward("Wrong");
            }
        }
        else                                                         //修改教师
        {
            boolean flag = dbuser.updateTeacher(stno,teacherform.getStname(),teacherform.getSdname());
            if(!flag)
            {
                session.setAttribute(Constants.Error_Message,"修改教师信息失败!");
                return mapping.findForward("Wrong");
            }
        }
        return mapping.findForward("TeacherList");
    }
}
```

（5）配置动作类 TeacherSaveAction

配置 TeacherSaveAction 动作类映射关系的代码应写在 struts-config.xml 文件的＜action-mappings＞元素对中。

配置代码如下：

```
< action path = "/bas/teachersaveaction"    type = "com.teach.actions.TeacherSaveAction"   name = "teacherform" scope = "request" validate = "false" >
    < forward name = "TeacherList"   path = "/bas/teacherlistaction.do"></forward>
</action>
```

（6）发布并测试程序

启动 Tomcat 服务器，将 TeacherTest 应用发布至服务器，打开浏览器，在地址栏中输入 http://localhost:8080/ TeacherTest，运行并测试是否能正确地插入或更新教师数据。

（7）修改 DBUser 类

在其中定义一个方法，方法为 checkTeacherDelete(String stno)，该方法的作用是在删除前检查该教师是否允许被删除，即该教师是否已被使用过，是则返回 true，否则返回 false。

实现代码如下：

```java
//检查教师是否已被使用
public boolean checkTeacherDelete(String stno)
{
```

项目六 教师测评系统(Struts模式)

```
        Connection con = null;
        boolean flag = false;

        try
        {
            con = dataSource.getConnection();
            Statement stat = con.createStatement();
            ResultSet result = stat.executeQuery("select * from sclass where stno = '" + stno + "'");
            if(result.next())
            {
                flag = true;
            }

            result = stat.executeQuery("select * from sccourse where stno = '" + stno + "'");
            if(result.next())
            {
                flag = true;
            }
        }
        catch(SQLException e)
        …//省略与CheckAdminUser(username,userpwd)方法中类同部分
        return flag;
}
```

(8) 修改 DBUser 类

在其中定义一个方法,方法为 deleteTeacher(String stno),该方法的作用是根据教师代码从数据库中删除该教师,成功返回 true,失败返回 false。

实现代码如下:

```
//删除一个教师
public boolean deleteTeacher(String stno)
    {
            Connection con = null;
            boolean flag = false;
            try
            {
                con = dataSource.getConnection();
                PreparedStatement pstat = con.prepareStatement("delete steacher where stno = ?");
                pstat.setString(1,stno);

                int row = pstat.executeUpdate();
                if(row == 1)
                    flag = true;
            }
            catch(SQLException e)
            …//省略与CheckAdminUser(username,userpwd)方法中类同部分
            return flag;
    }
```

(9) 新建并编制动作类文件 TeacherDeleteAction.java

该类的主要作用是处理数据删除请求。在该类中,先取出传递过来的教师代码,调用 checkTeacherDelete()方法检查该教师是否允许被删除,允许则调用 deleteTeacher()方法删除数据,删除数据成功后引导至 teacherlist.jsp 页面,删除数据失败时在 wrong.jsp 页面中显示错误信息。

实现代码如下:

```
package com.teach.actions;
```

```java
import com.teach.comms.Constants;
…//省略的包和类可参考 AdminLoginAction
public class TeacherDeleteAction extends Action
{
    public ActionForward execute (ActionMapping mapping, ActionForm form, HttpServletRequest request,
HttpServletResponse response) throws Exception
    {
        String stno = request.getParameter("stno");

        ServletContext context = servlet.getServletContext();
        DataSource dataSource = (DataSource)context.getAttribute("dbkey");
        DBUser dbuser = new DBUser(dataSource);

        HttpSession session = request.getSession();
        if(dbuser.checkTeacherDelete(stno))
        {
            session.setAttribute(Constants.Error_Message,"该教师已经被使用,不能删除!");
            return mapping.findForward("Wrong");
        }

        if (stno!= null && !stno.equals(""))              //删除时传递的参数
        {
            if(!dbuser.deleteTeacher(stno))
            {
                session.setAttribute(Constants.Error_Message,"删除教师记录失败!");
                return mapping.findForward("Wrong");
            }
        }
        return mapping.findForward("TeacherList");
    }
}
```

(10) 配置动作类 TeacherDeleteAction

配置 TeacherDeleteAction 动作类映射关系的代码应写在 struts-config.xml 文件的＜action-mappings＞元素对中。

配置代码如下：

```
<action path = "/bas/teacherdeleteaction" type = "com.teach.actions.TeacherDeleteAction" scope = "request" >
    <forward name = "TeacherList" path = "/bas/teacherlistaction.do"></forward>
</action>
```

(11) 发布并测试程序

启动 Tomcat 服务器，将 TeacherTest 应用发布至服务器，打开浏览器，在地址栏中输入 http://localhost:8080/TeacherTest，运行并测试是否能正确地删除教师数据。

任务四　编制课程基本情况维护模块

【任务描述】

本任务主要编制课程基本情况维护模块中的相关文件并测试运行，其编制方法与教师基本情况维护模块完全一致，可以在教师基本情况维护模块的基础上编制。课程基本情况维护模块主要是对课程情况表 scourse 中数据进行分页显示、添加、修改及删除操作。其运行示意图如图 6-17、图 6-18、图 6-19 所示。

项目六　教师测评系统(Struts模式)

图 6-17　课程基本情况列表页面

图 6-18　新增课程信息页面

图 6-19　修改课程信息页面

【任务分析】

1. 程序运行基本流程

图 6-20 所示为课程基本情况维护模块运行流程图。

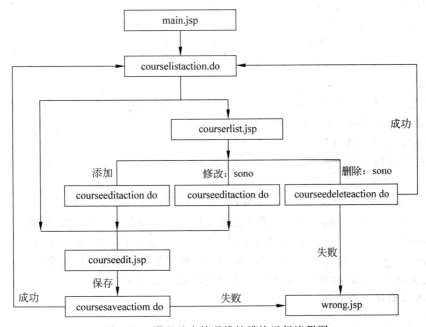

图 6-20　课程基本情况维护模块运行流程图

2. Action 映射表

课程基本情况维护模块中的 Action 映射表见表 6-17。

表 6-17　课程基本情况维护模块中的 Action 映射表

动作(Action)	入口	表单 bean(ActionForm)	出口
CourseListAction	main.jsp		courselist.jsp courseedit.jsp
CourseEditAction	courselist.jsp	CourseForm	courseedit.jsp
CourseSaveAction	courseedit.jsp	CourseForm	courselist.jsp wrong.jsp
CourseDeleteAction	courselist.jsp		courselist.jsp wrong.jsp
PageAction	courselist.jsp	DynaActionForm	courselist.jsp

【任务实施】

1. 新建并编制 JavaBean 类文件 Course.java

具体编制方法可参考 Teacher.java。

2. 修改 DBUser 类

在其中定义 6 个方法,分别为:

① getSCourse(),该方法的作用是取出所有课程数据,每个课程的数据存放在 Course 类型的一个对象中,所有课程数据存放在一个 List 类型的对象中,并返回出来。

② getSCourseBySono(String sono),该方法的作用是根据课程代码取出该课程数据,并存放在 Course 类型的一个对象中返回出来。

③ insertCourse(String soname),该方法的作用是保存课程数据至数据库,成功返回 true,失败返回 false,其中课程代码 sono 列的值(001、002、003…)是在程序中生成的。

④ updateCourse(String sono,String soname),该方法的作用是根据课程代码更新课程数据至数据库,成功返回 true,失败返回 false。

⑤ checkCourseDelete(String sono),该方法的作用是在删除前检查该课程是否允许被删除,即该课程是否已在选课表中被使用过,是则返回 true,否则返回 false。

⑥ deleteCourse(String sono),该方法的作用是根据课程代码从数据库中删除该课程,成功返回 true,失败返回 false。

具体编制时可参考教师基本情况维护模块中相关方法的定义。

3. 新建并编制表单 bean 类文件 CourseForm.java

具体编制方法可参考 TeacherForm.java。

4. 新建并编制动作类文件

新建四个动作类文件,分别为:

① CourseListAction.java　其中动作类的主要作用是处理课程数据列表显示请求;

② CourseEditAction.java　其中动作类的主要作用是处理课程数据编辑请求;

③ CourseSaveAction.java　其中动作类的主要作用是处理课程数据保存请求;

④ CourseDeleteAction.java　其中动作类的主要作用是处理课程数据删除请求。

具体编制时可参考教师基本情况维护模块中相关动作类的定义。

5. 新建并编制页面文件

新建两个页面文件,如图 6-17、图 6-18、图 6-19 所示,分别为:

① courselist.jsp　用于列表显示课程数据的页面文件；

② courseedit.jsp　用于编辑课程数据的页面文件。

具体编制时可参考教师基本情况维护模块中的相关页面。

6. 配置表单 bean 及动作 Action 类

需要在 struts-config.xml 文件的<form-beans>元素对中配置 CourseForm 表单 bean。

需要在 struts-config.xml 文件的<action-mappings>元素对中配置四个动作类 CourseListAction、CourseEditAction、CourseSaveAction、CourseDeleteAction 的映射关系。

具体配置方法可参考教师基本情况维护模块中表单 bean 及动作 Action 类的配置。

7. 修改主页面文件 main.jsp

将“课程数据维护”的超链接改为：

```
<a href = "courselistaction.do"  target = "inMain">课程数据维护</a>
```

其中,"inMain"为主框架右侧栏子窗体的 id 值。

8. 发布并测试程序

启动 Tomcat 服务器,将 TeacherTest 应用发布至服务器,打开浏览器,在地址栏中输入 http://localhost:8080/ TeacherTest,运行并测试课程基本情况维护模块。

任务五　编制班级基本情况维护模块

【任务描述】

本任务主要编制班级基本情况维护模块中的相关文件并测试运行,其编制方法与教师基本情况维护模块基本一致,可以在教师基本情况维护模块的基础上编制。班级基本情况维护模块主要是对班级情况表 sclass 中数据进行分页显示、添加、修改及删除操作。其运行示意图如图 6-21、图 6-22、图 6-23 所示。

班级情况表

序号	班级	年级	专业名称	班主任	所在系部	人数	备注
1	1011	10	财务会计	李继梅	财经管理系	54	
2	1012	10	财务会计	徐铭禧	财经管理系	53	
3	1013	10	财务会计	曹堃	财经管理系	54	
4	1021	10	国际贸易实务	王健	商贸系	51	
5	1022	10	国际贸易实务	邓云黪	商贸系	54	
6	1023	10	商务英语	王运奇	商贸系	46	
7	1024	10	商务英语	刘烨	商贸系	47	
8	1031	10	计算机网络技术	房艳萍	信息工程系	41	
9	1032	10	计算机网络技术	王登科	信息工程系	40	
10	1033	10	计算机网络技术	徐栋	信息工程系	42	
11	1034	10	应用化工技术	刘玲	信息工程系	44	
12	1041	10	动漫设计与制作	韦玮	动画艺术系	45	
13	1042	10	动漫设计与制作	浦晶晶	动画艺术系	46	
14	1043	10	广告设计与制作	马莉	动画艺术系	42	
15	1051	10	数控技术	杨国威	机电工程系	39	
16	1052	10	数控技术	肖雪武	机电工程系	38	
17	1053	10	模具设计与制造	杨晓俊	机电工程系	38	
18	1054	10	模具设计与制造	朱洪琪	机电工程系	38	
19	1055	10	汽车检测与维修技术	许杭建	机电工程系	36	
20	1056	10	汽车检测与维修技术	张长江	机电工程系	35	

共112条记录　第1页/共6页　GO　首页 上一页 下一页 末页　添 加　修 改　删 除

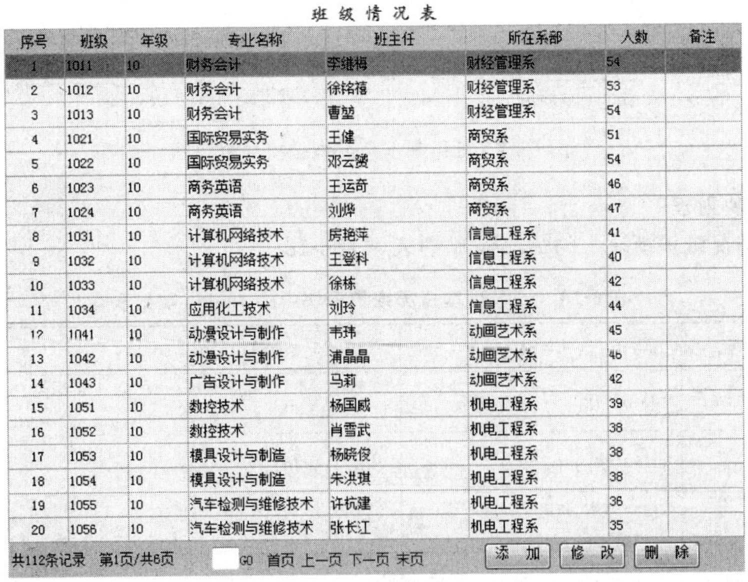

图 6-21　班级基本情况列表页面

图 6-22 新增班级信息页面

图 6-23 修改班级信息页面

【任务分析】

1. 程序运行基本流程

图 6-24 所示为班级基本情况维护模块流程图。

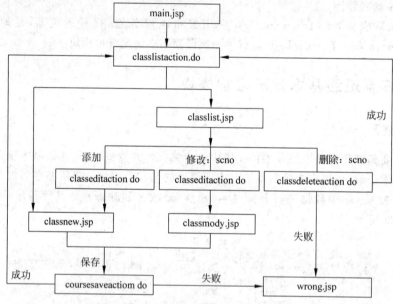

图 6-24 班级基本情况维护模块流程图

2. Action 映射表

班级基本情况维护模块中的 Action 映射表见表 6-18。

表 6-18 班级基本情况维护模块中的 Action 映射表

动作（Action）	入口	表单 bean（ActionForm）	出口
ClassListAction	main.jsp		classlist.jsp classnew.jsp
ClassEditAction	classlist.jsp	ClassForm	classnew.jsp classmody.jsp
ClassSaveAction	classnew.jsp classmody.jsp	ClassForm	classlist.jsp wrong.jsp
ClassDeleteAction	classlist.jsp		classlist.jsp wrong.jsp
PageAction	classlist.jsp	DynaActionForm	classlist.jsp
SearchAction	selectteacherbyname.jsp		selectteacherbyname.jsp

项目六 教师测评系统(Struts模式)

【任务实施】

1. 实现数据列表的分页显示

(1) 新建并编制 JavaBean 类文件 SClass.java

这是一个代表班级的实体 JavaBean 组件。其中的私有属性除了与班级情况表 sclass 中的列相同外,还要增加一个班主任姓名 stname,类中还要编写与属性配套的 get/set 方法。

实现代码如下:

```java
package com.teach.comms;
import java.io.Serializable;

public class SClass    implements Serializable
{
    private String scno;
    private String snno;
    private String scname;
    private String stno;
    private String stname;
    private String sdname;
    private int scpnumb;
    private String scnote;
    private String scout;

    public String getScno()
    {
        return scno;
    }
    public void setScno(String scno)
    {
        this.scno = scno;
    …//省略与其属性对应的 get 和 set 方法。
    }
}
```

(2) 修改 DBUser 类

在其中定义一个方法,方法为 getSClass(),该方法的作用是取出所有班级数据,取数时必须与教师情况表关联起来,取出班主任姓名。每个班级的数据存放在 SClass 类型的一个对象中,所有班级数据存放在一个 List 类型的对象中,并返回。

实现代码如下:

```java
public List getSClass()
    {
        Connection con = null;
        List   sClass = new ArrayList();
        SClass sclass = null;

        try
        {
            con = dataSource.getConnection();
            Statement stat = con.createStatement();
            ResultSet result = stat.executeQuery("select scno, snno, scname, sclass. stno, stname, sclass. sdname,scpnumb,scnote,scout from sclass,steacher where sclass.stno = steacher.stno order by scno");
            while(result.next())
            {
                sclass = new SClass();
                sclass.setScno(result.getString("scno"));
                sclass.setSnno(result.getString("snno"));
                sclass.setScname(result.getString("scname"));
                sclass.setStno(result.getString("stno"));
```

```
                sclass.setStname(result.getString("stname"));
                sclass.setSdname(result.getString("sdname"));
                sclass.setScpnumb(result.getInt("scpnumb"));
                sclass.setScnote(result.getString("scnote"));
                sclass.setScout(result.getString("scout"));

                sClass.add(sclass);
            }

        }
        catch(SQLException e)
        …//省略与 CheckAdminUser(username,userpwd)方法类同部分
        return sClass;
    }
```

（3）新建并编制动作类文件 ClassListAction.java

该类的主要作用是处理数据列表显示请求。在该类中调用 DBUser 类中的 getSClass()方法，取出所有班级数据，并存放至会话级变量 Search_AllList 中；如果没有班级数据，则引导至编辑新增页面 classnew.jsp 进行新增，否则进行分页处理，引导至 classlist.jsp 页面进行列表显示。

实现代码如下：

```
package com.teach.actions;

import java.util.List;
…//省略的包和类同 AdminLoginAction
public class ClassListAction extends Action
{
    public ActionForward execute(ActionMapping mapping, ActionForm form, HttpServletRequest request, HttpServletResponse response) throws Exception
    {
        ServletContext context = servlet.getServletContext();
        DataSource dataSource = (DataSource)context.getAttribute("dbkey");
        DBUser dbuser = new DBUser(dataSource);

        HttpSession session = request.getSession();
        List sClass = dbuser.getSClass();
        session.setAttribute(Constants.Search_AllList,sClass);

        if (sClass.size() == 0)
        {
            return  mapping.findForward("ClassNew");
        }
        else
        {
            PageControl pageControl = new PageControl();
            pageControl.setPageData(sClass,session,0);

            return (mapping.findForward("ClassList"));
        }
    }
}
```

（4）新建并编制 classlist.jsp 文件

如图 6-21 所示，编制方法与 teacherlist.jsp 页面编制方法一致。

实现代码如下：

```
<%@page contentType = "text/html;charset = utf-8" %>
<%@page import = "com.teach.comms.Constants,com.teach.comms.SClass,java.util.List" %>
```

项目六 教师测评系统(Struts模式)

```jsp
<%
    List sClass = (List)session.getAttribute(Constants.Search_PageList);

    Integer Records_All = (Integer)session.getAttribute(Constants.Records_All);
    Integer PageNo_Cur = (Integer)session.getAttribute(Constants.PageNo_Cur);
    Integer Pages_All = (Integer)session.getAttribute(Constants.Pages_All);

    int iRecords_All = Records_All.intValue();
    int iPageNo_Cur = PageNo_Cur.intValue();
    int iPages_All = Pages_All.intValue();
%>

<html>
    <head>
        <link href="../css/style_css.css" rel=stylesheet type="text/css">
        <script type="text/javascript" src="../js/ClientCom.js"></script>
        <script language="javascript">
            function add()                                               //添加
            {
                location.href = "classeditaction.do";
            }
            function modify()                                            //修改
            {
                if (document.all.keyvalue.value == "")
                    return false;

                location.href = "classeditaction.do?scno=" + document.all.keyvalue.value;
            }
            function Delete()                                            //删除
            {
                if (document.all.keyvalue.value == "")
                    return false;

                if (confirm("你是否要删除选中的记录数据?") == true)
                    location.href = "classdeleteaction.do?scno=" + document.all.keyvalue.value;

            }
            function Goto()
            {
                location.href = "pageaction.do?pageId=" + document.getElementById("pageId").value + "&forwardPage=ClassList";
            }
        </script>
    </head>
    <body>
        <table width="100%" height="100%">
            <tr>
                <td align="center" valign="top">
                    <input type="hidden" name="keyvalue" value="" />
                    <table width="650" bgcolor="#dec3b5" cellspacing="1" >
                        <caption>班 级 情 况 表</caption>
                        <tr height="20" align="center">
                            <td width="50"  bgcolor="#f3d5d5">序号</td>
                            <td width="50"  bgcolor="#f3d5d5">班级</td>
                            <td width="50"  bgcolor="#f3d5d5">年级</td>
                            <td width="120" bgcolor="#f3d5d5">专业名称</td>
                            <td width="120" bgcolor="#f3d5d5">班主任</td>
                            <td width="120" bgcolor="#f3d5d5">所在系部</td>
                            <td width="50"  bgcolor="#f3d5d5">人数</td>
                            <td bgcolor="#f3d5d5">备注</td>
                        </tr>
```

```jsp
<%
    if(sClass != null )
    {
        for(int i = 0;i < sClass.size();i++)
        {
            SClass sclass = (SClass)sClass.get(i);
%>
        <tr height = "20" style = "cursor:hand"  bgcolor = "#F1FCF7"
            id = "Row<% = i + 1 %>" title = "鼠标双击可以编辑该信息"
            onclick = "SetFocusIt(this,'<% = sclass.getScno() %>',1,keyvalue)"
            ondblclick = "modify()">
        <td  align = "center"><% = i + 1 + Constants.Rows_PerPapge * iPageNo_Cur%></td>
        <td><% = sclass.getScno() %></td>
        <td><% = sclass.getSnno() %></td>
        <td><% = sclass.getScname() %></td>
        <td><% = sclass.getStname() %></td>
        <td><% = sclass.getSdname() %></td>
        <td><% = sclass.getScpnumb() %></td>
        <td><% = sclass.getScnote() %></td>
        </tr>
<%
        if(i == 0)
        {
%>
            <script type = "text/javascript">
                SetFocusIt(Row1,'<% = sclass.getScno() %>',0,keyvalue)
            </script>
<%
        }
    }
}
%>
        </table>

        <table width = "650" bgcolor = "#f3d5d5">
        <tr height = "20" >
            <td>共<% = iRecords_All %>条记录   第<% = iPageNo_Cur + 1 %>页/共<% = iPages_All %>页</td>
            <td>
                <input type = "text"  name = "pageId" style = "width:25px;height:20px;"><a href = "javascript:Goto()">GO</a> 
                <a href = "pageaction.do?pageId = 1&forwardPage = ClassList">首页</a>
                <a href = "pageaction.do?pageId = <% = iPageNo_Cur %>&forwardPage = ClassList">上一页</a>
                <a href = "pageaction.do?pageId = <% = iPageNo_Cur + 2 %>&forwardPage = ClassList">下一页</a>
                <a href = "pageaction.do?pageId = <% = iPages_All %>&forwardPage = ClassList">末页</a>
            </td>
            <td> </td>
            <td>
                <input type = "button"  name = "addbutton"  value = "添  加"  onclick = "add()">
                <input type = "button"  name = "editbutton"  value = "修  改"  onclick = "modify()">
                <input type = "button"  name = "delbutton"  value = "删  除"  onclick = "Delete()">
            </td>
        </tr>
        </table>
    </td>
</tr>
</table>
```

```html
    </body>
</html>
```

(5) 配置动作类 ClassListAction

配置 ClassListAction 动作类映射关系的代码应写在＜action-mappings＞元素对中。

配置代码如下：

```xml
<action path = "/bas/classlistaction"    type = "com.teach.actions.ClassListAction"   scope = "request">
            <forward name = "ClassNew"    path = "/bas/classnew.jsp"/>
            <forward name = "ClassList"   path = "/bas/classlist.jsp"/>
</action>
```

(6) 修改主页面文件 main.jsp

将"班级数据维护"的超链接改为：

```html
<a href = "classlistaction.do"   target = "inMain">班级数据维护</a>
```

其中，"inMain"为主框架右侧栏子窗体的 id 值。

(7) 发布并测试程序

启动 Tomcat 服务器，将 TeacherTest 应用发布至服务器，打开浏览器，在地址栏中输入 http://localhost:8080/TeacherTest，运行并测试是否能正确地列出所有班级数据。

2. 实现数据的编辑

(1) 新建并编制表单 bean 类文件 ClassForm.java

表单 bean 类 ClassForm 中的属性与班级数据编辑页面中表单元素相对应，类中还有与属性配合的获取(getter)和设置(setter)属性的方法。

实现代码如下：

```java
package com.teach.forms;

import org.apache.struts.action.ActionForm;
public class ClassForm extends ActionForm
{
    private String scno;
    private String snno;
    private String scname;
    private String stno;
    private String stname;
    private String sdname;
    private int scpnumb;
    private String scnote;
    private String newmody;

    public String getScno()
    {
        return scno;
    }
    public void setScno(String scno)
    {
        this.scno = scno;
    }
    …//省略与其属性对应的 get 和 set 方法
}
```

(2) 修改 DBUser 类

在其中定义一个方法，方法为 getSClassByScno(String scno)，该方法的作用是根据班级代码取出该班级数据，并存放在 SClass 类型的一个对象中返回。

实现代码如下：

```java
public SClass getSClassByScno(String scno)
{
    Connection con = null;
    SClass sclass = null;

    try
    {
        con = dataSource.getConnection();
        Statement stat = con.createStatement();
        ResultSet result = stat.executeQuery("select scno, snno, scname, sclass.stno, stname, sclass.sdname, scpnumb, scnote from sclass, steacher where sclass.stno = steacher.stno and scno = '" + scno + "'");
        if(result.next())
        {
            sclass = new SClass();
            sclass.setScno(result.getString("scno"));
            sclass.setSnno(result.getString("snno"));
            sclass.setScname(result.getString("scname"));
            sclass.setStno(result.getString("stno"));
            sclass.setStname(result.getString("stname"));
            sclass.setSdname(result.getString("sdname"));
            sclass.setScpnumb(result.getInt("scpnumb"));
            sclass.setScnote(result.getString("scnote"));
        }
    }
    catch(SQLException e)
    …//省略与CheckAdminUser(username,userpwd)方法类同部分
    return sclass;
}
```

（3）新建并编制动作类文件ClassEditAction.java

该类的主要作用是处理数据编辑请求。在该类中，先取出传递过来的班级代码，如果为空，则是添加，如果不为空，则是修改；添加时直接引导至classnew.jsp页面，修改时调用DBUser类中的getSClassByScno()方法，取出该班级数据，存放在ClassForm类的对象中，并引导至classmody.jsp页面。

实现代码如下：

```java
package com.teach.actions;

import com.teach.comms.DBUser;
…//省略的包和类可参考AdminLoginAction
public class ClassEditAction extends Action
{
    public ActionForward execute(ActionMapping mapping, ActionForm form,
            HttpServletRequest request, HttpServletResponse response) throws Exception
    {
        ClassForm classeditform = (ClassForm) form;
        String scno = request.getParameter("scno");

        ServletContext context = servlet.getServletContext();
        DataSource dataSource = (DataSource)context.getAttribute("dbkey");
        DBUser dbuser = new DBUser(dataSource);

        if (scno == null || scno.equals(""))
        {
            return mapping.findForward("ClassNew");
        }
```

项目六 教师测评系统(Struts模式)

```java
        else
        {
            SClass sclass = dbuser.getSClassByScno(scno);
            if(sclass!= null)
            {
                classeditform.setScno(sclass.getScno());
                classeditform.setSnno(sclass.getSnno());
                classeditform.setScname(sclass.getScname());
                classeditform.setStno(sclass.getStno());
                classeditform.setStname(sclass.getStname());
                classeditform.setSdname(sclass.getSdname());
                classeditform.setScpnumb(sclass.getScpnumb());
                classeditform.setScnote(sclass.getScnote());
            }
            return mapping.findForward("ClassMody");
        }
    }
}
```

(4) 修改 DBUser 类

在其中定义一个方法,方法为 searchDatasByTxt(String code,String keyTxt),该方法的作用是根据输入的关键字去模糊查询匹配数据记录,并将结果数据组织成 XML 形式返回。

实现代码如下:

```java
//对话框查询匹配数据记录的方法
//keyTxt-输入的匹配关键字
//code-区别教师、课程的代码
public String searchDatasByTxt(String code,String keyTxt)
{
    Connection con = null;
    StringBuffer datas = new StringBuffer();
    try
    {
        con = dataSource.getConnection();
        Statement stat = con.createStatement();
        datas.append("<datas>");

        if(code.equals("sono"))
        {
            String strSql = "select sono,soname from scouresulte where soname like '%" + keyTxt + "%'";
            ResultSet result = stat.executeQuery(strSql);
            while(result.next())
            {
                datas.append("<data>" + result.getString("sono") + result.getString("soname") + "</data>");
            }
        }
        else if(code.equals("stno"))
        {
            String strSql = "select stno,stname from steacher where stname like '%" + keyTxt + "%'";
            ResultSet result = stat.executeQuery(strSql);
            while(result.next())
            {
                datas.append("<data>" + result.getString("stno") + result.getString("stname") + "</data>");
            }
        }
        datas.append("</datas>");
    }
    catch(SQLException e)
    … //省略与 CheckAdminUser(username,userpwd)方法类同部分
    return datas.toString();
}
```

(5) 新建并编制动作类文件 SearchAction.java

该类的主要作用是处理对话框发出的模糊查询教师或课程的请求。在该类中,先取出传递过来的 code,即用来区分教师或课程的 stno、sono,keytxt 即输入的匹配关键字,然后调用 searchDatasByTxt()方法进行查询,将查询结果返回给客户端页面。

实现代码如下:

```java
package com.teach.comms;

import javax.servlet.ServletContext;
…//省略的包和类可参考 AdminLoginAction
public class SearchAction extends Action
{
    public ActionForward execute (ActionMapping mapping, ActionForm form, HttpServletRequest request, HttpServletResponse response) throws Exception
    {
        String code = request.getParameter("code");
        String keyTxt = request.getParameter("keytxt");

        ServletContext context = servlet.getServletContext();
        DataSource dataSource = (DataSource)context.getAttribute("dbkey");
        DBUser dbuser = new DBUser(dataSource);

        String datas = dbuser.searchDatasByTxt(code,keyTxt);
        response.setContentType("application/xml;charset=utf-8");
        response.getWriter().write(datas);                       //向客户端页面返回值

        return null;
    }
}
```

(6) 新建并编制 selectteacherbyname.jsp 文件

如图 6-25 所示,该页面文件用来生成一个用于选择班主任的对话框,班主任应该是教师中的一位。在对话框的文本框中输入教师的姓或名后,程序中采用 Ajax 技术按模糊匹配规则查询所有满足条件的教师,将其列出在对话框的列表框中,选择后返回。

图 6-25 教师选择对话框

项目六 教师测评系统(Struts模式)

实现代码如下：

```jsp
<%@page contentType="text/html;charset=utf-8"%>
<html>
    <head>
        <title>选择教师</title>
        <link href="../css/style_css.css" rel=stylesheet type="text/css">
        <script language="javascript" src="../js/ClientCom.js"></script>
        <script language="javascript">
            //确定
            function selectIt()
            {
                var selObj = document.getElementById("sel");
                //是否选择教师
                if(selObj.selectedIndex<0)
                    return;

                //取出所选项的代码及名称
                var val = selObj.options[selObj.selectedIndex].value;
                var txt = selObj.options[selObj.selectedIndex].text;

                //将代码及名称连接后返回
                window.returnValue = val + "@" + txt;

                //关闭对话框
                window.close();
            }

            //取消
            function Exit()
            {
                window.returnValue = '';
                window.close();

            }
        </script>
    </head>
    <body bgcolor="#ece9d8" onLoad="document.all.txt.focus()">
        <table width="100%" height="100%">
            <tr>
                <td align="center" valign="middle">
                    <table width="250">
                        <tr align="center">
                            <td><input id="txt" style="width:250;" type="text" onkeyup="search('stno');" onkeydown="downArrow();"></td>
                        </tr>
                        <tr align="center">
                            <td>
                                <select id="sel" style="width:250;height:300;" multiple ondblclick="selectIt()" onkeydown="if(event.keyCode==13) selectIt();"></select>
                            </td>
                        </tr>
                        <tr align="center">
                            <td>
                                <input type="button" value="确 定" onclick="selectIt()" />    <input type="button" value="取 消" onclick="Exit()"/>
                            </td>
                        </tr>
                    </table>
                </td>
            </tr>
        </table>
    </body>
</html>
```

```
</html>
```

（7）修改 JavaScript 程序文件 ClientCom.js

页面文件 selectteacherbyname.jsp 中用到的一些函数是在本文件中定义的，因此要将下列函数加入到文件中。

实现代码如下：

```javascript
//1 获取输入值,并开始使用 XMLHttpRequest 对象与服务器进行通信
function search( code)
{
    var keyValue = document.getElementById("txt").value;
    if(keyValue.length > 0)
    {
        var xmlrequest = createRequest();
        xmlrequest.onreadystatechange = getReadyStateHandler(xmlrequest,update);
        xmlrequest.open("post","searchaction.do?code = " + code,true);
        //设置 HTTP 头部信息
        xmlrequest.setRequestHeader("Content - Type","application/x - www - form - urlencoded");
        //向 Action 中传递参数
        xmlrequest.send("keytxt = " + keyValue);
    }
    else
    {
        //清除输入内容后,清空列表框
        var sel = document.getElementById("sel");
        sel.length = 0;
    }
}
//2 创建 XMLHttpRequest 对象
function createRequest()
{
    var xmlrequest = false;
    if (window.XMLHttpRequest)
    {
        xmlrequest = new XMLHttpRequest();
    }
    else if(window.ActiveXObject)
    {
        try
        {
            xmlrequest = new ActiveXObject("Msxml2.XMLHTTP");
        }
        catch(e1)
        {
            try
            {
                xmlrequest = new ActiveXObject("Microsoft.XMLHTTP");
            }
            catch(e2)
            {
                xmlrequest = false;
            }
        }
    }
    return xmlrequest;
}
//3 服务器返回状态值处理函数
function getReadyStateHandler(req,responseXmlHandler)
{
    return function()
    {
        if(req.readyState == 4)
```

```
            {
                if(req.status == 200)
                {
                    //相当于调用 update(req.responseXML)
                    responseXmlHandler(req.responseXML);
                }
                else
                {
                    //出错
                    alert("HTTP error " + req.status + ":" + req.statusText);
                }
            }
        }
    }
    //4 从返回值: req.responseXML 中分离出数据,并设置至页面域中
    function update(respXML)    //cartXML = req.responseXML
    {
        //只有一对标签对
        var datas = respXML.getElementsByTagName("datas")[0];
        //可能有多对标签对
        var data  = datas.getElementsByTagName("data");

        var sel = document.getElementById("sel");
        sel.length = 0;

        for (var i = 0;i < data.length; i++)
        {
            ndValue = data[i].firstChild.nodeValue;

            var code = ndValue.substring(0,3);
            var name = ndValue.substring(3,ndValue.length);
            //创建一个 option,并设置其值
            var oOption = document.createElement("option");
            oOption.value = code;
            oOption.text = name;
            //将一个 option 项加入至列表框中
            sel.options.add(oOption);
        }
    }
    //5
    function downArrow()
    {
        var sel = document.getElementById("sel");
        e = event.keyCode;
        //使用下箭头时,提示层获得焦点
        if(e == 40)
            sel.focus();
    }
```

(8) 新建并编制 classnew.jsp 及 classmody.jsp 文件

如图 6-22、图 6-23 所示的页面,编制的基本方法与编制 teacheredit.jsp 页面方法一致。班主任是通过"选择…"按钮打开对话框后选取的,文件 classnew.jsp 与 classmody.jsp 的区别在于,提交表单时,执行保存动作传递的参数不同,设定的参数名为"action",值分别是"new"和"mody",用于区分新增和修改,班级修改时班级号 scno 不能修改。classmody.jsp 可将 classnew.jsp 另存后进行修改。

classnew.jsp 实现代码如下:

```
<%@page contentType = "text/html;charset = utf - 8" %>
<%@taglib uri = "http://struts.apache.org/tags - html" prefix = "html" %>
<html>
```

```
<head>
    <link href="../css/style_css.css" rel=stylesheet type="text/css">
    <title>班级信息编辑修改</title>
    <script language="javascript" src="../js/ClientCom.js"></script>
    <script type="text/javascript">
        function checkdata()
        {
            var scno = document.classform.scno.value;
            if (scno=="" || scno.length!=4 || !isNumberic(scno) )
            {
                alert("班级号不能为空,必须输入,并且只能是4位数字!");
                document.classform.scno.focus();
                return false;
            }

            var snno = document.classform.snno.value;
            if (snno=="" || snno.length!=2 || !isNumberic(snno) )
            {
                alert("年级号不能为空,必须输入,并且只能是2位数字!");
                document.classform.snno.focus();
                return false;
            }

            var scname = document.classform.scname.value;
            if (scname=="" )
            {
                alert("专业名称不能为空,必须输入!");
                document.classform.scname.focus();
                return false;
            }

            var stno = document.classform.stno.value;
            if (stno=="" )
            {
                alert("班主任不能为空,必须选择!");
                document.classform.stno.focus();
                return false;
            }

            var sdname = document.classform.sdname.value;
            if (sdname=="" )
            {
                alert("所在系部不能为空,必须选择!");
                document.classform.sdname.focus();
                return false;
            }

            return true;
        }

        function isNumberic( str )
        {
            var len = str.length;
            for (var i=0;i<len;i++)
                if ( str.charAt(i)<'0' || str.charAt(i)>'9' )
                    return false;
            return true;
        }

        //按姓或名模糊查询教师
        function SelectTeacher()
        {
            //打开模式对话框
```

项目六 教师测评系统(Struts模式)

```
            var rtn = window.showModalDialog("selectteacherbyname.jsp",""," dialogwidth:280px;
dialogheight:400px;center:yes;status:no;scroll:no;help:no");
            if(rtn == '' )
            return;

            //取返回值
            document.all.stno.value = rtn.split("@")[0];
            document.all.stname.value = rtn.split("@")[1];
        }

        function Exit()
        {
            history.go(-1);
        }
    </script>
</head>
<body onload = "document.classform.scno.focus()" >
    <script for = document event = onkeydown language = "javascript">
        if (event.keyCode == 13)
        {
            var objType = window.event.srcElement.type;
            if(objType == "submit")
                return checkdata();
            else
                event.keyCode = 9;

        }
    </script>
<table width = "100%" height = "100%" >
    <tr align = "center" valign = "middle">
        <td>
<html:form  action = "/bas/classsaveaction?action = new" >
<table  bgcolor = "#F1FCF7" >
            <tr>
                <td colspan = "2" align = "center">添加班级信息</td>
            </tr>
            <tr>
                <td align = "center">班     级</td><td><html:text property = "scno"/></td><td></td>
            </tr>
            <tr>
                <td align = "center">年     级</td><td><html:text property = "snno"/></td><td></td>
            </tr>
            <tr>
                <td align = "center">专业名称</td><td><html:text property = "scname"/></td><td></td>
            </tr>
            <tr>
                <td align = "center">班  主  任</td><td><html:text property = "stname" readonly = "true"/></td><td><input type = "button" value = "选择…" onclick = "SelectTeacher()"></td>
                <html:hidden property = "stno"/>
            </tr>
            <tr>
                <td align = "center">所在系部</td><td><html:text property = "sdname"/></td><td></td>
            </tr>
            <tr>
                <td align = "center">班级人数</td><td><html:text property = "scpnumb"/></td><td></td>
            </tr>
            <tr>
                <td align = "center">备     注</td><td><html:text property = "scnote"/></td><td></td>
            </tr>
            <tr>
```

```
                            <td></td><td align="center"><html:submit value="保 存"   onclick="return CheckData()"/>
     <html:button property="exit" value="取 消"  onclick="Exit()"/></td><td></td>
                            </tr>
                        </table>
                </html:form>
            </td>
        </tr>
    </table>
  </body>
</html>
```

(9) 配置表单 bean 及动作 Action 类

配置 ClassForm 表单 bean 的代码应写在 struts-config.xml 文件的＜form-beans＞元素对中。
配置代码如下：

```
<form-bean name="classform"      type="com.teach.forms.ClassForm"/>
```

配置 ClassEditAction、SearchAction 动作类映射关系的代码应写在 struts-config.xml 文件的＜action-mappings＞元素对中。
配置代码如下：

```
<action path="/bas/classeditaction"      type="com.teach.actions.ClassEditAction"   name="classform"
scope="request"   validate="false">
         <forward name="ClassNew"    path="/bas/classnew.jsp"/>
         <forward name="ClassMody"   path="/bas/classmody.jsp"/>
</action>
<action path="/bas/searchaction"   type="com.teach.comms.SearchAction"/>
```

(10) 发布并测试程序

请注意，为了在测试时能正确地显示编辑页面，可以先将 classnew.jsp 和 classmody.jsp 页面表单属性 action 值中的 classsaveaction 改为 classeditaction，这是因为 classsaveaction 尚未编制和配置。启动 Tomcat 服务器，将 TeacherTest 应用发布至服务器，打开浏览器，在地址栏中输入 http://localhost:8080/TeacherTest，运行并测试班级记录增加及修改页面是否能正确地显示出来，是否能正确地打开对话框选择班主任。

3．实现数据的保存、删除

(1) 修改 DBUser 类

在其中定义一个方法，方法为 insertClass(String scno, String snno, String scname, String stno, int scpnumb, String scnote, String sdname)，该方法的作用是保存班级数据至数据库，表中 scout 列的默认值为"1"，表示该班现在允许进行测评，保存成功返回 true，失败返回 false。

在其中再定义一个方法，方法为 updateClass(String scno, String snno, String scname, String stno, int scpnumb, String scnote, String sdname)，该方法的作用是根据班级代码更新班级数据至数据库，成功返回 true，失败返回 false。

实现代码如下：

```
//插入一个新班级
    public boolean insertClass(String scno,String snno,String scname,String stno,int scpnumb,String scnote,
String sdname)
        {
            Connection con = null;
            boolean flag = false;
            try
             {
                con = dataSource.getConnection();
```

```
                PreparedStatement pstat = con.prepareStatement("insert into sclass values(?,?,?,?,?,?,?,?)");
                pstat.setString(1,scno);
                pstat.setString(2,snno);
                pstat.setString(3,scname);
                pstat.setString(4,stno);
                pstat.setInt(5,scpnumb);
                pstat.setString(6,scnote);
                pstat.setString(7,"0");
                pstat.setString(8,sdname);

            int row = pstat.executeUpdate();
            if(row == 1)
                flag = true;
        …//省略与CheckAdminUser(username,userpwd)方法中类同部分
            return  flag;
    }

//更新一个班级信息
public boolean updateClass(String scno, String snno, String scname, String stno, int scpnumb, String scnote, String sdname)
    {
            Connection con = null;
            boolean flag = false;
            try
            {
                con = dataSource.getConnection();

                PreparedStatement pstat = con.prepareStatement("update sclass set snno = ?, scname = ?, stno = ?, scpnumb = ?, scnote = ?, sdname = ? where scno = ?");
                pstat.setString(1,snno);
                pstat.setString(2,scname);
                pstat.setString(3,stno);
                pstat.setInt(4,scpnumb);
                pstat.setString(5,scnote);
                pstat.setString(6,sdname);
                pstat.setString(7,scno);

                int row = pstat.executeUpdate();
                if(row == 1)
                    flag = true;
            }
            catch(SQLException e)
            …//省略与CheckAdminUser(username,userpwd)方法中类同部分
            return  flag;
    }
```

（2）新建并编制动作类文件ClassSaveAction.java

该类的主要作用是数据保存请求。在该类中，先取出参数"action"的值，如果是"new"，则是添加，调用insertClass()方法插入数据，如果是"mody"，则是修改，调用updateClass()方法更新数据。插入或更新数据成功后引导至classlist.jsp页面，插入或更新数据失败时在wrong.jsp页面中显示错误信息。

实现代码如下：

```
package com.teach.actions;

import com.teach.comms.Constants;
…//省略的包和类可参考AdminLoginAction
public class ClassSaveAction extends Action
{
    public ActionForward execute(ActionMapping mapping,ActionForm form,
            HttpServletRequest request,HttpServletResponse response) throws Exception
```

```
            {
                ClassForm classeditform = (ClassForm) form;
                String action = request.getParameter("action");

                ServletContext context = servlet.getServletContext();
                DataSource dataSource = (DataSource)context.getAttribute("dbkey");
                DBUser dbuser = new DBUser(dataSource);

                HttpSession session = request.getSession();
                if (action.equals("new"))
                {
                    if(!insertClass(classeditform,dbuser))
                    {
                        session.setAttribute(Constants.Error_Message,"添加班级信息失败!");
                        return mapping.findForward("Wrong");
                    }
                }
                else if(action.equals("mody"))
                {
                    if(!updateClass(classeditform,dbuser))
                    {
                        session.setAttribute(Constants.Error_Message,"修改班级信息失败!");
                        return mapping.findForward("Wrong");
                    }
                }

                return mapping.findForward("ClassList");
            }
            private boolean insertClass(ClassForm classeditform,DBUser dbuser)
            {
                String scno = (classeditform.getScno()).trim();
                String snno = (classeditform.getSnno()).trim();
                String scname = (classeditform.getScname()).trim();
                String stno = (classeditform.getStno()).trim();
                String sdname = (classeditform.getSdname()).trim();
                int scpnumb = classeditform.getScpnumb();
                String scnote = (classeditform.getScnote()).trim();

                return dbuser.insertClass(scno,snno,scname,stno,scpnumb,scnote,sdname);
            }
            private boolean updateClass(ClassForm classeditform,DBUser dbuser)
            {
                String scno = (classeditform.getScno()).trim();
                String snno = (classeditform.getSnno()).trim();
                String scname = (classeditform.getScname()).trim();
                String stno = (classeditform.getStno()).trim();
                String sdname = (classeditform.getSdname()).trim();
                int scpnumb = classeditform.getScpnumb();
                String scnote = (classeditform.getScnote()).trim();

                return dbuser.updateClass(scno,snno,scname,stno,scpnumb,scnote,sdname);
            }
        }
```

(3) 配置动作类 ClassSaveAction

配置 ClassSaveAction 动作类映射关系的代码应写在 struts-config.xml 文件的 <action-mappings> 元素对中。

配置代码如下：

```
<action path = "/bas/classsaveaction"        type = "com.teach.actions.ClassSaveAction" name = "classform"
    scope = "request" validate = "false" >
```

```xml
    <forward name = "ClassList"    path = "/bas/classlistaction.do"></forward>
</action>
```

(4) 发布并测试程序

启动 Tomcat 服务器,将 TeacherTest 应用发布至服务器,打开浏览器,在地址栏中输入 http://localhost:8080/TeacherTest,运行并测试是否能正确地插入或更新班级数据。

(5) 修改 DBUser 类

在其中定义一个方法,方法为 checkClassDelete(String scno),该方法的作用是在删除前检查该班级是否允许被删除,即该班级是否已被使用过,是则返回 true,否则返回 false。

实现代码如下:

```java
public boolean checkClassDelete(String scno)
{
        Connection con = null;
        boolean flag = false;

        try
        {
            con = dataSource.getConnection();
            Statement stat = con.createStatement();
            ResultSet result = stat.executeQuery("select * from sccourse where scno = '" + scno + "'");
            if(result.next())
            {
                    flag = true;
            }
        }
        catch(SQLException e)
        …//省略部分同 CheckAdminUser(username,userpwd)方法中该部分
        return flag;
}
```

(6) 修改 DBUser 类

在其中定义一个方法,方法为 deleteClass(String scno),该方法的作用是根据班级号从数据库中删除该班级,成功返回 true,失败返回 false。

实现代码如下:

```java
public boolean deleteClass(String scno)
    {
            Connection con = null;
            boolean flag = false;
            try
            {
                con = dataSource.getConnection();
                PreparedStatement pstat = con.prepareStatement("delete sclass where scno = ?");
                pstat.setString(1,scno);
                int row = pstat.executeUpdate();
                if(row == 1)
                 flag = true;
            }
            catch(SQLException e)
            …//省略部分同 CheckAdminUser(username,userpwd)方法中该部分
            return   flag;

}
```

(7) 新建并编制动作类文件 ClassDeleteAction.java

该类的主要作用是处理数据删除请求。在该类中,先取出传递过来的班级代码,调用

checkClassDelete()方法检查该班级是否允许被删除,允许则调用deleteClass()方法删除数据,删除数据成功后引导至classlist.jsp页面,删除数据失败时在wrong.jsp页面中显示错误信息。

实现代码如下:

```java
package com.teach.actions;

import com.teach.comms.*;
…//省略的包和类可参考AdminLoginAction
public class ClassDeleteAction extends Action
{
    public ActionForward execute(ActionMapping mapping, ActionForm form, HttpServletRequest request,
HttpServletResponse response) throws Exception
    {
        String scno = request.getParameter("scno");

        ServletContext context = servlet.getServletContext();
        DataSource dataSource = (DataSource)context.getAttribute("dbkey");
        DBUser dbuser = new DBUser(dataSource);

        HttpSession session = request.getSession();

        if(dbuser.checkClassDelete(scno))
        {
            session.setAttribute(Constants.Error_Message,"该班级已经有选课记录,不能删除!");
            return mapping.findForward("Wrong");
        }

        if (scno!= null && !scno.equals(""))
        {
            if(!dbuser.deleteClass(scno))
            {
                session.setAttribute(Constants.Error_Message,"删除班级记录失败!");
                return mapping.findForward("Wrong");
            }
        }

        return mapping.findForward("ClassList");
    }
}
```

(8) 配置动作类 ClassDeleteAction

配置 ClassDeleteAction 动作类映射关系的代码应写在 struts-config.xml 文件的＜action-mappings＞元素对中。

配置代码如下:

```xml
<action path="/bas/classdeleteaction"    type="com.teach.actions.ClassDeleteAction" scope="request">
    <forward name="ClassList"    path="/bas/classlistaction.do"></forward>
</action>
```

(9) 发布并测试程序

启动 Tomcat 服务器,将 TeacherTest 应用发布至服务器,打开浏览器,在地址栏中输入 http://localhost:8080/TeacherTest,运行并测试是否能正确地删除班级数据。

任务六　编制班级科目维护模块

【任务描述】

本任务主要编制班级科目维护模块中的相关文件并测试运行,其编制方法与班级基本情况维护模块基本一致，可以在班级基本情况维护模块的基础上编制。班级科目维护模块主要是对班级选课表 sccourse 中数据进行列表显示、添加、修改及删除操作。其运行示意图如图 6-26、图 6-27、图 6-28、图 6-29 所示。

图 6-26 班级选择下拉列表

图 6-28 新增班级科目页面

图 6-29 修改班级科目页面

【任务分析】

1. 程序运行基本流程

图 6-30 所示为班级科目维护模块运行流程图。

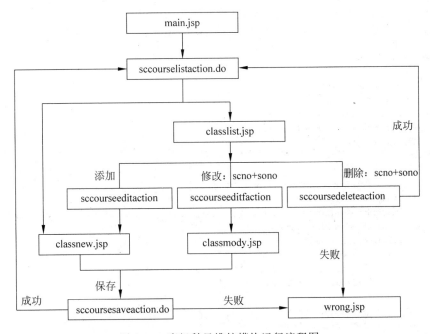

图 6-30 班级科目维护模块运行流程图

2. Action 映射表

班级科目维护模块中的 Action 映射表见表 6-19。

表 6-19 班级科目维护模块中的 Action 映射表

动作（Action）	入口	表单 bean（ActionForm）	出口
ScCourseListAction	main.jsp		sccourselist.jsp sccoursenew.jsp
ScCourseEditAction	sccourselist.jsp	ScCourseForm	sccoursenew.jsp sccoursemody.jsp
ScCourseSaveAction	sccoursenew.jsp sccoursemody.jsp	ScCourseForm	sccourselist.jsp wrong.jsp
ScCourseDeleteAction	sccourselist.jsp		sccourselist.jsp wrong.jsp
SearchAction	selectteacherbyname.jsp selectcoursebyname.jsp		selectteacherbyname.jsp selectcoursebyname.jsp

【任务实施】

1. 实现数据的列表显示

（1）修改 DBUser 类

在其中定义一个方法，方法为 getScnoAll()，该方法的作用是取出所有的班级代号 scno 及对应的班主任姓名 stname，取数时用班级情况表与教师情况表关联起来，并按年级、班级号排序。定义三个 Vector 类型的对象，分别为 rtn、scnoVector 和 stnameVector，取出的班级代号 scno 及对应的班主任姓名 stname 按顺序分别存放在 scnoVector、stnameVector 两个元素中，将这两个元素放入 rtn 中返回。

实现代码如下：

```java
public Vector getScnoAll()
{
    Connection con = null;
    Vector rtn = new Vector();
    Vector scnoVector = new Vector();
    Vector stnameVector = new Vector();
    String scno, stname;
    try
    {
        con = dataSource.getConnection();
        Statement stat = con.createStatement();
        ResultSet result = stat.executeQuery("select  scno,stname from sclass,steacher where sclass.stno = steacher.stno order by snno,scno");
        while(result.next())
        {
            scno = result.getString("scno");
            stname = result.getString("stname");

            scnoVector.add(scno);
            stnameVector.add(stname);
        }
    }
    catch(SQLException e)
    …//省略部分同 CheckAdminUser(username,userpwd)方法中该部分
```

项目六 教师测评系统(Struts模式)

```
        return rtn;
}
```

(2) 修改动作类文件 AdminLoginAction.java

AdminLoginAction 类用于处理系统管理员的登录请求。在登录成功后,调用 getScnoAll() 方法,用一个 Vector 类型的对象接收该方法的返回值,并将两个元素中的值分别保存至 Constants 类的 Bas_Scnoall 及 Bas_Stnameall 变量中。

在 if…else…结构的 else 分支中加入以下代码:

```
Vector rtn;
Vector scnoVector;
Vector stnameVector;
rtn = dbuser.getScnoAll();
scnoVector = (Vector)rtn.get(0);
stnameVector = (Vector)rtn.get(1);

session.setAttribute(Constants.Bas_ScnoAll,scnoVector);
session.setAttribute(Constants.Bas_StnameAll,stnameVector);
```

(3) 修改主页面文件 main.jsp

在页面代码的开始处,将保存在 Constants 类的 Bas_Scnoall 及 Bas_Stnameall 变量中的数据取出,实现代码如下:

```
<%
…
Vector scnoVector = (Vector)session.getAttribute(Constants.Bas_ScnoAll);
Vector stnameVector = (Vector)session.getAttribute(Constants.Bas_StnameAll);
%>
```

如图 6-26 所示,将取出的数据设置至选择班级下拉列表框中,下拉列表框中各项的值为要执行的选课动作(sccourselist.do),并将班级号作为参数,下拉列表框中各项要显示的字符串为班级号及班主任姓名,并为下拉列表框的选项改变事件(onchange)定义动作,班级科目数据维护所在 <tr> 标签对的实现代码如下:

```
<tr height = "25">
    <td align = "center">
        <fieldset style = "width:150px;">
            <legend align = "center">班级科目数据维护</legend>
            <select name = "select1" style = "width:120px;" onchange = "selectit(this.options[this.selectedIndex].value)">
                <option value = "">---请选择班级---</option>
                <%
                    for(int i = 0;i<scnoVector.size();i++)
                    {
                        String scno = (String)scnoVector.get(i);
                        String stname = (String)stnameVector.get(i);
                %>
                <option value = "sccourselistaction.do?scno=<% = scno %>"><% = (scno + "  " + stname) %></option>
                <%
                    }
                %>
            </select>
        </fieldset>
    </td>
</tr>
```

其中,selectit 是一个 JavaScript 函数,在页面开始处定义,实现代码如下:

```
<script type = "text/javascript">
        var j = 0;
        function selectit(str)
        {
            if(str.length > 0)
            {
                j++;
                var str1 = str + "&j = " + j.toString();
                window.open(str1,'inMain');
            }
        }
</script>
```

在选项改变时,将当前班级的选课记录取出并显示框架的右侧栏中。

(4) 发布并测试程序

启动 Tomcat 服务器,将 TeacherTest 应用发布至服务器,打开浏览器,在地址栏中输入 http://localhost:8080/TeacherTest,运行并测试是否能正确地在下拉列表框中将班级号及班主任姓名显示出来。

(5) 新建并编制 JavaBean 类文件 ScCourse.java

这是一个代表班级选课的实体 JavaBean 组件。其中的私有属性除了与班级选课表 sccourse 中的列相同外,还要增加课程名称 soname 及任课教师姓名 stname,类中还要编写与属性配套的 get/set 方法。

实现代码如下:

```
package com.teach.comms;

import java.io.Serializable;
public class ScCourse   implements Serializable
{
    private String scno;
    private String sono;
    private String soname;
    private String stno;
    private String stname;

    public String getScno()
    {
        return scno;
    }
    public void setScno(String scno)
    {
        this.scno = scno;
    }

    …//省略与其余属性对应的 get 和 set 方法
}
```

(6) 修改 DBUser 类

在其中定义一个方法,方法为 getScCourse(String scno),该方法的作用是取出某一个班级的选课数据,取数时必须与课程情况表及教师情况表关联起来,取出课程名称及任课教师姓名。班级的每条选课数据存放在 ScCourse 类型的一个对象中,班级的所有选课数据存放在一个 List 类型的对象中,并返回。

实现代码如下:

```
public List getScCourse(String scno)
{
```

```
        Connection con = null;
        List    scCourse = new ArrayList();
        ScCourse sccourse = null;
        try
        {
            con = dataSource.getConnection();
            Statement stat = con.createStatement();
            ResultSet result = stat.executeQuery("select scno, sccourse.sono, soname, sccourse.stno, stname
from sccourse, scourse, steacher where sccourse.sono = scourse.sono and sccourse.stno = steacher.stno and scno =
'" + scno + "'");
            while(result.next())
            {
                sccourse = new ScCourse();
                sccourse.setScno(result.getString("scno"));
                sccourse.setSono(result.getString("sono"));
                sccourse.setSoname(result.getString("soname"));
                sccourse.setStno(result.getString("stno"));
                sccourse.setStname(result.getString("stname"));

                scCourse.add(sccourse);
            }

        }
        catch(SQLException e)
        …//省略部分同 CheckAdminUser(username,userpwd)方法中该部分
        return scCourse;
}
```

(7) 新建并编制动作类文件 ScCourseListAction.java

该类的主要作用是处理数据列表显示请求。在该类中，先取出传入的班级号，调用 DBUser 类中的 getScCourse() 方法，取出当前班级的选课数据，并存放至 session 级变量 Search_AllList 中，将当前班级号存放至 session 级变量 Bas_Sccourse_Curscno 中，引导至 sccourselist.jsp 页面。

实现代码如下：

```
package com.teach.actions;

import java.util.List;
…//省略的包和类可参考 AdminLoginAction
public class ScCourseListAction extends Action
{
    public ActionForward execute (ActionMapping mapping, ActionForm form, HttpServletRequest request,
HttpServletResponse response) throws Exception
    {

        String scno = request.getParameter("scno");

        ServletContext context = servlet.getServletContext();
        DataSource dataSource = (DataSource)context.getAttribute("dbkey");
        DBUser dbuser = new DBUser(dataSource);
        HttpSession session = request.getSession();
        if(scno == null || scno.equals(""))   //新增、修改、删除后重取数据
        {
            scno = (String)session.getAttribute(Constants.Bas_ScCourse_CurScno);
        }
        else
        {
            session.setAttribute(Constants.Bas_ScCourse_CurScno, scno);
        }

        List sScCourse = dbuser.getScCourse(scno);
        if (sScCourse.size() == 0)
```

```
        {
            return mapping.findForward("ScCourseNew");
        }
        else
        {
            session.setAttribute(Constants.Search_AllList,sScCourse);
            return (mapping.findForward("ScCourseList"));
        }
    }
}
```

(8) 新建并编制 sccourselist.jsp 文件

如图 6-27 所示,编制方法与 teacherlist.jsp 页面编制方法基本一致,但由于班级的选课记录一般不会超过 20 条,此页面不需要实现分页显示。

实现代码如下:

```
<%@page contentType="text/html;charset=utf-8" %>
<%@page import="com.teach.comms.Constants,com.teach.comms.ScCourse,java.util.List" %>
<%
    List scCourse = (List)session.getAttribute(Constants.Search_AllList);
    String CurScno = (String)session.getAttribute(Constants.Bas_ScCourse_CurScno);
%>
<html>
    <head>
        <link href="../css/style_css.css" rel=stylesheet type="text/css">
        <script type="text/javascript" src="../js/ClientCom.js"></script>
        <script type="text/javascript">
            function add()                                          //添加
            {
                location.href="sccourseeditaction.do";
            }
            function modify()                                       //修改
            {
                if(document.all.keyvalue.value=="")
                    return false;

                location.href="sccourseeditaction.do?scono="+document.all.keyvalue.value;
            }
            function Delete()                                       //删除
            {
                if(document.all.keyvalue.value=="")
                    return false;

                if(confirm("你是否要删除选中的记录数据?")==true)
                    location.href="sccoursedeleteaction.do?scono="+document.all.keyvalue.value;
            }
        </script>
    </head>
    <body>
        <table width="100%" height="100%">
            <tr>
                <td align="center" valign="top">
                    <input type="hidden" name="keyvalue" value="" />
                    <table width="650" bgcolor="#dec3b5" cellspacing="1" >
                        <caption><b><font face="华文新魏" color=#d71345>[<%=CurScno%>]</font></b>班级科目情况表</caption>
                        <tr height="20" align="center">
                            <td width="50"  bgcolor="#f3d5d5">序号</td>
                            <td width="200"  bgcolor="#f3d5d5">课程名称</td>
                            <td bgcolor="#f3d5d5">任课教师</td>
                        </tr>
```

项目六 教师测评系统(Struts模式)

```
                    <%
                       if(scCourse != null )
                       {
                          for(int i = 0;i < scCourse.size();i++)
                          {
                             ScCourse sccourse = (ScCourse)scCourse.get(i);
                    %>
                    <tr height = "20" style = "cursor:hand"  bgcolor = "#F1FCF7"
                        id = "Row<% = i + 1%>" title = "鼠标双击可以编辑该信息"
                        onclick = " SetFocusIt ( this, '<% = sccourse. getScno ( ) % > <% = sccourse.getSono()%>',1,keyvalue)"
                        ondblclick = "modify()">
                       <td align = "center"><% = i + 1 %></td>
                       <td><% = sccourse.getSoname()%></td>
                       <td><% = sccourse.getStname()%></td>
                    </tr>
                    <%
                       if(i == 0)
                       {
                    %>
                    <script type = "text/javascript">
                           SetFocusIt( Row1, '<% = sccourse. getScno ( ) % > <% = sccourse.getSono()%>',0,keyvalue)
                    </script>
                    <%
                          }
                       }
                    }
                    %>
                 </table>
                 <table width = "650" bgcolor = "#f3d5d5">
                    <tr height = "20">
                       <td> </td>
                       <td>
                          <input type = "button"   name = "addbutton"   value = "添  加" onclick = "add()">
                          <input type = "button"   name = "editbutton"  value = "修  改" onclick = "modify()">
                          <input type = "button"   name = "delbutton"   value = "删  除" onclick = "Delete()">
                       </td>
                    </tr>
                 </table>
              </td>
           </tr>
        </table>
     </body>
</html>
```

(9) 配置动作类 ScCourseListAction

配置 ScCourseListAction 动作类映射关系的代码应写在<action-mappings>元素对中。

配置代码如下：

```
<action path = "/bas/sccourselistaction"    type = "com.teach.actions.ScCourseListAction"   scope = "request">
     <forward name = "ScCourseNew"    path = "/bas/sccourseedit.do"/>
     <forward name = "ScCourseList"   path = "/bas/sccourselist.jsp"/>
</action>
```

(10) 发布并测试程序

启动 Tomcat 服务器，将 TeacherTest 应用发布至服务器，打开浏览器，在地址栏中输入 http://localhost:8080/TeacherTest，运行并测试是否能正确地列出当前班级的所选课程及任课教师数据。

2. 实现数据的编辑

(1) 新建并编制表单 bean 类文件 ScCourseForm.java

表单 bean 类 ScCourseForm 中的属性及方法与 ScCourse 类中一致，区别是本类要从 ActionForm 类继承。

实现代码如下：

```java
package com.teach.forms;

import org.apache.struts.action.ActionForm;
public class ScCourseForm extends ActionForm
{
    private String scno;
    private String sono;
    private String soname;
    private String stno;
    private String stname;
    private String newmody;

    public String getScno()
    {
        return scno;
    }
    public void setScno(String scno)
    {
        this.scno = scno;
    }

    …//省略与其余属性对应的 get 和 set 方法
}
```

(2) 修改 DBUser 类

在其中定义一个方法，方法为 getScCourseByScono（String scono），该方法的作用是根据班级号及课程代码取出该班级的一条选课数据，并存放在 ScCourse 类型的一个对象中返回。

实现代码如下：

```java
public ScCourse getScCourseByScono(String scono)
    {
        Connection con = null;
        ScCourse sccourse = null;
        //分离班级号及课程代码
        String scno = scono.substring(0,4);
        String sono = scono.substring(4,7);
        try
        {
            con = dataSource.getConnection();
            Statement stat = con.createStatement();
            ResultSet result = stat.executeQuery("select scno, sccourse.sono, soname, sccourse.stno, stname from sccourse, scourse, steacher where sccourse.sono = scourse.sono and sccourse.stno = steacher.stno and scno = '" + scno + "' and sccouresulte.sono = '" + sono + "'");
            if(result.next())
            {
                sccourse = new ScCourse();
                sccourse.setScno(result.getString("scno"));
                sccourse.setSono(result.getString("sono"));
                sccourse.setSoname(result.getString("soname"));
                sccourse.setStno(result.getString("stno"));
                sccourse.setStname(result.getString("stname"));
            }
```

项目六 教师测评系统(Struts模式)

```
        }
        catch(SQLException e)
        …//省略部分同 CheckAdminUser(username,userpwd)方法中该部分
        return sccourse;
}
```

（3）新建并编制动作类文件 ScCourseEditAction.java

该类的主要作用是处理数据编辑请求。在该类中，先取出传递过来的班级号及课程代码，如果为空，则是添加，如果不为空，则是修改；添加时先从会话级变量 ScCourse_CurScno 中取出当前班级号，并设置至表单 ScCourseForm 类的对象中，引导至 sccoursenew.jsp 页面；修改时调用 DBUser 类中的 getScCourseByScono()方法，取出该班级的一条选课数据，存放在 ScCourseForm 类的对象中，并引导至 sccoursemody.jsp 页面。

实现代码如下：

```
package com.teach.actions;

import com.teach.comms.*;
…//省略的包和类可参考 AdminLoginAction
public class ScCourseEditAction extends Action
{
    public ActionForward execute (ActionMapping mapping, ActionForm form, HttpServletRequest request, HttpServletResponse response) throws Exception
    {
        ScCourseForm sccourseform = (ScCourseForm) form;
        String scono = request.getParameter("scono");

        ServletContext context = servlet.getServletContext();
        DataSource dataSource = (DataSource)context.getAttribute("dbkey");
        DBUser dbuser = new DBUser(dataSource);
        HttpSession session = request.getSession();

        String scno = (String)session.getAttribute(Constants.Bas_ScCourse_CurScno);
        if ( (dbuser.getScCourse(scno).size()) == 0 || (scono == null || scono.equals("")) )
        {
            sccourseform.setScno(scno);
            return mapping.findForward("ScCourseNew");
        }
        else
        {
            ScCourse sccourse = dbuser.getScCourseByScono(scono);
            if(sccourse!= null)
            {
                sccourseform.setScno(sccourse.getScno());
                sccourseform.setSono(sccourse.getSono());
                sccourseform.setSoname(sccourse.getSoname());
                sccourseform.setStno(sccourse.getStno());
                sccourseform.setStname(sccourse.getStname());
            }
            return mapping.findForward("ScCourseMody");
        }
    }
}
```

（4）新建并编制 selectcoursebyname.jsp 文件

如图 6-31 所示，该页面文件用来生成一个用于选择课程的对话框。在对话框的文本框中输入文本后，程序中采用 Ajax 技术按模糊匹配规则查询所有名称中包含该文本的课程，将其列出在对话框的列表框中，选择后返回。编制时只要将 selectteacherbyname.jsp 另存为 selectcoursebyname.jsp 后将 onkeyup="search('stno');"代码中的 stno 改为 sono 即可。

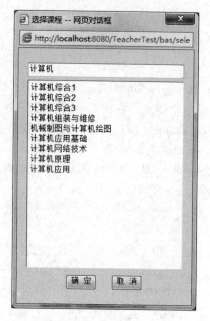

图 6-31 课程选择对话框

(5) 新建并编制 sccoursenew.jsp 及 sccoursemody.jsp 文件

如图 6-28、图 6-29 所示,编制的基本方法与编制 sccoursenew.jsp、sccoursemody.jsp 页面方法一致。班主任和课程是通过"选择…"按钮打开对话框后选取的,页面文件 sccoursenew.jsp 与 sccoursemody.jsp 的区别在于,提交表单时,执行保存动作传递的参数不同,设定的参数名为 "action",值分别是"new"和"mody",用于区分新增和修改,修改时班级所选课程不能修改,只能先删除原来选错的记录后再重新选取。sccoursemody.jsp 可将 sccoursenew.jsp 另存后进行修改。

sccoursenew.jsp 实现代码如下:

```jsp
<%@page contentType="text/html;charset=utf-8"%>
<%@taglib uri="http://struts.apache.org/tags-html" prefix="html"%>
<html>
    <head>
        <link href="../css/style_css.css" rel=stylesheet type="text/css">
        <title>班级选课信息编辑修改</title>
        <script language="javascript" src="../js/ClientCom.js"></script>
        <script type="text/javascript">
            function checkdata()
            {
                var sono = document.sccourseform.sono.value;
                if (sono=="" || sono.length!=3 || !isNumberic(sono))
                {
                    alert("课程不能为空,必须选择!");
                    document.sccourseform.sono.focus();
                    return false;
                }
                var stno = document.sccourseform.stno.value;
                if (stno=="" || stno.length!=3 || !isNumberic(stno))
                {
                    alert("任课教师不能为空,必须选择!");
                    document.sccourseform.stno.focus();
                    return false;
                }
                return true;
            }
```

项目六 教师测评系统(Struts模式)

```javascript
function isNumberic( str )
{
    var len = str.length;
    for (var i = 0;i<len;i++)
        if ( str.charAt(i)<'0' || str.charAt(i)>'9' )
            return false;
    return true;
}

//输入课程名称模糊查询选择课程
function SelectCourse()
{
    //打开模式对话框
    var  rtn = window.showModalDialog ( "selectcoursebyname.jsp","","dialogwidth:280px;dialogheight:400px;center:yes;status:no;scroll:no;help:no");
    if(rtn == '' )
        return;

    //取返回值
    document.all.sono.value = rtn.split("@")[0];
    document.all.soname.value = rtn.split("@")[1];
}

//输入教师姓名模糊查询选择教师
function SelectTeacher()
{
    //打开模式对话框
    var  rtn = window.showModalDialog ( "selectteacherbyname.jsp","","dialogwidth:280px;dialogheight:400px;center:yes;status:no;scroll:no;help:no");
    if(rtn == '' )
        return;

    //取返回值
    document.all.stno.value = rtn.split("@")[0];
    document.all.stname.value = rtn.split("@")[1];
}

function Exit()
{
  history.go(-1);
}
</script>
</head>
<body onload = "document.sccourseform.sono.focus()" >
  <table  width = "100%" height = "100%"  >
    <tr align = "center" valign = "middle">
    <td>

    <html:form  action = "/bas/sccoursesaveaction.do?action = new"  >
      <table  bgcolor = "#F1FCF7" >
        <tr>
          <td colspan = "2" align = "center">添加班级选课信息</td>
        </tr>
        <tr>
          <td align = "center">课程名称</td><td><html:text property = "soname" readonly = "true"/></td><td><input type = "button" value = "选择…" onclick = "SelectCourse()"></td>
        </tr>
        <tr>
          <td align = "center">任课教师</td><td><html:text property = "stname" readonly = "true"/></td><td><input type = "button" value = "选择…" onclick = "SelectTeacher()"></td>
        </tr>
            <html:hidden property = "scno"/>
            <html:hidden property = "sono"/>
            <html:hidden property = "stno"/>
```

```
                </tr>
                <tr>
                    <td></td><td align="center"><html:submit value="保 存" onclick="return checkdata()"/>     <html:button property="exit" value="取 消" onclick="Exit()"/></td><td></td>
                </tr>
            </table>
        </html:form>
        </td>
    </tr>
    </table>
    </body>
</html>
```

(6) 配置表单 bean 及动作 Action 类

配置 ScCourseForm 表单 bean 的代码应写在 struts-config.xml 文件的＜form-beans＞元素对中。

配置代码如下：

```
<form-bean name="sccourseform"    type="com.teach.forms.ScCourseForm"/>
```

配置 ScCourseEditAction 动作类映射关系的代码应写在 struts-config.xml 文件的＜action-mappings＞元素对中。

配置代码如下：

```
<action path="/bas/sccourseeditaction"    type="com.teach.actions.ScCourseEditAction"    name="sccourseform"    scope="request" validate="false">
        <forward name="ScCourseNew"    path="/bas/sccoursenew.jsp"/>
        <forward name="ScCourseMody"    path="/bas/sccoursemody.jsp"/>
</action>
```

(7) 发布并测试程序

请注意，为了在测试时能正确地显示编辑页面，可以先将 sccoursenew.jsp 和 sccoursemody.jsp 页面表单属性 action 值中的 sccoursesaveaction 改为 sccourseeditaction，这是因为 sccoursesaveaction 尚未编制和配置。启动 Tomcat 服务器，将 TeacherTest 应用发布至服务器，打开浏览器，在地址栏中输入 http://localhost:8080/TeacherTest，运行并测试班级选课记录增加及修改页面是否能正确地显示出来，是否能正确地打开对话框选择班主任及课程。

3. 实现数据的保存、删除

(1) 修改 DBUser 类

在其中定义一个方法，方法为 insertScCourse(String scno,String sono,String stno)，该方法的作用是保存班级选课数据至数据库，保存成功返回 true，失败返回 false。

在其中再定义一个方法，方法为 updateScCourse(String scno,String sono,String stno)，该方法的作用是根据班级号及课程代码更新班级选课数据至数据库，成功返回 true，失败返回 false。

实现代码如下：

```
//插入一个新选课信息
public boolean insertScCourse(String scno,String sono,String stno)
{
            Connection con = null;
            boolean flag = false;

            try
            {
                con = dataSource.getConnection();
```

项目六 教师测评系统（Struts模式）

```java
                PreparedStatement pstat = con.prepareStatement("insert into sccourse values(?,?,?)");
                pstat.setString(1,scno);
                pstat.setString(2,sono);
                pstat.setString(3,stno);

                int row = pstat.executeUpdate();
                if(row == 1)
                    flag = true;
            }
            catch(SQLException e)
            …//省略部分同 CheckAdminUser(username,userpwd)方法中该部分
            return   flag;
        }

//更新一条选课信息
public boolean updateScCourse(String scno,String sono,String stno)
    {
            Connection con = null;
            boolean flag = false;

            try
            {
                con = dataSource.getConnection();

                PreparedStatement stat = con.prepareStatement("update sccourse set stno = ? where scno = ? and sono = ?");

                stat.setString(1,stno);
                stat.setString(2,scno);
                stat.setString(3,sono);

                int row = stat.executeUpdate();
                if(row == 1)
                    flag = true;
            }
            catch(SQLException e)
            …//省略部分同 CheckAdminUser(username,userpwd)方法中该部分
            return   flag;
    }
}
```

（2）新建并编制动作类文件 ScCourseSaveAction.java

该类的主要作用是理数据保存请求。在该类中，先取出参数 action 的值，如果是 new，则是添加，调用 insertScCourse 方法插入数据，如果是 mody，则是修改，调用 updateScCourse 方法更新数据。插入或更新数据成功后引导至 sccourselist.jsp 页面，插入或更新数据失败时在 wrong.jsp 页面中显示错误信息。

实现代码如下：

```java
package com.teach.actions;

import com.teach.comms.*;
…//省略的包和类可参考 AdminLoginAction
public class ScCourseSaveAction extends Action
{
    public ActionForward execute(ActionMapping mapping, ActionForm form, HttpServletRequest request, HttpServletResponse response) throws Exception
    {
        ScCourseForm sccourseeditform = (ScCourseForm)form;
        String action = request.getParameter("action");

        ServletContext context = servlet.getServletContext();
        DataSource dataSource = (DataSource)context.getAttribute("dbkey");
```

```
            DBUser dbuser = new DBUser(dataSource);

            HttpSession session = request.getSession();
            String scno = sccourseeditform.getScno();
            String sono = sccourseeditform.getSono();
            String stno = sccourseeditform.getStno();
            if (action.equals("new"))
            {
                boolean flag = dbuser.insertScCourse(scno,sono,stno);
                if(!flag)
                {
                    session.setAttribute(Constants.Error_Message,"添加班级科目信息失败!");
                    return mapping.findForward("Wrong");
                }
            }
            else if(action.equals("mody"))
            {
                boolean flag = dbuser.updateScCourse(scno,sono,stno);
                if(!flag)
                {
                    session.setAttribute(Constants.Error_Message,"修改班级科目信息失败!");
                    return mapping.findForward("Wrong");
                }
            }

            return mapping.findForward("ScCourseList");
        }
    }
```

（3）配置动作类 ScCourseSaveAction

配置 ScCourseSaveAction 动作类映射关系的代码应写在 struts-config.xml 文件的＜action-mappings＞元素对中。

配置代码如下：

```
<action path = "/bas/sccoursesaveaction"          type = "com.teach.actions.ScCourseSaveAction"    name = "sccourseform"   scope = "request" validate = "false" >
         <forward name = "ScCourseList"          path = "/bas/sccourselistaction.do"></forward>
</action>
```

（4）发布并测试程序

启动 Tomcat 服务器，将 TeacherTest 应用发布至服务器，打开浏览器，在地址栏中输入 http://localhost:8080/ TeacherTest,运行并测试是否能正确地插入或更新班级选课数据。

（5）修改 DBUser 类

在其中定义一个方法，方法为 deleteScCourse(String scono),该方法的作用是根据班级号及课程代码从数据库中删除该班级的一个选课数据,成功返回 true,失败返回 false。

实现代码如下：

```
public boolean deleteScCourse(String scono)
    {
        Connection con = null;
        boolean flag = false;
        String scno = scono.substring(0,4);
        String sono = scono.substring(4,7);

        try
        {
            con = dataSource.getConnection();
            PreparedStatement stat = con.prepareStatement("delete from sccourse where scno = ? and sono = ?");
```

```
                stat.setString(1,scno);
                stat.setString(2,sono);

                int row = stat.executeUpdate();
                if(row == 1)
                    flag = true;
            }
            catch(SQLException e)
            …//省略部分同 CheckAdminUser(username,userpwd)方法中该部分
            return flag;
}
```

(6) 新建并编制动作类文件 ScCourseDeleteAction.java

该类的主要作用是处理数据删除请求。在该类中，先取出传递过来的班级号及课程代码，调用 deleteScCourse()方法删除数据，删除数据成功后引导至 sccourselist.jsp 页面，删除数据失败时在 wrong.jsp 页面中显示错误信息。

实现代码如下：

```
package com.teach.actions;

import com.teach.comms.Constants;
…//省略部分同 CheckAdminUser(Username,userpuce)方法中该部分
public class ScCourseDeleteAction extends Action
{
    public ActionForward execute(ActionMapping mapping, ActionForm form, HttpServletRequest request,
HttpServletResponse response) throws Exception
    {
        String scono = request.getParameter("scono");

        ServletContext context = servlet.getServletContext();
        DataSource dataSource = (DataSource)context.getAttribute("dbkey");
        DBUser dbuser = new DBUser(dataSource);

        HttpSession session = request.getSession();
        if (scono!= null && !scono.equals(""))
        {
            if(!dbuser.deleteScCourse(scono))
            {
                session.setAttribute(Constants.Error_Message,"删除班级科目记录失败!");
                return mapping.findForward("Wrong");
            }
        }
        return mapping.findForward("ScCourseList");
    }
}
```

(7) 配置动作类 ScCourseDeleteAction

配置 ScCourseDeleteAction 动作类映射关系的代码应写在 struts-config.xml 文件的＜action-mappings＞元素对中。

配置代码如下：

```
<action path = "/bas/sccoursedeleteaction"    type = "com.teach.actions.ScCourseDeleteAction" scope = "request">
        <forward name = "ScCourseList"    path = "/bas/sccourselistaction.do"></forward>
</action>
```

(8) 发布并测试程序

启动 Tomcat 服务器，将 TeacherTest 应用发布至服务器，打开浏览器，在地址栏中输入 http://localhost:8080/ TeacherTest，运行并测试是否能正确地删除班级选课数据。

任务七　编制用户情况维护模块

【任务描述】

本任务主要编制系统用户情况维护模块中的相关文件并测试运行,其编制方法与教师基本情况维护模块基本一致,可以在教师基本情况维护模块的基础上编制。系统用户情况维护模块主要是对系统用户表 suser 中数据进行列表显示、添加、修改及删除操作。其运行示意图如图 6-32、图 6-33、图 6-34 所示。

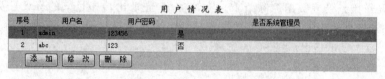

图 6-32　系统用户列表页面

图 6-33　新增用户信息页面　　　　图 6-34　修改用户信息页面

【任务分析】

1. 程序运行基本流程

图 6-35 所示为系统用户模块运行流程图。

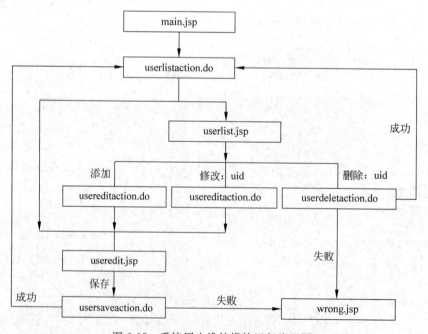

图 6-35　系统用户维护模块运行流程图

2. Action 映射表

系统用户维护模块中的 Action 映射表见表 6-20。

表 6-20 系统用户维护模块中的 Action 映射表

动作(Action)	入口	表单 bean(ActionForm)	出口
UserListAction	main.jsp		userlist.jsp
UserEditAction	userlist.jsp	UserForm	useredit.jsp
UserSaveAction	useredit.jsp	UserForm	userlist.jsp wrong.jsp
UserDeleteAction	userlist.jsp		userlist.jsp wrong.jsp

【任务实施】

1. 实现数据的列表显示

(1) 新建并编制 JavaBean 类文件 Suser.java

这是一个代表系统用户的实体 JavaBean 组件。其中的私有属性除了与系统用户表 suser 中的列相同外，类中还要编写与属性配套的 get/set 方法。

实现代码如下：

```java
package com.teach.comms;

import java.io.Serializable;
public class Suser implements Serializable
{
    int uid;
    private String username;
    private String userpwd;
    private int isadmin;

    public int getUid()
    {
        return uid;
    }
    public void setUid(int uid)
    {
        this.uid = uid;
    }
    …//省略与其余属性对应的 get 和 set 方法
}
```

(2) 修改 DBUser 类

在其中定义一个方法，方法为 getSuser()，该方法的作用是取出所有用户数据，每个用户的数据存放在 Suser 类型的一个对象中，所有用户数据存放在一个 List 类型的对象中，并返回。

实现代码如下：

```java
public List getSUser()
    {
        Connection con = null;
        List sSuser = new ArrayList();
        Suser suser = null;
        try
```

```java
        {
            con = dataSource.getConnection();
            Statement stat = con.createStatement();
            ResultSet result = stat.executeQuery("select * from suser order by uid");
            while(result.next())
            {
                suser = new Suser();
                suser.setUid(result.getInt("uid"));
                suser.setUsername(result.getString("username"));
                suser.setUserpwd(result.getString("userpwd"));
                suser.setIsadmin(result.getInt("isadmin"));

                sSuser.add(suser);
            }
        }
        catch(SQLException e)
        …//省略部分同 CheckAdminUser(username,userpwd)方法中该部分
        return sSuser;
}
```

(3) 新建并编制动作类文件 UserListAction.java

该类的主要作用是处理数据列表显示请求。在该类中,调用 DBUser 类中的 getSuser() 方法,取出所有用户数据,并存放至会话级变量 Search_AllList 中,并引导至 userlist.jsp 页面。

实现代码如下:

```java
package com.teach.actions;

import java.util.List;
……//省略的包和类可参考 AdminLoginAction
public class UserListAction extends Action
{
    public ActionForward execute (ActionMapping mapping, ActionForm form, HttpServletRequest request, HttpServletResponse response) throws Exception
    {
        ServletContext context = servlet.getServletContext();
        DataSource dataSource = (DataSource)context.getAttribute("dbkey");
        DBUser dbuser = new DBUser(dataSource);

        HttpSession session = request.getSession();
        List sSuser = dbuser.getSUser();
        if (sSuser.size() == 0)
        {
            return mapping.findForward("UserEdit");
        }
        else
        {
            session.setAttribute(Constants.Search_AllList,sSuser);

            return (mapping.findForward("UserList"));
        }
    }
}
```

(4) 新建并编制 userlist.jsp 文件

如图 6-32 所示页面,编制方法与 teacherlist.jsp 页面编制方法基本一致,但由于系统用户数一般不会超过 20 个,此页面不需要实现分页显示。

实现代码如下:

```jsp
<%@page contentType="text/html;charset=utf-8"%>
<%@page import="com.teach.comms.Constants,com.teach.comms.Suser,java.util.List"%>
```

```jsp
<%
    List sUser = (List)session.getAttribute(Constants.Search_AllList);
%>
<html>
    <head>
        <link href = "../css/style_css.css" rel = stylesheet type = "text/css">
        <script type = "text/javascript" src = "../js/ClientCom.js"></script>
        <script type = "text/javascript">
        function add()              //添加
        {
         location.href = "usereditaction.do";
        }
        function modify()           //修改
        {
         if (document.all.keyvalue.value == "")
              return false;
location.href = "usereditaction.do?uid = " + document.all.keyvalue.value;
        }
        function Delete()           //删除
        {
            if (document.all.keyvalue.value == "")
                return false;

            if (confirm("你是否要删除选中的记录数据?") == true)
location.href = "userdeleteaction.do?uid = " + document.all.keyvalue.value;

        }
        </script>
    </head>
    <body>
        <table width = "100%" height = "100%">
           <tr>
             <td align = "center" valign = "top">
                <input type = "hidden" name = "keyvalue" value = "" />
                 <table width = "650"  bgcolor = "#dec3b5" cellspacing = "1" >
                      <caption>用 户 情 况 表</caption>
                      <tr height = "20" align = "center" >
                         <td width = "50"   bgcolor = "#f3d5d5">序号</td>
                         <td width = "120"  bgcolor = "#f3d5d5">用户名</td>
                         <td width = "120"  bgcolor = "#f3d5d5">用户密码</td>
                         <td bgcolor = "#f3d5d5">是否系统管理员</td>
                      </tr>
        <%
            if(sUser != null )
            {
                for(int i = 0;i < sUser.size();i++)
                {
                    Suser suser = (Suser)sUser.get(i);
        %>
                <tr  height = "20" style = "cursor:hand"  bgcolor = "#F1FCF7"
                    id = "Row<% = i+1 %>" title = "鼠标双击可以编辑该信息"
                    onclick = "SetFocusIt(this,'<% = suser.getUid() %>',1,keyvalue)"
                    ondblclick = "modify()">
                    <td  align = "center"><% = i+1 %></td>
                    <td><% = suser.getUsername() %></td>
                    <td><% = suser.getUserpwd() %></td>
                    <td><% = (suser.getIsadmin() == 1)?"是":"否" %></td>
                </tr>
        <%
            if(i == 0)
               {
        %>
```

```
                    <script type = "text/javascript">
                        SetFocusIt(Row1,'<% = suser.getUid() %>',0,keyvalue)
                    </script>
        <%
                }
            }
        }
        %>
            </table>
            <table width = "650" bgcolor = "#f3d5d5">
                <tr height = "20">
                    <td> </td>
                    <td>
                        <input type = "button"   name = "addbutton"    value = "添 加"   onclick = "add()">
                        <input type = "button"   name = "editbutton"   value = "修 改"   onclick = "modify()">
                        <input type = "button"   name = "delbutton"    value = "删 除"   onclick = "Delete()">
                    </td>
                </tr>
            </table>
        </td>
    </tr>
</table>
</body>
</html>
```

(5) 配置动作类 UserListAction

配置 UserListAction 动作类映射关系的代码应写在<action-mappings>元素对中。
配置代码如下：

```
<action path = "/bas/userlistaction"        type = "com.teach.actions.UserListAction" scope = "request">
        <forward name = "UserEdit"          path = "/bas/useredit.jsp"/>
        <forward name = "UserList"          path = "/bas/userlist.jsp"/>
</action>
```

(6) 修改主页面文件 main.jsp

将"系统用户数据维护"的超链接改为：

`<a href = "userlistaction.do" target = "inMain">系统用户数据维护</a>`

其中，"inMain"为主框架右侧栏子窗体的 id 值。

(7) 发布并测试程序

启动 Tomcat 服务器，将 TeacherTest 应用发布至服务器，打开浏览器，在地址栏中输入 http://localhost:8080/TeacherTest，运行并测试是否能正确地列出系统用户数据。

2. 实现数据的编辑

(1) 新建并编制表单 bean 类文件 UserForm.java

表单 bean 类 UserForm 中的属性及方法与 Suser 类中一致，区别是本类要从 ActionForm 类继承。
实现代码如下：

```java
package com.teach.forms;

import org.apache.struts.action.ActionForm;
public class UserForm extends ActionForm
{
    int uid;
    private String username;
    private String userpwd;
    private int isadmin;
```

项目六 教师测评系统(Struts模式)

```java
    public int getUid()
    {
        return uid;
    }
    public void setUid(int uid)
    {
        this.uid = uid;
    }
    …//省略与其余属性对应的get和set方法
}
```

(2) 修改DBUser类

在其中定义一个方法,方法为getSuserByUid(int uid),该方法的作用是根据用户ID号(uid)取出该用户数据,并存放在Suser类型的一个对象中返回。

实现代码如下:

```java
public Suser getSuserByUid(int uid)
{
    Connection con = null;
    Suser suser = null;
    try
    {
        con = dataSource.getConnection();
        Statement stat = con.createStatement();
        ResultSet result = stat.executeQuery("select * from suser where uid = " + String.valueOf(uid));
        if(result.next())
        {
            suser = new Suser();
            suser.setUid(result.getInt("uid"));
            suser.setUsername(result.getString("username"));
            suser.setUserpwd(result.getString("userpwd"));
            suser.setIsadmin(result.getInt("isadmin"));
        }
    }
    catch(SQLException e)
    …//省略部分同CheckAdminUser(username,userpwd)方法中该部分
    return suser;
}
```

(3) 新建并编制动作类文件UserEditAction.java

该类的主要作用是处理数据编辑请求。在该类中,先取出传递过来的用户ID号(uid),如果为空,则是添加;如果不为空,则是修改。添加时直接引导至useredit.jsp页面,修改时调用DBUser类中的getSuserByUid(int uid)方法,取出该用户数据,存放在UserForm类的对象中,并引导至useredit.jsp页面。

实现代码如下:

```java
package com.teach.actions;

import com.teach.comms.*;
…//省略的包和类可参考AdminLoginAction
public class UserEditAction extends Action
{
    public ActionForward execute( ActionMapping mapping, ActionForm form, HttpServletRequest request,
    HttpServletResponse response) throws Exception
    {
        UserForm userform = (UserForm) form;
        String suid = request.getParameter("uid");

        ServletContext context = servlet.getServletContext();
```

```java
        DataSource dataSource = (DataSource)context.getAttribute("dbkey");
        DBUser dbuser = new DBUser(dataSource);
        if(suid!= null && !suid.equals(""))
        {
            int uid = Integer.parseInt(suid);
            Suser suser = dbuser.getSuserByUid(uid);
            if(suser!= null)
            {
                userform.setUid(suser.getUid());
                userform.setUsername(suser.getUsername());
                userform.setUserpwd(suser.getUserpwd());
                userform.setIsadmin(suser.getIsadmin());
            }
        }
        return mapping.findForward("UserEdit");
    }
}
```

(4) 新建并编制 useredit.jsp 文件

如图 6-33、图 6-34 所示,编制的基本方法与编制 teacheredit.jsp 页面方法一致。
实现代码如下:

```jsp
<%@page contentType="text/html;charset=utf-8"%>
<%@taglib uri="http://struts.apache.org/tags-html" prefix="html"%>
<html>
    <head>
        <link href="../css/style_css.css" rel=stylesheet type="text/css">
        <script type="text/javascript" src="../js/ClientCom.js"></script>
        <title>用户信息编辑修改</title>
        <script type="text/javascript">
            function checkdata()
            {
                var username = document.userform.username.value;
                if (username == "")
                {
                    alert("用户名不能为空,必须输入!");
                    document.userform.username.focus();
                    return false;
                }

                var userpwd = document.userform.userform.value;
                if (userpwd == "")
                {
                    alert("用户密码不能为空,必须输入!");
                    document.userform.userpwd.focus();
                    return false;
                }
                return true;
            }

            function Exit()
            {
                history.go(-1);
            }
        </script>
    </head>

    <body onload="document.userform.username.focus()">
        <table width="100%" height="100%">
            <tr align="center" valign="middle">
                <td>
                    <html:form action="/bas/usersaveaction.do">
```

```
                <table bgcolor="#F1FCF7">
                    <tr>
                        <td colspan="2" align="center">用户信息编辑</td>
                    </tr>
                    <tr>
                        <td align="center">用 户 名</td><td><html:text property="username"/></td>
                        <html:hidden property="uid"/>
                    </tr>
                    <tr>
                        <td align="center">用户密码</td><td><html:text property="userpwd"/></td>
                    </tr>
                    <tr>
                        <td align="center">是否系统管理员</td>
                        <td>
                            <html:radio property="isadmin" value="1">是</html:radio>
                            <html:radio property="isadmin" value="0">否</html:radio>
                        </td>
                    </tr>
                    <tr>
                        <td></td><td align="center"><html:submit value="保 存" onclick="return checkdata()"/>
     <html:button property="exit" value="取 消" onclick="Exit()"/></td>
                    </tr>
                </table>
            </html:form>
        </td>
    </tr>
</table>
</body>
</html>
```

(5) 配置表单 bean 及动作 Action 类

配置 UserForm 表单 bean 的代码应写在 struts-config.xml 文件的<form-beans>元素对中。

配置代码如下：

```
<form-bean name="userform"    type="com.teach.forms.UserForm"/>
```

配置 UserEditAction 动作类映射关系的代码应写在 struts-config.xml 文件的<action-mappings>元素对中。

配置代码如下：

```
<action path="/bas/usereditaction"         type="com.teach.actions.UserEditAction"   name="userform"
        scope="request">
            <forward name="UserEdit"    path="/bas/useredit.jsp"/>
</action>
```

(6) 发布并测试程序

请注意，为了在测试时能正确地显示编辑页面，可以先将 suseredit.jsp 页面表单属性 action 值中的 usersaveaction 改为 usereditaction，这是因为 usersaveaction 尚未编制和配置。启动 Tomcat 服务器，将 TeacherTest 应用发布至服务器，打开浏览器，在地址栏中输入 http://localhost:8080/TeacherTest，运行并测试系统用户数据是否能正确地显示出来。

3．实现数据的保存、删除

(1) 修改 DBUser 类

在其中定义一个方法，方法为 insertSuser(String username,String userpwd,int isadmin)，该方法的作用是保存用户数据至数据库，成功返回 true，失败返回 false，其中用户 ID 号(uid)在表中该列是自动增长的，不需要提供值。

在其中再定义一个方法，方法为 updateSuser(int uid,String username,String userpwd,int isadmin)，该

方法的作用是根据用户 ID 号(uid)更新用户数据至数据库,成功返回 true,失败返回 false。

实现代码如下:

```java
//插入一个新用户
    public boolean insertSuser(String username,String userpwd,int isadmin)
    {
        Connection con = null;
        boolean flag = false;

        try
        {
            con = dataSource.getConnection();
            PreparedStatement pstat = con.prepareStatement("insert into suser values(?,?,?)");
            pstat.setString(1,username);
            pstat.setString(2,userpwd);
            pstat.setInt(3,isadmin);

            int row = pstat.executeUpdate();
            if(row == 1)
                flag = true;
        }
        catch(SQLException e)
        …//省略部分同 CheckAdminUser(username,userpwd)方法中该部分
        return   flag;
}

//更新一个用户信息
public boolean updateSuser(int uid,String username,String userpwd,int isadmin)
    {
        Connection con = null;
        boolean flag = false;
        try
        {
             con = dataSource.getConnection();
             PreparedStatement pstat = con.prepareStatement("update suser set username = ?,userpwd = ?,isadmin = ? where uid = ?");
            pstat.setString(1,username);
            pstat.setString(2,userpwd);
            pstat.setInt(3,isadmin);
             pstat.setInt(4,uid);

            int row = pstat.executeUpdate();
            if(row == 1)
                flag = true;
        }
        catch(SQLException e)
        …//省略部分同 CheckAdminUser(username,userpwd)方法中该部分
        return   flag;
}
```

(2) 新建并编制动作类文件 UserSaveAction.java

该类的主要作用是理数据保存请求。在该类中,先取出表单中的用户 ID 号(uid),如果为空或 0,则是添加,调用 insertSuser()方法插入数据;否则,则是修改,调用 updateSuser()方法更新数据。插入或更新数据成功后引导至 userlist.jsp 页面,插入或更新数据失败时在 wrong.jsp 页面中显示错误信息。思考一下,如何确保当前用户不被修改?

实现代码如下:

```java
package com.teach.actions;

import com.teach.comms.Constants;
…//省略的包和类可参考 AdminLoginAction
```

项目六 教师测评系统(Struts模式)

```java
public class UserSaveAction extends Action
{
    public ActionForward execute(ActionMapping mapping, ActionForm form, HttpServletRequest request,
HttpServletResponse response) throws Exception
    {
        UserForm userform = (UserForm) form;
        int uid = userform.getUid();

        ServletContext context = servlet.getServletContext();
        DataSource dataSource = (DataSource)context.getAttribute("dbkey");
        DBUser dbuser = new DBUser(dataSource);
        HttpSession session = request.getSession();
        String username = userform.getUsername();
        String userpwd = userform.getUserpwd();
        int isadmin = userform.getIsadmin();

        if ((new Integer(uid)) == null || uid == 0)
        {
            boolean flag = dbuser.insertSuser(username,userpwd,isadmin);
            if(!flag)
            {
                session.setAttribute(Constants.Error_Message,"添加用户失败!");
                return mapping.findForward("Wrong");
            }

        }
        else
        {
            boolean flag = dbuser.updateSuser(uid,username,userpwd,isadmin);
            if(!flag)
            {
                session.setAttribute(Constants.Error_Message,"修改用户失败!");
                return mapping.findForward("Wrong");
            }
        }
        return mapping.findForward("UserList");
    }
}
```

(3) 配置动作类 UserSaveAction

配置 UserSaveAction 动作类映射关系的代码应写在 struts-config.xml 文件的<action-mappings>元素对中。

配置代码如下:

```xml
<action path = "/bas/usersaveaction"     type = "com.teach.actions.UserSaveAction" name = "userform"
scope = "request" validate = "false" >
    <forward name = "UserList"      path = "/bas/userlistaction.do"></forward>
</action>
```

(4) 发布并测试程序

启动 Tomcat 服务器,将 TeacherTest 应用发布至服务器,打开浏览器,在地址栏中输入 http://localhost:8080/ TeacherTest,运行并测试是否能正确地插入或更新用户数据。

(5) 修改 DBUser 类

在其中定义一个方法,方法为 deleteSuser(int uid),该方法的作用是根据用户 ID 号(uid)从数据库中删除该用户,成功返回 true,失败返回 false。

实现代码如下:

```java
public boolean deleteSuser(int uid)
    {
```

```
            Connection con = null;
            boolean flag = false;

                try
                {
                    con = dataSource.getConnection();
                    PreparedStatement pstat = con.prepareStatement("delete suser  where uid = ?");
                    pstat.setInt(1,uid);

                    int row = pstat.executeUpdate();
                    if(row == 1)
                        flag = true;
                }
                catch(SQLException e)
                …//省略部分同 CheckAdminUser(username,userpwd)方法中该部分
                return  flag;
        }
```

(6) 新建并编制动作类文件 UserDeleteAction.java

该类的主要作用是处理数据删除请求。在该类中,先取出传递过来的用户代码,调用 deleteSuser()方法删除数据,删除数据成功后引导至 suserlist.jsp 页面,删除数据失败时在 wrong.jsp 页面中显示错误信息。思考一下,如何确保当前用户不被删改?

实现代码如下:

```
package com.teach.actions;

import com.teach.comms.*;
…//省略的包和类可参考 AdminLoginAction
public    class UserDeleteAction extends Action
{
      public ActionForward execute(ActionMapping mapping, ActionForm form, HttpServletRequest request, HttpServletResponse response) throws Exception
    {
        int  uid = Integer.parseInt(request.getParameter("uid"));

        ServletContext context = servlet.getServletContext();
        DataSource dataSource = (DataSource)context.getAttribute("dbkey");
        DBUser dbuser = new DBUser(dataSource);
        HttpSession session = request.getSession();

        if ((new Integer(uid))!= null && uid!= 0)
        {
            boolean flag = dbuser.deleteSuser(uid);
            if(!flag)
            {
                session.setAttribute(Constants.Error_Message,"删除用户记录失败!");
                return mapping.findForward("Wrong");
            }
        }
        return mapping.findForward("UserList");
    }
}
```

(7) 配置动作类 UserDeleteAction

配置 UserDeleteAction 动作类映射关系的代码应写在 struts-config.xml 文件的 <action-mappings> 元素对中。

配置代码如下:

```
<action path = "/bas/userdeleteaction"        type = "com.teach.actions.UserDeleteAction" scope = "request" >
        < forward name = "UserList"           path = "/bas/userlistaction.do"></forward>
</action>
```

（8）发布并测试程序

启动 Tomcat 服务器，将 TeacherTest 应用发布至服务器，打开浏览器，在地址栏中输入 http://localhost:8080/ TeacherTest，运行并测试是否能正确地删除用户数据。

任务八　编制系统初始化模块

【任务描述】

本任务主要编制系统初始化模块中的相关文件并测试运行。系统初始化分两种情况，一种是在使用系统之前将上一次的测评数据全部清除，即删除表 scyn、styn 中的数据；另外一种是在使用系统之前将系统数据库存中的数据全部清除，只剩系统用户表中的系统管理员记录。单击主页面中"系统初始化"超链接，出现如图 6-36 所示页面，在单击相应按钮后，执行相应操作（系统化确认如图 6-37 所示），操作失败后，转至 wrong.jsp 页面，并显示"系统初始化失败！"信息，操作成功后，转至 info.jsp 页面，并显示"系统初始化成功！"信息，如图 6-38 所示。

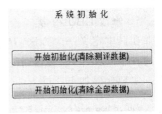

图 6-36　系统初始化页面

【任务分析】

图 6-37　系统初始化确认对话框

图 6-38　显示系统初始化成功信息的页面

1. 程序运行基本流程

图 6-39 所示为系统初始化模块运行流程图。

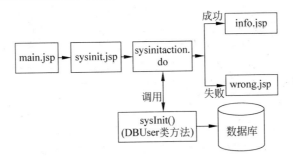

图 6-39　系统初始化模块运行流程图

2. Action 映射表

系统初始化模块中的 Action 映射表见表 6-21。

表 6-21 系统初始化模块中的 Action 映射表

动作(Action)	入口	表单 bean(ActionForm)	出口
SysInitAction	sysinit.jsp		info.jsp wrong.jsp

【任务实施】

1. 修改 DBUser 类

在其中定义一个方法,方法为 sysInit(String parm),该方法的作用是根据参数执行系统初始化的两种不同操作,成功返回 true,失败返回 false。

实现代码如下:

```java
public boolean sysInit(String parm)
{
    Connection con = null;
    boolean flag = false;
    try
    {
        con = dataSource.getConnection();
        Statement stat = con.createStatement();

        stat.executeUpdate("delete from stynsum");
        stat.executeUpdate("delete from scynsum");
        stat.executeUpdate("delete from styn");
        stat.executeUpdate("delete from scyn");

        if(parm.equals("2"))
        {
            stat.executeUpdate("delete from sccourse");
            stat.executeUpdate("delete from sclass");
            stat.executeUpdate("delete from scourse");
            stat.executeUpdate("delete from steacher");
            stat.executeUpdate("delete from suser where username!='admin'");
        }
        flag = true;
    }
    catch(SQLException e)
    …//省略部分同 CheckAdminUser(username,userpwd)方法中该部分
    return flag;
}
```

2. 新建并编制 info.jsp 文件

前面编制的 wrong.jsp 页面文件是用来显示操作失败信息的,如图 6-38 所示。本页面文件用来显示操作成功的提示信息,提示信息事先是保存在会话级变量 Constants.Show_Message 中的,然后在页面中取出并显示,两个页面文件的编制方法基本一致。

实现代码如下:

```jsp
<%@page contentType="text/html;charset=utf-8" %>
<%@page import="com.teach.comms.Constants" %>
<%
    String Show_Message = (String)session.getAttribute(Constants.Show_Message);
%>
```

```html
<html>
  <head>
    <title>教师教育教学情况测评系统</title>
    <link href = "../css/style_css.css" rel = stylesheet type = "text/css">
  </head>

  <body>
    <table width = "100%" height = "100%">
    <tr>
        <td align = "center" valign = "top">
          <div align = center>
            <table bgcolor = "#f1fcf7">
              <tr height = "30">
                <td width = "400" align = "center" bgcolor = "#dbc2b0"><b>信息显示页面</b></td>
              </tr>
              <tr>
                <td align = "center" bgcolor = "#f5efe7">
                  <font size = "5" color = "#ff0000"><% = Show_Message %></font>
                </td>
              </tr>
            </table>
            <hr width = "400" color = "#dbc2b0" size = "2">
          </div>
        </td>
      </tr>
    </table>
  </body>
</html>
```

3. 配置全局转发

与 wrong.jsp 页面一致,要从多个处理中转至 info.jsp 页面,在 Struts 中,也可以通过在 struts-config.xml 配置全局转发来实现。

配置代码如下:

```xml
<global - forwards>
    <forward name = "Wrong"    path = "/bas/wrong.jsp"/>
    <forward name = "Info"     path = "/bas/info.jsp"/>
</global - forwards>
```

4. 新建并编制动作类文件 SysInitAction.java

SysInitAction 类的主要作用是处理系统初始化请求。在该动作类中,先取出页面中传入的参数 parm,然后调用 sysInit()方法执行系统初始化。执行成功后,并引导至 info.jsp 页面,显示"系统初始化成功!"信息,失败后在 wrong.jsp 页面中显示错误信息。

实现代码如下:

```java
package com.teach.actions;
import org.apache.struts.action.*;
…//省略的包和类可参考 AdminLoginAction
public class SysInitAction extends Action
{
    public ActionForward execute (ActionMapping mapping, ActionForm form, HttpServletRequest request, HttpServletResponse response) throws Exception
    {
        String parm = request.getParameter("parm");

        ServletContext context = servlet.getServletContext();
        DataSource dataSource = (DataSource)context.getAttribute("dbkey");
        DBUser dbuser = new DBUser(dataSource);
```

```
HttpSession session = request.getSession();
if(parm.equals("1"))
{
    if(!dbuser.sysInit(parm))
    {
        session.setAttribute(Constants.Error_Message,"系统初始化失败!");
        return (mapping.findForward("Wrong"));
    }
}
else if(parm.equals("2"))
{
    if(!dbuser.sysInit(parm))
    {
        session.setAttribute(Constants.Error_Message,"系统初始化失败!");
        return (mapping.findForward("Wrong"));
    }
}
session.setAttribute(Constants.Show_Message,"系统初始化成功!");
return (mapping.findForward("Info"));
}
```

5．新建并编制 sysinit.jsp 文件

如图 6-36 所示，页面上有两个按钮，单击"开始初始化（清除测评数据）"按钮时执行 SysInitAction 动作类，传入的参数为"1"；单击"开始初始化（清除全部数据）"按钮时也是执行 SysInitAction 动作类，传入的参数为"2"。

实现代码如下：

```
<%@page contentType="text/html;charset=utf-8"%>
<%@taglib uri="http://struts.apache.org/tags-html" prefix="html"%>
<html>
    <head>
        <link href="../css/style_css.css" rel=stylesheet type="text/css">
        <title>系统初始化</title>
        <script type="text/javascript">
            function Start_Init1()
            {
                if (confirm("系统初始化将清除数据库中所有的测评数据!你确认要开始初始化?") == true)
                location.href = "sysinitaction.do?parm=1";
            }
            function Start_Init2()
            {
                if (confirm("系统初始化将清除数据库中除用户以外所有的数据!你确认要开始初始化?") == true)
                location.href = "sysinitaction.do?parm=2";
            }
        </script>
    </head>
    <body >
        <table  width="100%" height="100%"  border="0" cellspacing="0" cellpadding="0" style="border-collapse: collapse" bordercolor="#111111">
            <tr align="center" valign="middle">
                <td>
                    <table  bgcolor="#f1fcf7">
                        <tr>
                            <td colspan="3" align="center">系 统 初 始 化</td>
                        </tr>
                        <tr  height="20">
                            <td width="100"> </td><td> </td><td> </td>
                        </tr>
                        <tr  height="20">
                            <td width="100"> </td><td> </td><td> </td>
```

项目六 教师测评系统(Struts模式)

```
        </tr>
        <tr>
          <td colspan = "3" align = "center"><html:button property = "init1" value = "开始初始化
(清除测评数据)" onclick = "Start_Init1()"/></td>
        </tr>
        <tr height = "20">
          <td width = "100"> </td><td> </td><td> </td>
        </tr>
        <tr>
          <td colspan = "3" align = "center"><html:button property = "init2" value = "开始初始化
(清除全部数据)" onclick = "Start_Init2()"/></td>
        </tr>
        <tr height = "20">
          <td width = "100"> </td><td> </td><td> </td>
        </tr>
      </table>
    </td>
  </tr>
</table>
</body>
</html>
```

6. 配置动作类

配置 SysInitAction 动作类映射关系的代码应写在 struts-config.xml 文件的＜action-mappings＞元素对中。

配置代码如下：

```
<action path = "/bas/sysinitaction"    type = "com.teach.actions.SysInitAction"    scope = "request"
/>
```

7. 修改主页面文件 main.jsp

将"系统初始化"的超链接改为：

```
<a href = "sysinit.jsp" target = "inMain">系统初始化</a>
```

其中，"inMain"为主框架右侧栏子窗体的 id 值。

8. 发布并测试程序

启动 Tomcat 服务器,将 TeacherTest 应用发布至服务器,打开浏览器,在地址栏中输入 http://localhost:8080/TeacherTest,运行并测试系统管理子系统的系统初始化模块。

任务九 编制开放或禁止测评系统模块

【任务描述】

本任务主要编制开放或禁止测评系统模块中的相关文件并测试运行。开放测评系统,是指允许使用学生测评子系统,即将 sclass 表中 scout 列的值设置为"1";禁止测评系统,是指不允许使用学生测评子系统,即将 sclass 表中 scout 列的值设置为"0"。单击主页面中"开放或禁止测评系统"超链接,出现如图 6-40 所示页面,在单击按钮后,执行相应操作(开发测评系统确认见图 6-41),操作失败后,转至 wrong.jsp 页面,并显示"开放测评系统失败!"或"禁止测评系统失败!"信息;操作成功后,转至 info.jsp 页面,并显示"已开放测评系统!"(见图 6-42)或"已禁止测评系统!"信息(见图 6-43 和图 6-44)。

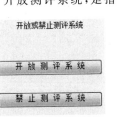

图 6-40 开放或禁止测评系统页面

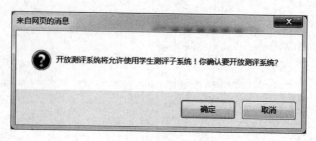

图 6-41　开放测评系统确认对话框

图 6-42　显示成功开放测评系统信息的页面

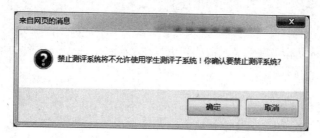

图 6-43　禁止测评系统确认对话框

图 6-44　显示成功禁止测评系统信息的页面

【任务分析】

1．程序运行基本流程

图 6-45 所示为开放或禁止测评系统模块运行流程图。

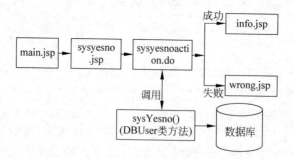

图 6-45　开放或禁止测评系统模块运行流程图

2．Action 映射表

开放或禁止测评系统模块中的 Action 映射表见表 6-22。

项目六 教师测评系统(Struts模式)

表 6-22 开放或禁止测评系统模块中的 Action 映射表

动作(Action)	入口	表单 bean(ActionForm)	出口
SysYesnoAction	sysyesno.jsp		info.jsp wrong.jsp

【任务实施】

1. 修改 DBUser 类

在其中定义一个方法,方法为 sysYesno(String parm),该方法的作用是根据参数执行开放或禁止测评系统两种不同的操作,成功返回 true,失败返回 false。

实现代码如下:

```java
public boolean sysYesno(String parm)
{
    Connection con = null;
    boolean flag = false;
    try
    {
        con = dataSource.getConnection();
        Statement stat = con.createStatement();

        if(parm.equals("1"))
        {
            stat.executeUpdate("update sclass set scout = '1'");
        }
        if(parm.equals("2"))
        {
            stat.executeUpdate("update sclass set scout = '0'");
        }
        flag = true;
    }
    catch(SQLException e)
    …//省略部分同 CheckAdminUser(username,userpwd)方法中该部分
    return flag;
}
```

2. 新建并编制动作类文件 SysYesnoAction.java

SysYesnoAction 类的主要作用是处理系统初始化请求。在该动作类中,先取出页面中传入的参数 parm,然后调用 sysYesno()方法执行开放或禁止测评系统两种不同的操作。执行成功后,并引导至 info.jsp 页面,显示"已成功开放测评系统!"、"已成功禁止测评系统!"信息;失败后在 wrong.jsp 页面中显示错误信息。

实现代码如下:

```java
package com.teach.actions;

import com.teach.comms.*;
…//省略的包和类可参考 AdminLoginAction
public class SysYesnoAction extends Action
{
    public ActionForward execute(ActionMapping mapping, ActionForm form, HttpServletRequest request, HttpServletResponse response) throws Exception
    {
        String parm = request.getParameter("parm");

        ServletContext context = servlet.getServletContext();
```

```java
            DataSource dataSource = (DataSource)context.getAttribute("dbkey");
            DBUser dbuser = new DBUser(dataSource);

            HttpSession session = request.getSession();
            if(parm.equals("1"))
            {
                if(!dbuser.sysYesno(parm))
                {
                    session.setAttribute(Constants.Error_Message,"开放测评系统失败!");
                    return (mapping.findForward("Wrong"));
                }
                session.setAttribute(Constants.Show_Message,"已成功开放测评系统!");
            }
            else if(parm.equals("2"))
            {
                if(!dbuser.sysYesno(parm))
                {
                    session.setAttribute(Constants.Error_Message,"禁止测评系统失败!");
                    return (mapping.findForward("Wrong"));
                }
                session.setAttribute(Constants.Show_Message,"已成功禁止测评系统!");
            }
            return (mapping.findForward("Info"));
        }
    }
```

3. 新建并编制 sysyesno.jsp 文件

如图 6-40 所示,页面上有两个按钮,单击"开放测评系统"按钮时执行 SysYesnoAction 动作类,传入的参数为"1";单击"禁止测评系统"按钮时也是执行 SysYesnoAction 动作类,传入的参数为"2"。

4. 配置动作类

配置 SysYesnoAction 动作类映射关系的代码应写在 struts-config.xml 文件的＜action-mappings＞元素对中。

配置代码如下:

```xml
<action path="/bas/sysyesnoaction"   type="com.teach.actions.SysYesnoAction" scope="request"
/>
```

5. 修改主页面文件 main.jsp

将"开放或禁止测评系统"的超链接改为:

```html
<a href="sysyesno.jsp" target="inMain">开放或禁止测评系统</a>
```

其中,"inMain"为主框架右侧栏子窗体的 id 值。

6. 发布并测试程序

启动 Tomcat 服务器,将 TeacherTest 应用发布至服务器,打开浏览器,在地址栏中输入 http://localhost:8080/TeacherTest,运行并测试系统管理子系统的开放禁止测评系统模块。

任务十　编制汇总班主任测评数据模块

【任务描述】

本任务主要编制汇总班主任测评数据模块中的相关文件并测试运行。汇总班主任测评数据,是指将班主任教育情况测评表 scyn 中的数据按要求汇总至班主任教育情况测评汇总表 scynsum

项目六 教师测评系统(Struts模式)

中,以便于查询统计。单击主页面中"汇总班主任测评数据"超链接,出现如图6-46所示页面,在单击相关按钮后,执行相应操作。操作失败后,转至wrong.jsp页面,并显示"汇总班主任测评数据失败!"信息,操作成功后,转至info.jsp页面,并显示"已成功汇总班主任测评数据!"信息(见图4-67)。

图6-46 汇总班主任测评数据页面　　图6-47 显示成功汇总班主任测评数据信息的页面

【任务分析】

1. 程序运行基本流程

图6-48所示为汇总班主任测评数据模块运行流程图。

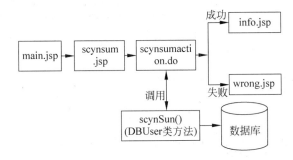

图6-48 汇总班主任测评数据模块运行流程图

2. Action映射表

汇总班主任测评数据模块中的Action映射表见表6-23。

表6-23 汇总班主任测评数据模块中的Action映射表

动作(Action)	入口	表单bean(ActionForm)	出口
ScynsumAction	scynsum.jsp		info.jsp wrong.jsp

【任务实施】

1. 修改DBUser类

在其中定义一个方法,方法为scynSum(),该方法的作用是将班主任测评情况表scyn中的数据汇总至班主任测评情况汇总表scynsum中,汇总前先删除班主任测评情况汇总表scynsum中的数据,成功返回true,失败返回false。

实现代码如下:

```
public boolean scynSum()
{
    Connection con = null;
    String strSql1,strSql2,strSql3;
    int i = 0;
    boolean flag = false;
    String scno,snno,stno,stname;
```

```java
        int num, itorder;
        double itsum, it1, it2, it3, it4, it5, it6, it7, it8, it9, it10;

        try
        {
            con = dataSource.getConnection();

            strSql1 = "delete from scynsum";
            Statement stat1 = con.createStatement();
            stat1.executeUpdate(strSql1);

            strSql2 = "  select scno,count( * )as num,snno,stno,stname,"
                    + "  sum(item1)/count( * ) as it1,sum(item2)/count( * ) as it2,sum(item3)/count( * ) as it3,sum(item4)/count( * ) as it4,sum(item5)/count( * )   as it5,"
                    + "  sum(item6)/count( * ) as it6,sum(item7)/count( * ) as it7,sum(item8)/count( * ) as it8,sum(item9)/count( * ) as it9,sum(item10)/count( * ) as it10,"
                    + "  sum(item1)/count( * ) + sum(item2)/count( * ) + sum(item3)/count( * ) + sum(item4)/count( * ) + sum(item5)/count( * ) + "
                    + "  sum(item6)/count( * ) + sum(item7)/count( * ) + sum(item8)/count( * ) + sum(item9)/count( * ) + sum(item10)/count( * ) as itsum"
                    + "  from scyn group by scno,snno,stno,stname   having datalength(rtrim(scno)) = 4"
                    + "  order by sum(item1)/count( * ) + sum(item2)/count( * ) + sum(item3)/count( * ) + sum(item4)/count( * ) + sum(item5)/count( * ) + "
                    + "  sum(item6)/count( * ) + sum(item7)/count( * ) + sum(item8)/count( * ) + sum(item9)/count( * ) + sum(item10)/count( * ) desc";

            Statement stat2 = con.createStatement();
            ResultSet result = stat2.executeQuery(strSql2);
            while(result.next())
            {
                i++;
                snno = result.getString("snno");
                scno = result.getString("scno");
                stno = result.getString("stno");
                stname = result.getString("stname");
                num = result.getInt("num");
                itsum = result.getDouble("itsum");
                itorder = i;
                it1 = result.getDouble("it1");
                it2 = result.getDouble("it2");
                it3 = result.getDouble("it3");
                it4 = result.getDouble("it4");
                it5 = result.getDouble("it5");
                it6 = result.getDouble("it6");
                it7 = result.getDouble("it7");
                it8 = result.getDouble("it8");
                it9 = result.getDouble("it9");
                it10 = result.getDouble("it10");

                strSql3 = "insert into scynsum values('" + scno + "','"
                        + snno + "','"
                        + stno + "','"
                        + stname + "',"
                        + String.valueOf(num) + ","
                        + String.valueOf(itsum) + ","
                        + String.valueOf(itorder) + ","
                        + String.valueOf(it1) + ","
                        + String.valueOf(it2) + ","
                        + String.valueOf(it3) + ","
                        + String.valueOf(it4) + ","
                        + String.valueOf(it5) + ","
                        + String.valueOf(it6) + ","
                        + String.valueOf(it7) + ","
                        + String.valueOf(it8) + ","
                        + String.valueOf(it9) + ","
                        + String.valueOf(it10)
```

```
            + ")";
        Statement stat3 = con.createStatement();
        stat3.executeUpdate(strSql3);
    }
    flag = true;
}
catch(SQLException e)
…//省略部分同 CheckAdminUser(username,userpwd)方法中该部分
return flag;
}
```

2. 新建并编制动作类文件 ScynsumAction.java

ScynsumAction 类的主要作用是处理汇总班主任测评数据的请求。在该动作类中,调用 scynSum()方法汇总班主任测评数据,执行成功后,并引导至 info.jsp 页面,显示"已成功汇总班主任测评数据!"信息,失败后在 wrong.jsp 页面中显示错误信息。

实现代码如下:

```
package com.teach.actions;

import com.teach.comms.*;
…//省略的包和类可参考 AdminLoginAction
public class ScynsumAction extends Action
{
    public ActionForward execute(ActionMapping mapping, ActionForm form, HttpServletRequest request, HttpServletResponse response) throws Exception
    {
        ServletContext context = servlet.getServletContext();
        DataSource dataSource = (DataSource)context.getAttribute("dbkey");
        DBUser dbuser = new DBUser(dataSource);

        HttpSession session = request.getSession();
        if(!dbuser.scynSum())
        {
            session.setAttribute(Constants.Error_Message,"汇总班主任测评数据失败!");
            return (mapping.findForward("Wrong"));
        }

        session.setAttribute(Constants.Show_Message,"已成功汇总班主任测评数据!");
        return (mapping.findForward("Info"));
    }
}
```

3. 新建并编制 scynsum.jsp 文件

如图 6-46 所示,页面上有一个按钮,单击"开始汇总班级测评数据…"按钮时通过 script 代码中的 scynsum()函数执行 ScynsumAction 动作类。页面的具体编制可参考系统初始化模块中页面编制方法。

4. 配置动作类

配置 ScynsumAction 动作类映射关系的代码应写在 struts-config.xml 文件的＜action-mappings＞元素对中。

配置代码如下:

```
<action path="/bas/scynsumaction"  type="com.teach.actions.ScynsumAction"  scope="request"
/>
```

5. 修改主页面文件 main.jsp

将"汇总班级测评数据"的超链接改为:

```
<a href = "scynsum.jsp" target = "inMain">汇总班级测评数据</a>
```

其中,"inMain"为主框架右侧栏子窗体的 id 值。

6. 发布并测试程序

启动 Tomcat 服务器,将 TeacherTest 应用发布至服务器,打开浏览器,在地址栏中输入 http://localhost:8080/TeacherTest,运行并测试系统管理子系统的汇总班级测评数据模块。

任务十一　编制汇总任课教师测评数据模块

【任务描述】

本任务主要编制汇总任课教师测评数据模块中的相关文件并测试运行。汇总任课教师测评数据,是指将教师教育情况测评表 styn 中的数据按要求汇总至教师教育情况测评汇总表 stynsum 中,以便于查询统计。单击主页面中"汇总任课教师测评数据"超链接,出现如图 6-49 所示页面,在单击相关按钮后,执行相应操作,操作失败后,转至 wrong.jsp 页面,并显示"汇总任课教师测评数据失败!"信息;操作成功后,转至 info.jsp 页面,并显示"已成功汇总任课教师测评数据!"信息,如图 6-50 所示。

图 6-49　汇总任课教师测评数据页面　　　图 6-50　显示成功汇总任课教师测评数据信息的页面

【任务分析】

1. 程序运行基本流程

图 6-51 所示为汇总任课教师测评数据模块运行流程图。

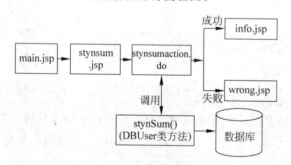

图 6-51　汇总任课教师测评数据模块运行流程图

2. Action 映射表

汇总任课教师测评数据模块中的 Action 映射表见表 6-24。

表 6-24　汇总任课教师测评数据模块中的 Action 映射表

动作(Action)	入口	表单 bean(ActionForm)	出口
StynsumAction	stynsum.jsp		info.jsp wrong.jsp

项目六 教师测评系统(Struts模式)

【任务实施】

1. 修改 DBUser 类

在其中定义一个方法,方法为 stynSum(),该方法的作用是将教师测评情况表 styn 中的数据汇总至教师测评情况汇总表 stynSum 中,汇总前先删除教师测评情况汇总表 stynsum 中的数据,成功返回 true,失败返回 false。

实现代码如下:

```java
public boolean stynSum()
{
        Connection con = null;
        String strSql1,strSql2,strSql3;
        int i = 0;
        boolean flag = false;
        String scno,snno,stno,stname,sono,soname;
        int num,itorder;
        double itsum,it1,it2,it3,it4,it14,it5,it6,it7,it8,it58,it9,it10,it11,it12,it912;

    try
    {
        con = dataSource.getConnection();
        Statement stat1 = con.createStatement();
        Statement stat2 = con.createStatement();
        Statement stat3 = con.createStatement();

        strSql1 = "delete from stynsum";
        stat1.executeUpdate(strSql1);

        strSql2 = "  select stno,count(*) as num,scno,snno,stname,sono,soname,"
            + " sum(item1)/count(*) as it1,sum(item2)/count(*) as it2,sum(item3)/count(*) as it3,sum(item4)/count(*) as it4,sum(item1)/count(*) + sum(item2)/count(*) + sum(item3)/count(*) + sum(item4)/count(*) as it14,"
            + " sum(item5)/count(*) as it5,sum(item6)/count(*) as it6,sum(item7)/count(*) as it7,sum(item8)/count(*) as it8,sum(item5)/count(*) + sum(item6)/count(*) + sum(item7)/count(*) + sum(item8)/count(*) as it58,"
            + " sum(item9)/count(*) as it9,sum(item10)/count(*) as it10,sum(item11)/count(*) as it11,sum(item12)/count(*) as it12,sum(item9)/count(*) + sum(item10)/count(*) + sum(item11)/count(*) + sum(item12)/count(*) as it912,"
            + "  sum(item1)/count(*) + sum(item2)/count(*) + sum(item3)/count(*) + sum(item4)/count(*) + sum(item5)/count(*) +"
            + "  sum(item6)/count(*) + sum(item7)/count(*) + sum(item8)/count(*) + sum(item9)/count(*) + sum(item10)/count(*) + sum(item11)/count(*) + sum(item12)/count(*) as itsum"
            + " from styn "
            + " group by scno,sono,stno,snno,stname,soname"
            + " order by sum(item1)/count(*) + sum(item2)/count(*) + sum(item3)/count(*) + sum(item4)/count(*) + sum(item5)/count(*) +"
            + " sum(item6)/count(*) + sum(item7)/count(*) + sum(item8)/count(*) + sum(item9)/count(*) + sum(item10)/count(*) + sum(item11)/count(*) + sum(item12)/count(*) desc";

        ResultSet result = stat2.executeQuery(strSql2);
        while(result.next())
        {
            i++;
            snno = result.getString("snno");
            sono = result.getString("sono");
            scno = result.getString("scno");
            stno = result.getString("stno");
            stname = result.getString("stname");
            soname = result.getString("soname");
```

```
                    num = result.getInt("num");
                    itsum = result.getDouble("itsum");
                    itorder = i;
                    it1 = result.getDouble("it1");
                    it2 = result.getDouble("it2");
                    it3 = result.getDouble("it3");
                    it4 = result.getDouble("it4");
                    it14 = result.getDouble("it14");
                    it5 = result.getDouble("it5");
                    it6 = result.getDouble("it6");
                    it7 = result.getDouble("it7");
                    it8 = result.getDouble("it8");
                    it58 = result.getDouble("it58");
                    it9 = result.getDouble("it9");
                    it10 = result.getDouble("it10");
                    it11 = result.getDouble("it11");
                    it12 = result.getDouble("it12");
                    it912 = result.getDouble("it912");

                    strSql3 = "insert into stynsum values('" + scno + "','"
                        + sono + "','"
                        + stno + "','"
                        + snno + "','"
                        + stname + "','"
                        + soname + "','"
                        + String.valueOf(num) + ","
                        + String.valueOf(itorder) + ","
                        + String.valueOf(itsum) + ","
                        + String.valueOf(it1) + ","
                        + String.valueOf(it2) + ","
                        + String.valueOf(it3) + ","
                        + String.valueOf(it4) + ","
                        + String.valueOf(it14) + ","
                        + String.valueOf(it5) + ","
                        + String.valueOf(it6) + ","
                        + String.valueOf(it7) + ","
                        + String.valueOf(it8) + ","
                        + String.valueOf(it58) + ","
                        + String.valueOf(it9) + ","
                        + String.valueOf(it10) + ","
                        + String.valueOf(it11) + ","
                        + String.valueOf(it12) + ","
                        + String.valueOf(it912)
                        + ")";
                    stat3.executeUpdate(strSql3);
                }
                flag = true;
            }
            catch(SQLException e)
            …//省略部分同CheckAdminUser(username,userpwd)方法中该部分
            return flag;
        }
```

2．新建并编制动作类文件StynsumAction.java

StynsumAction类的主要作用是处理汇总教师测评数据的请求。其编制方法与ScynsumAction编制方法完全相同。只需将ScynsumAction.java另存为StynsumAction.java后，修改两处，一是将要调用的DBUser类中汇总数据的方法由scynSum()改为"stynSum()"，二是将相关文字由"班级"改为"教师"。

3．新建并编制stynsum.jsp文件

如图6-49所示的页面,其编制方法与scynsum.jsp编制方法完全相同。只需将scynsum.jsp

另存为 stynsum.jsp 后,修改两处,一是将页面单击按钮要执行的动作类由"scynsumaction.do"改为"stynsumaction.do",二是将相关文字由"班级"改为"教师"。

4. 配置动作类

配置 StynsumAction 动作类映射关系的代码应写在 struts-config.xml 文件的＜action-mappings＞元素对中。

配置代码如下:

```
<action path = "/bas/stynsumaction"  type = "com.teach.actions.StynsumAction"  scope = "request" />
```

5. 修改主页面文件 main.jsp

将"汇总教师测评数据"的超链接改为:

```
<a href = "stynsum.jsp" target = "inMain">汇总教师测评数据</a>
```

其中,"inMain"为主框架右侧栏子窗体的 id 值。

6. 发布并测试程序

启动 Tomcat 服务器,将 TeacherTest 应用发布至服务器,打开浏览器,在地址栏中输入 http://localhost:8080/TeacherTest,运行并测试系统管理子系统的汇总教师测评数据模块。

任务十二 编制测评班主任模块的学生登录部分

【任务描述】

本任务主要编制学生测评子系统中测评班主任模块的学生登录部分的相关文件并测试运行。测评班主任模块的学生登录部分主要是要在学生登录时取出待测班级及班主任的相关数据。单击首页面中"学生测评子系统"超链接,出现如图 6-52 所示的登录页面。

图 6-52 学生测评子系统登录页面

输入学号和密码后,单击"登录"按钮,如果学号前 4 位所对应的班级不存在,或者已禁止测评系统,显示如图 6-53 所示出错信息。

如果该学生已经测评过,并且已全部测评完毕,则显示如图 6-54 所示提示信息;否则,进入如图 6-55 所示测评子系统的主页面。

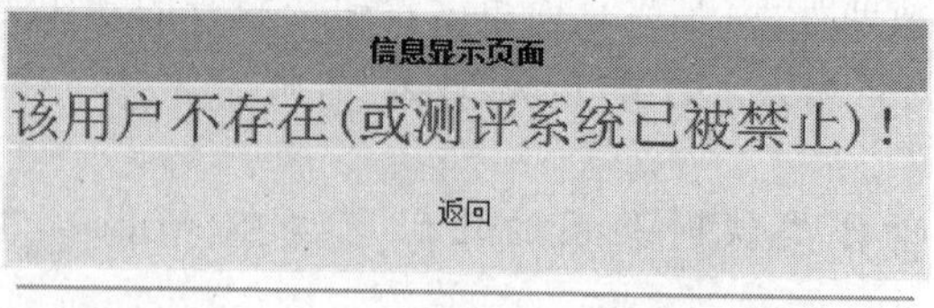

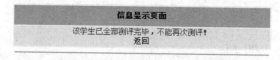

图 6-53 显示登录失败信息的页面

图 6-54 显示学生已全部测评完毕信息的页面

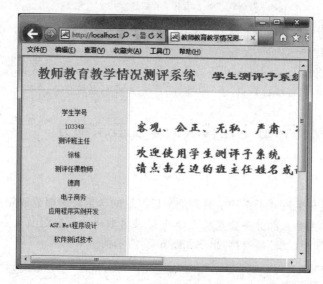

图 6-55 学生测评子系统主页面

【任务分析】

1. 程序运行基本流程

图 6-56 所示为学生测评子系统登录模块运行流程图。

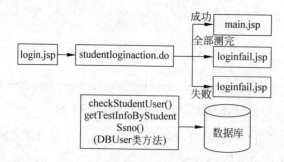

图 6-56 学生测评子系统登录模块运行流程图

2. Action 映射表

学生测评子系统登录模块中的 Action 映射表见表 6-25。

项目六 教师测评系统(Struts模式)

表 6-25 学生测评子系统登录模块中的 Action 映射表

动作(Action)	入口	表单 bean(ActionForm)	出口
StudentLoginAction	login.jsp	LoginForm	main.jsp loginfail.jsp

【任务实施】

1. 新建并编制 login.jsp 文件

如图 6-52 所示,该页面为学生测评时的登录页面。只需将系统管理子系统中的 login.jsp 页面复制到 test 文件夹中,将页面进行修改。一是修改页面上的文字,二是修改验证学号和密码是否合法的 JavaScript 代码,三是修改表单中的 action 属性的值。

实现代码如下:

```jsp
<%@page contentType="text/html;charset=utf-8"%>
<html>
    <head>
        <title>教师教育教学情况测评系统---测评子系统</title>
        <link href="../css/style_css.css" rel=stylesheet type="text/css">
        <script language="javascript">
            function checkdata()
            {
                var username = document.loginform.username.value;
                if (username == "" || username.length!=6 || !isNumberic(username))
                {
                    alert("学号不能为空,长度必须是6位,只能是数字!");
                    document.loginform.username.focus();
                    return false;
                }
                var userpwd = document.loginform.userpwd.value;
                if (userpwd == "")
                {
                    alert("密码不能为空!");
                    document.loginform.userpwd.focus();
                    return false;
                }
                if (username!=userpwd)
                {
                    alert("密码不正确!");
                    document.loginform.userpwd.focus();
                    return false;
                }
                return true;
            }
            function isNumberic(str)
            {
                var len = str.length;
                for (var i=0;i<len;i++)
                    if (str.charAt(i)<'0' || str.charAt(i)>'9')
                        return false;
                return true;
            }
        </script>
    </head>
    <body onload="document.loginform.username.focus()">
        <table width="100%" height="100%">
            <tr>
                <td align="center" valign="middle">
```

```html
<form action = "studentloginaction.do"  method = "post"  name = "loginform">
    <table  bgcolor = "#f1fcf7">
        <tr>
            <td colspan = "2" align = "center">学 生 登 录</td>
        </tr>
        <tr>
            <td align = "center">学   号  </td>
            <td><input type = "text"  name = "username" style = "width:125px; autocomplete = "off"></td>
        </tr>
        <tr>
            <td align = "center">密   码  </td>
            <td><input type = "password"  name = "userpwd" style = "width:125px; autocomplete = "off"></td>
        </tr>
        <tr>
            <td> </td>
            <td align = "center"><input type = "submit"  value = "登 录"  onclick = "return checkdata()"></td>
        </tr>
    </table>
</form>
        </td>
    </tr>
</table>
</body>
</html>
```

2. 修改 DBUser 类

在其中定义一个方法,方法为 checkStudentUser(String username),参数为用户名,即学生的学号。该方法的作用是根据传入的学号,判断该学号的前 4 位所对应的班级是否存在,并且测评子系统是否开放(scout 的值为 1),如果班级存在并且测评系统开放,返回一布尔值 true,否则返回 false。

实现代码如下:

```java
public boolean checkStudentUser(String username)
{
    Connection con = null;
    boolean flag = false;
    try
    {
        con = dataSource.getConnection();
        Statement stat = con.createStatement();
        ResultSet result = stat.executeQuery("select * from sclass where scout = '1' and scno = '" + username.substring(0,4) + "'");
        if ( result.next())
        {
            flag = true;
        }
    }
    catch(SQLException e)
    … //省略部分同 CheckAdminUser(username,userpwd)方法中该部分
    return flag;
}
```

3. 修改 DBUser 类

在其中定义一个方法,方法为 getTestInfoByStudentSsno(String ssno),参数为登录的用户名,即学生的学号。该方法的作用是根据输入的学号,获取该学生所在班级的相关信息、是否对班主任测评过的信息、测评任课教师的相关信息。在本模块中只需从班级情况表、教师情况表中获取该学生所在班级的相关信息、从班主任教育情况测评表中获取是否对班主任测评过的信息。

项目六 教师测评系统（Struts模式）

实现代码如下：

```java
public Vector getTestInfoByStudentSsno(String ssno)
{
    Connection con = null;
    
    String stno,stname = "",snno,sono,soname = "";
    boolean flag = false;
    
    Vector rtn;
    rtn = new Vector();
    
    rtn0.add(0," * * * ");              //是否已全部测评过
    rtn0.add(1,ssno);                    //学号
    rtn0.add(2,ssno.substring(0,4));     //班级号
    
    try
    {
        con = dataSource.getConnection();
        //1
        //取年级代码、班主任代码、姓名
        Statement stat1 = con.createStatement();
        ResultSet result1 = stat1.executeQuery("select snno,sclass.stno,stname from sclass,steacher where sclass.stno = steacher.stno and scno = '" + ssno.substring(0,4) + "'");
        if ( result1.next())
        {
            snno = result1.getString("snno");
            stno = result1.getString("stno");
            stname = result1.getString("stname");
            
            rtn0.add(3,snno);
            rtn0.add(4,stno);
            rtn0.add(5,stname);
        }
        result1.close();
        stat1.close();
        
        //是否已测评过班主任的信息
        Statement stat2 = con.createStatement();
        ResultSet result2 = stat2.executeQuery( "select stno from scyn where ssno = '" + ssno + "'");
        if ( result2.next())
        {
            flag = true;                        //已测评过
            stno = result2.getString("stno");
            rtn0.add(6,stno);                   //已测过
        }
        else
        {
            rtn0.add(6," * * * ");              //未测过
        }
        result2.close();
        stat2.close();
        
        //3
        if(flag)                                //已全部测评过
            rtn0.set(0,"yes");
        
    }
    catch(SQLException e)
    … //省略部分同 CheckAdminUser(username,userpwd)方法中该部分
    return rtn;
}
```

4. 新建并编制动作类文件 StudentLoginAction.java

StudentLoginAction 类的主要作用是处理学生的登录请求。在该动作类中，先取出表单 bean 中的用户名及密码，调用 DBUser 类中的 checkStudentUser() 方法验证该学生用户的合法性，若不合法则转至 loginfail.jsp 页面，并显示错误信息。合法则调用 DBUser 类中的 getTestInfoByStudentSsno() 方法，获取该学生是否已全部测评过、待测评班级及班主任的相关信息，将返回值存放在一个 Vector 类型的对象 rtn 中，并将 rtn 对象中的第一个元素的值取出并放入 Vector 类型的对象 rtn0 中；判断 rtn0 中的第一个元素值是否等于"yes"，相等则说明已测评过，转至 loginfail.jsp 页面，显示提示信息；未测评过则取出 rtn0 中后面 6 个元素的值的存入常量类 Constants 中定义的 6 个会话变量中，转至主页面 main.jsp。

实现代码如下：

```java
package com.teach.actions;

import java.util.Vector;
…//省略的包和类可参考 AdminLoginAction
public class StudentLoginAction extends Action
{
    public ActionForward execute(ActionMapping mapping, ActionForm form, HttpServletRequest request, HttpServletResponse response) throws Exception
    {
        LoginForm loginform = (LoginForm) form;
        String username = loginform.getUsername();
        String userpwd = loginform.getUserpwd();

        ServletContext context = servlet.getServletContext();
        DataSource dataSource = (DataSource)context.getAttribute("dbkey");

        ActionMessages errors = new ActionMessages();
        HttpSession session = request.getSession();

        DBUser dbuser = new DBUser(dataSource);
        if (!dbuser.checkStudentUser(username))
        {
            errors.add(ActionMessages.GLOBAL_MESSAGE, new ActionMessage("errors.studentLoginFail"));
            if (!errors.isEmpty())
            {
                saveErrors(request, errors);
            }
            return mapping.findForward("LoginFail");
        }
        else
        {
            Vector rtn = dbuser.getTestInfoByStudentSsno(username);

            //1
            //取是否已全部测评过及待测评班级及班主任的相关信息
            Vector rtn0 = (Vector)rtn.get(0);

            if( (rtn0.get(0)).equals("yes"))
            {
                errors.add(ActionMessages.GLOBAL_MESSAGE, new ActionMessage("errors.allTested"));
                if (!errors.isEmpty())
                {
                    saveErrors(request, errors);
                }
                return mapping.findForward("AllTested");
            }
            else
            {
                session.setAttribute(Constants.Test_CurSsno, rtn0.get(1));
```

项目六 教师测评系统(Struts模式)

```
        session.setAttribute(Constants.Test_CurScno,rtn0.get(2));
        session.setAttribute(Constants.Test_CurSnno,rtn0.get(3));
        session.setAttribute(Constants.Test_CurManagerNo,rtn0.get(4));
        session.setAttribute(Constants.Test_CurManagerName,rtn0.get(5));
        session.setAttribute(Constants.Test_ManagerNo,rtn0.get(6));

            return mapping.findForward("Success");                        //登录成功
        }
    }
  }
}
```

5. 新建并编制 first.jsp 文件

如图 6-57 所示,该页面仅用来在登录成功后在框架右侧栏中显示欢迎信息。

> 客观、公正、无私、严肃、准确地评价老师是你成熟的表现!
>
> 欢迎使用学生测评子系统
> 请点击左边的班主任姓名或课程名称进行测评!

图 6-57 登录成功后显示欢迎信息的页面

实现代码如下:

```jsp
<%@page contentType = "text/html;charset = utf-8" %>
<html>

  <body>
    <br>
    <br>
    <p>
    <b><font face = "华文新魏" color = "#ff0000" size = "5">客观、公正、无私、严肃、准确地评价老师是你成熟的表现!</font></b>
    <br>
    <br>
    <b><font face = "华文新魏" color = "#996633" size = "5">欢迎使用学生测评子系统
    <br>
    请点击左边的班主任姓名或课程名称进行测评!</font></b>
    <br>
  </body>
</html>
```

6. 新建并编制 main.jsp 文件

如图 6-55 所示,该页面为系统主框架页面,其编制方法与系统管理子系统中的 main.jsp 页面基本一致。只不过左侧栏中的超链接为待测班主任及待测课程的超链接。

实现代码如下:

```jsp
<%@page contentType = "text/html;charset = utf-8" %>
<%@page import = "com.teach.comms.Constants,java.util.Vector" %>

<%
    String ssno = (String)session.getAttribute(Constants.Test_CurSsno);
    String stname = (String)session.getAttribute(Constants.Test_CurManagerName);
    if(ssno == null || ssno.equals(""))
    {
        response.sendRedirect("login.jsp");
    }
%>
<html>
  <head>
    <title>教师教育教学情况测评系统 -- -学生测评子系统</title>
    <link href = "../css/main.css" rel = "stylesheet" type = "text/css" />
```

```html
<link href="../css/style_css.css" rel="stylesheet" type="text/css" />
</head>
<body>
    <div id="main">
        <div id="top">
            <table width="100%" height="100%">
                <tr align="center" valign="middle">
                    <td align="left">
                           
                        <font face="宋体" color="#FF0000" size="5"><b>教师教育教学情况测评系统</b></font>
                           
                        <font face="隶书" color="#FF0000" size="5"><b>学生测评子系统</b></font>
                    </td>
                </tr>
            </table>
        </div>
        <div id="mid">
            <div id="left">

                <table width="100%" cellspacing="1">
                    <tr height="25">
                        <td align="center"> </td>
                    </tr>
                    <tr height="25">
                        <td align="center">学生学号</td>
                    </tr>
                    <tr height="25">
                        <td align="center"><font color="#0000cc"><%=ssno%></font></td>
                    </tr>
                    <tr height="25">
                        <td align="center">测评班主任</td>
                    </tr>
                    <tr height="25">
                        <td align="center"><a href=" " target="inMain"></a></td>
                    </tr>

                    <tr height="25">
                        <td align="center">测评任课教师</td>
                    </tr>
                    <tr height="25">
                        <td align="center">
                            <a href="" target="inMain"></a>
                        </td>
                    </tr>
                </table>

            </div>
            <div id="right">
                <iframe src="first.jsp" name="inMain" id="inMain" width="100%" height="100%" frameborder=0 scrolling="yes"></iframe>
            </div>
        </div>
        <div id="end">
            <table width="100%" height="100%">
                <tr>
                    <td align="center" valign="middle">
                        <table>
                            <tr>
                                <td>
                                    版权所有 2009 XXX 高等职业技术学校
                                </td>
                            </tr>
                        </table>
                    </td>
```

```
                    </tr>
                </table>
        </div>
      </div>
    </body>
</html>
```

7．配置动作类 StudentLoginAction

配置 StudentLoginAction 动作类映射关系的代码应写在 struts-config.xml 文件的＜action-mappings＞元素对中。

配置代码如下：

```
< action path = "/test/studentloginaction"   type = " com. teach. actions. StudentLoginAction"   name =
"loginform" scope = "request" validate = "false"  >
        <forward name = "Success"    path = "/test/main.jsp"/>
        <forward name = "LoginFail"      path = "/bas/loginfail.jsp"/>
        <forward name = "AllTested"      path = "/bas/loginfail.jsp"/>
</action>
```

8．发布并测试程序

启动 Tomcat 服务器，将 TeacherTest 应用发布至服务器，打开浏览器，在地址栏中输入 http://localhost:8080/ TeacherTest，运行并测试学生测评子系统中测评班主任部分的登录模块。

任务十三　编制测评班主任模块

【任务描述】

本任务主要编制测评班主任模块的相关文件并测试运行。单击主页面中“班主任”姓名超链接后，如果该班主任已被测评过，出现如图 6-58 所示错误信息页面，如果未被测评过，出现如图 6-59 所示测评页面。选择各项(各项中必须选择一个)并且填入意见后，单击“提交”按钮，如保存成功，则出现如图 6-60 所示信息页面。

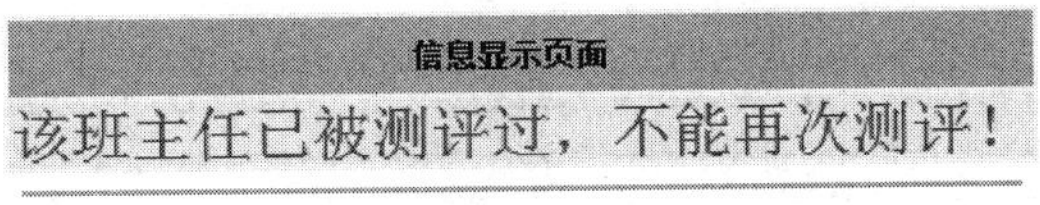

图 6-58　显示已测评信息的页面

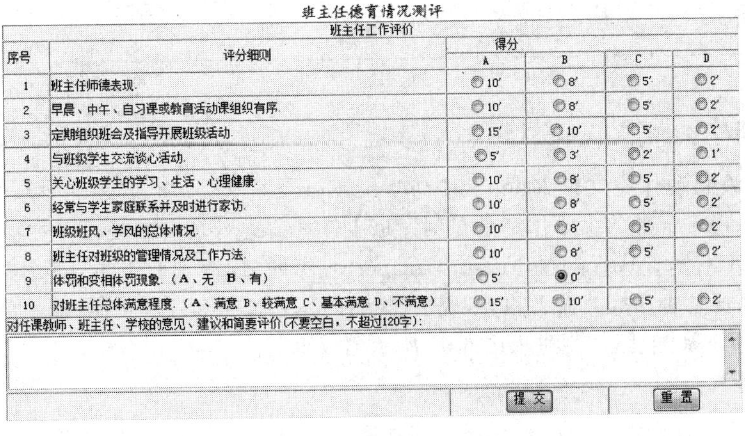

班主任德育情况测评

序号	评分细则	得分 A	B	C	D
1	班主任师德表现.	10′	8′	5′	2′
2	早晨、中午、自习课或教育活动课组织有序.	10′	8′	5′	2′
3	定期组织班会及指导开展班级活动.	15′	10′	5′	2′
4	与班级学生交流谈心活动.	5′	3′	2′	1′
5	关心班级学生的学习、生活、心理健康.	10′	8′	5′	2′
6	经常与学生家庭联系并及时进行家访.	10′	8′	5′	2′
7	班级班风、学风的总体情况.	10′	8′	5′	2′
8	班主任对班级的管理情况及工作方法.	10′	8′	5′	2′
9	体罚和变相体罚现象.（A、无　B、有）	5′	0′		
10	对班主任总体满意程度.（A、满意 B、较满意 C、基本满意 D、不满意）	15′	10′	5′	2′

对任课教师、班主任、学校的意见、建议和简要评价(不要空白，不超过120字)：

提交　重置

图 6-59　测评班主任的页面

信息显示页面
保存测评数据成功！

图 6-60 显示测评成功信息的页面

【任务分析】

1. 程序运行基本流程

图 6-61 所示为测评班主任模块运行流程图。

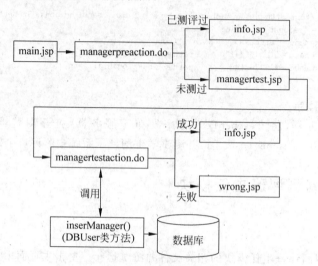

图 6-61 测评班主任模块运行流程图

2. Action 映射表

测评班主任模块中的 Action 映射表见表 6-26。

表 6-26 测评班主任模块中的 Action 映射表

动作（Action）	入口	表单 bean（ActionForm）	出口
ManagerPreAction	main.jsp		managertest.jsp info.jsp
ManagerTestAction	managertest.jsp	ManagerTestForm	info.jsp wrong.jsp

【任务实施】

1. 新建并编制动作类文件 ManagerPreAction.java

ManagerPreAction 类的主要作用是在测评班主任之前进行预判断处理。在该动作类中，先获取会话级变量 Test_CurManagerNo、Test_ManagerNo 中的值，存入两个 String 类型的变量 stno1、stno2 中，判断两值是否相等。如果相等，说明已测评过，转至 info.jsp 页面，显示如图 6-58 所示的信息；否则，说明未被测评过，则进入图 6-59 所示测评页面。

实现代码如下：

```
package com.teach.actions;
```

项目六 教师测评系统(Struts模式)

```
import com.teach.comms.Constants;
…//省略的包和类可参考 AdminLoginAction
public class ManagerPreAction extends Action
{
    public ActionForward execute(ActionMapping mapping, ActionForm form, HttpServletRequest request,
HttpServletResponse response) throws Exception
    {
        HttpSession session = request.getSession();

        String stno1 = (String)session.getAttribute(Constants.Test_CurManagerNo);
        String stno2 = (String)session.getAttribute(Constants.Test_ManagerNo);

        if(stno1.equals(stno2))
        {
            session.setAttribute(Constants.Show_Message,"该班主任已被测评过,不能再次测评!");
            return mapping.findForward("Info");
        }
        else
            return (mapping.findForward("ManagerTest"));
    }
}
```

2．新建并编制 managertest.jsp 文件

如图 6-59 所示,该页面为学生对班主任进行测评的页面。可先用嵌套表格编制页面,表格上下、左右居中,上有一个表单,表单中有 10 组单选按钮,每一组单选按钮的 property 的值为 rb1、rb2…,value 的值即为该项目分数;书写意见的多项编辑框的 property 的值为 sctext,有 3 行 90 列。提交表单时,用 JavaScript 代码验证每一组单选按钮必须选择一个。

实现代码如下:

```
<%@page contentType="text/html;charset=utf-8"%>
<%@taglib uri="http://struts.apache.org/tags-html" prefix="html"%>

<html>
    <head>
        <title>教师教育教学情况测评系统---测评子系统</title>
        <link href="../css/style_css.css" rel=stylesheet type="text/css">
        <script language="javascript">
        function checkdata()
        {
            var rb1 = document.getElementsByName("rb1");
            var rb2 = document.getElementsByName("rb2");
            var rb3 = document.getElementsByName("rb3");
            var rb4 = document.getElementsByName("rb4");
            var rb5 = document.getElementsByName("rb5");
            var rb6 = document.getElementsByName("rb6");
            var rb7 = document.getElementsByName("rb7");
            var rb8 = document.getElementsByName("rb8");
            var rb9 = document.getElementsByName("rb9");
            var rb10 = document.getElementsByName("rb10");

            if (rb1[0].checked==false && rb1[1].checked==false && rb1[2].checked==false && rb1[3].checked==false )
            {
                alert("第[1]项你一个都没有选择!"); return false;
            }
            if (rb2[0].checked==false && rb2[1].checked==false && rb2[2].checked==false && rb2[3].checked==false )
            {
                alert("第[2]项你一个都没有选择!"); return false;
            }
            if (rb3[0].checked==false && rb3[1].checked==false && rb3[2].checked==false && rb3[3].
```

```
            checked == false  )
                    {
                      alert("第[3]项你一个都没有选择!");  return false;
                    }
                  if (rb4[0].checked == false && rb4[1].checked == false && rb4[2].checked == false && rb4[3].checked == false  )
                    {
                      alert("第[4]项你一个都没有选择!");  return false;
                    }
                  if (rb5[0].checked == false && rb5[1].checked == false && rb5[2].checked == false && rb5[3].checked == false  )
                    {
                      alert("第[5]项你一个都没有选择!");  return false;
                    }
                  if (rb6[0].checked == false && rb6[1].checked == false && rb6[2].checked == false && rb6[3].checked == false  )
                    {
                      alert("第[6]项你一个都没有选择!");  return false;
                    }
                  if (rb7[0].checked == false && rb7[1].checked == false && rb7[2].checked == false && rb7[3].checked == false  )
                    {
                      alert("第[7]项你一个都没有选择!");  return false;
                    }
                  if (rb8[0].checked == false && rb8[1].checked == false && rb8[2].checked == false && rb8[3].checked == false  )
                    {
                      alert("第[8]项你一个都没有选择!");  return false;
                    }
                  if (rb9[0].checked == false && rb9[1].checked == false  )
                    {
                      alert("第[9]项你一个都没有选择!");  return false;
                    }
                  if (rb10[0].checked == false && rb10[1].checked == false && rb10[2].checked == false && rb10[3].checked == false  )
                    {
                      alert("第[10]项你一个都没有选择!");  return false;
                    }
                }
            </script>
       </head>

       <body>
            <table width="100%" height="100%" >
                <tr align="center" valign="middle">
                    <td>

                        <html:form action="/test/managertestaction.do"  method="post">
                          <table  border="1" bgcolor="#F1FCF7" cellspacing="0">
                              <caption>班主任德育情况测评</caption>
                              <tr>
                                  <td colspan="6" align="center">班主任工作评价</td>
                              </tr>
                              <tr>
                                  <td rowspan="2" width="30"   align="center">序号</td>
                                  <td rowspan="2"              align="center">评分细则</td>
                                  <td colspan="4" width="120"  align="center">得分</td>
                              </tr>
                              <tr>
                                  <td align="center">A</td>
                                  <td align="center">B</td>
                                  <td align="center">C</td>
                                  <td align="center">D</td>
```

```
        </tr>
        <tr>
          <td align="center">1</td>
          <td>班主任师德表现.</td>
          <td align="center"><html:radio property="rb1" value="10">10'</html:radio></td>
          <td align="center"><html:radio property="rb1" value="8">8'</html:radio></td>
          <td align="center"><html:radio property="rb1" value="5">5'</html:radio></td>
          <td align="center"><html:radio property="rb1" value="2">2'</html:radio></td>
        </tr>
        <tr>
          <td align="center">2</td>
          <td>早晨、中午、自习课或教育活动课组织有序.</td>
          <td align="center"><html:radio property="rb2" value="10">10'</html:radio></td>
          <td align="center"><html:radio property="rb2" value="10">8'</html:radio></td>
          <td align="center"><html:radio property="rb2" value="5">5'</html:radio></td>
          <td align="center"><html:radio property="rb2" value="2">2'</html:radio></td>
        </tr>
        <tr>
          <td align="center">3</td>
          <td>定期组织班会及指导开展班级活动.</td>
          <td align="center"><html:radio property="rb3" value="15">15'</html:radio></td>
          <td align="center"><html:radio property="rb3" value="10">10'</html:radio></td>
          <td align="center"><html:radio property="rb3" value="5">5'</html:radio></td>
          <td align="center"><html:radio property="rb3" value="2">2'</html:radio></td>
        </tr>
        <tr>
          <td align="center">4</td>
          <td>与班级学生交流谈心活动.</td>
          <td align="center"><html:radio property="rb4" value="5">5'</html:radio></td>
          <td align="center"><html:radio property="rb4" value="3">3'</html:radio></td>
          <td align="center"><html:radio property="rb4" value="2">2'</html:radio></td>
          <td align="center"><html:radio property="rb4" value="1">1'</html:radio></td>
        </tr>
        <tr>
          <td align="center">5</td>
          <td>关心班级学生的学习、生活、心理健康.</td>
          <td align="center"><html:radio property="rb5" value="10">10'</html:radio></td>
          <td align="center"><html:radio property="rb5" value="8">8'</html:radio></td>
          <td align="center"><html:radio property="rb5" value="5">5'</html:radio></td>
          <td align="center"><html:radio property="rb5" value="2">2'</html:radio></td>
        </tr>
        <tr>
          <td align="center">6</td>
          <td>经常与学生家庭联系并及时进行家访.</td>
          <td align="center"><html:radio property="rb6" value="10">10'</html:radio></td>
          <td align="center"><html:radio property="rb6" value="8">8'</html:radio></td>
          <td align="center"><html:radio property="rb6" value="5">5'</html:radio></td>
          <td align="center"><html:radio property="rb6" value="2">2'</html:radio></td>
        </tr>
        <tr>
          <td align="center">7</td>
          <td>班级班风、学风的总体情况.</td>
          <td align="center"><html:radio property="rb7" value="10">10'</html:radio></td>
          <td align="center"><html:radio property="rb7" value="8">8'</html:radio></td>
          <td align="center"><html:radio property="rb7" value="5">5'</html:radio></td>
          <td align="center"><html:radio property="rb7" value="2">2'</html:radio></td>
        </tr>
          <tr>
            <td align="center">8</td>
            <td>班主任对班级的管理情况及工作方法.</td>
            <td align="center"><html:radio property="rb8" value="10">10'</html:radio></td>
            <td align="center"><html:radio property="rb8" value="8">8'</html:radio></td>
            <td align="center"><html:radio property="rb8" value="5">5'</html:radio></td>
```

```
            <td align="center"><html:radio property="rb8" value="2">2'</html:radio></td>
         </tr>
         <tr>
            <td align="center">9</td>
            <td>体罚和变相体罚现象.(A、无 B、有)</td>
            <td align="center"><html:radio property="rb9" value="5">5'</html:radio></td>
            <td align="center"><html:radio property="rb9" value="0">0'</html:radio></td>
            <td align="center"> </td>
            <td align="center"> </td>
         </tr>
         <tr>
            <td align="center">10</td>
            <td>对班主任总体满意程度.(A、满意 B、较满意 C、基本满意 D、不满意)</td>
            <td align="center"><html:radio property="rb10" value="15">15'</html:radio></td>
            <td align="center"><html:radio property="rb10" value="10">10'</html:radio></td>
            <td align="center"><html:radio property="rb10" value="5">5'</html:radio></td>
            <td align="center"><html:radio property="rb10" value="2">2'</html:radio></td>
         </tr>
         <tr>
            <td colspan="6">对任课教师、班主任、学校的意见、建议和简要评价(不要空白,不超过 120 字):</td>
         </tr>
         <tr>
            <td colspan="6"><textarea rows="3" cols="90" name="sctext"></textarea></td>
         </tr>
         <tr>
            <td colspan="2"> </td>
            <td align="center" colspan="2" align="center"><html:submit property="Submit" value="提 交" onclick="return checkdata()" /></td>
            <td align="center" colspan="2" align="center"><html:reset property="Reset" value="重 置"/></td>
         </tr>
       </table>
     </html:form>
    </td>
  </tr>
 </table>
 </body>
</html>
```

3. 配置动作类 ManagerPreAction

配置 ManagerPreAction 动作类映射关系的代码应写在 struts-config.xml 文件的<action-mappings>元素对中。

配置代码如下:

```
<action path="/test/managerpreaction"    type="com.teach.actions.ManagerPreAction"    scope="request">
       <forward name="ManagerTest"    path="/test/managertest.jsp"/>
</action>
```

4. 修改主页面文件 main.jsp

将"测评班主任"下面的超链接改为:

```
<a href="managerpreaction.do" target="inMain"><%=stname%></a>
```

其中,"inMain"为主框架右侧栏子窗体的 id 值。

5. 发布并测试程序

启动 Tomcat 服务器,将 TeacherTest 应用发布至服务器,打开浏览器,在地址栏中输入 http://localhost:8080/TeacherTest,运行并测试是否能正确地打开测评班主任的页面。

项目六 教师测评系统(Struts模式)

6. 修改 DBUser 类

在其中定义一个方法,方法为 insertManager(String ssno,String scno,String snno,String stno,String stname,int r1,int r2,int r3,int r4,int r5,int r6,int r7,int r8,int r9,int r10,String sctext)。该方法的作用是保存对班主任的测评数据至数据库,成功返回 true,失败返回 false,其中用户 ID 号(scid)列的值(1,2,3…)是自动生成的。

实现代码如下:

```java
public boolean insertManager(String ssno, String scno, String snno, String stno, String stname, int r1, int r2, int r3, int r4, int r5, int r6, int r7, int r8, int r9, int r10, String sctext)
{
    Connection con = null;
    boolean flag = false;
    try
    {
        con = dataSource.getConnection();
        PreparedStatement stat = con.prepareStatement("insert into scyn values(?,?,?,?,?,?,?,?,?,?,?,?,?,?,?,?)");

        stat.setString(1,ssno);
        stat.setString(2,scno);
        stat.setString(3,snno);
        stat.setString(4,stno);
        stat.setString(5,stname);
        stat.setInt(6,r1);
        stat.setInt(7,r2);
        stat.setInt(8,r3);
        stat.setInt(9,r4);
        stat.setInt(10,r5);
        stat.setInt(11,r6);
        stat.setInt(12,r7);
        stat.setInt(13,r8);
        stat.setInt(14,r9);
        stat.setInt(15,r10);
        stat.setString(16,sctext);

        int row = stat.executeUpdate();
        if(row == 1)
            flag = true;
    }
    catch(SQLException e)
    …//省略部分同 CheckAdminUser(username,userpwd)方法中该部分
    return flag;
}
```

7. 新建并编制表单 bean 类文件 ManagerTestForm.java

该表单 bean 类中的属性应与页面 managertest.jsp 表单中域的名称一致,有 10 个 int 型属性(rb1,rb2…),一个 String 型属性 sctext,类中还有与属性配合的获取(getter)和设置(setter)属性的方法。

实现代码如下:

```java
package com.teach.forms;

import org.apache.struts.action.ActionForm;
public class ManagerTestForm extends ActionForm
{
    private int rb1;
    private int rb2;
    private int rb3;
```

```
        private int rb4;
        private int rb5;
        private int rb6;
        private int rb7;
        private int rb8;
        private int rb9;
        private int rb10;
        private String sctext;

        public int getRb1()
        {
            return rb1;
        }
        public void setRb1(int rb1)
        {
            this.rb1 = rb1;
        }
        …//省略与其余属性对应的get 和set 方法
    }
```

8．新建并编制动作类文件 ManagerTestAction.java

ManagerPreAction.java 类的主要作用是处理测评数据的保存请求。在该类中，先将表单 bean 中的测评数据取出放入 int 型变量 rb1 至 rb10 中，将意见、建议文本放入 String 类型变量 sctext 中；再取出会话级变量中的学号、班级、年级、班主任代码、班主任姓名、是否已测班主任相关数据；如果该班主任已经被测评过，则显示图 6-58 的出错信息；否则调用 insertManager()方法插入数据，保存成功则在 info.jsp 页面中显示"保存测评数据成功！"信息，插入数据失败时在 wrong.jsp 页面中显示错误信息。

实现代码如下：

```
package com.teach.actions;

import com.teach.comms.*;
…//省略的包和类可参考 AdminLoginAction
public class ManagerTestAction extends Action
{
    public ActionForward execute(ActionMapping mapping, ActionForm form, HttpServletRequest request, HttpServletResponse response) throws Exception
    {
        ManagerTestForm managertestform = (ManagerTestForm) form;
        int rb1 = managertestform.getRb1();
        …
        int rb10 = managertestform.getRb10();
        String sctext = managertestform.getSctext();

        if(sctext == null)
        {
            sctext = "";
        }
        ServletContext context = servlet.getServletContext();
        DataSource dataSource = (DataSource)context.getAttribute("dbkey");
        HttpSession session = request.getSession();

        String ssno = (String)session.getAttribute(Constants.Test_CurSsno );
        String scno = (String)session.getAttribute(Constants.Test_CurScno );
        String snno = (String)session.getAttribute(Constants.Test_CurSnno );
        String stno = (String)session.getAttribute(Constants.Test_CurManagerNo );
        String stname = (String)session.getAttribute(Constants.Test_CurManagerName );

        DBUser dbuser = new DBUser(dataSource);
```

项目六 教师测评系统(Struts模式)

```
        String stnoyesno = (String)session.getAttribute(Constants.Test_ManagerNo );
        if(stnoyesno.equals(stno))
        {
            session.setAttribute(Constants.Show_Message,"该班主任已被测评过,不能再次测评!");
            return mapping.findForward("Info");
        }
        boolean flag = dbuser.insertManager(ssno,scno,snno,stno,stname,rb1,rb2,rb3,rb4,rb5,rb6,rb7,rb8,
    rb9,rb10,sctext);
        if(!flag)                    //插入记录失败
        {
            session.setAttribute(Constants.Error_Message,"保存测评数据失败!");
            return mapping.findForward("Wrong");
        }
        else                         //插入记录成功
        {
            session.setAttribute(Constants.Test_ManagerNo,stno);
            session.setAttribute(Constants.Show_Message,"保存测评数据成功!");

            return (mapping.findForward("Info"));
        }
    }
}
```

9. 配置表单 bean 及动作 Action 类

配置 ManagerTestForm 表单 bean 的代码应写在 struts-config.xml 文件的＜form-beans＞元素对中。

配置代码如下：

```
<form-bean name="managertestform"    type="com.teach.forms.ManagerTestForm" />
```

配置 ManagerTestAction 动作类映射关系的代码应写在 struts-config.xml 文件的＜action-mappings＞元素对中。

配置代码如下：

```
<action path="/test/managertestaction"     type="com.teach.actions.ManagerTestAction"
    name="managertestform"    scope="request" validate="false"   />
```

10. 发布并测试程序

启动 Tomcat 服务器,将 TeacherTest 应用发布至服务器,打开浏览器,在地址栏中输入 http://localhost:8080/TeacherTest,运行并测试学生测评子系统的测评班主任模块。

任务十四　编制测评任课教师模块的学生登录部分

【任务描述】

本任务主要编制学生测评子系统中测评任课教师模块的学生登录部分相关文件并测试运行。测评任课教师模块的学生登录部分主要是要在学生登录时取出待测课程及任课教师的相关数据。成功登录后,在主页面左侧栏将出现如图 6-62 所示的页面。

【任务分析】

1. 程序运行基本流程

图 6-63 所示为学生测评子系统登录模块运行流程图。

图 6-62　显示全部待测课程(任课教师)页面

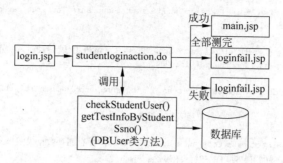

图 6-63 学生测评子系统登录模块运行流程图

2. Action 映射表

学生测评子系统登录模块中的 Action 映射表见表 6-27。

表 6-27 学生测评子系统登录模块中的 Action 映射表

动作(Action)	入口	表单 bean(ActionForm)	出口
StudentLoginAction	login.jsp	LoginForm	main.jsp loginfail.jsp

【任务实施】

1. 修改 DBUser 类

方法 getTestInfoByStudentSsno(String ssno) 在前面模块中已经定义过，获取的是待测班级及班主任的相关信息，现在只需对其进行修改，以获取该学生对任课教师的测评信息。可以根据学生所在班级从班级课程情况表、课程情况表中获取待测的课程代码、课程名称、教师代码、教师姓名，根据学号从教师教学情况测评表中获取已测评过的课程代码、教师代码。

修改后该方法实现的代码如下：

```java
public Vector getTestInfoByStudentSsno(String ssno)
{
    Connection con = null;
    String stno,stname = "", snno,sono,soname = "";
    boolean flag = false;
    int i,j;

    Vector rtn;
    rtn = new Vector();

    Vector rtn0,rtn1,rtn2,rtn3,rtn4,rtn5,rtn6;
    rtn0 = new Vector();
    rtn1 = new Vector();
    rtn2 = new Vector();
    rtn3 = new Vector();
    rtn4 = new Vector();
    rtn5 = new Vector();
    rtn6 = new Vector();

    rtn0.add(0,"***");              //是否已全部测评过
    rtn0.add(1,ssno);                //学号
    rtn0.add(2,ssno.substring(0,4)); //班级号

    try
    {
```

```java
            con = dataSource.getConnection();
            //1
            //取年级代码、班主任代码、姓名
            Statement stat1 = con.createStatement();
            ResultSet result1 = stat1.executeQuery("select snno,sclass.stno,stname from sclass,steacher where sclass.stno = steacher.stno and scno = '" + ssno.substring(0,4) + "'");
            if ( result1.next())
            {
                snno = result1.getString("snno");
                stno = result1.getString("stno");
                stname = result1.getString("stname");

                rtn0.add(3,snno);
                rtn0.add(4,stno);
                rtn0.add(5,stname);
            }
            result1.close();
            stat1.close();

            //是否已测评过班主任的信息
            Statement stat2 = con.createStatement();
            ResultSet result2 = stat2.executeQuery( "select stno from scyn where ssno = '" + ssno + "'");
            if ( result2.next())
            {
                flag = true;              //已测评过
                stno = result2.getString("stno");
                rtn0.add(6,stno);         //已测过
            }
            else
            {
                rtn0.add(6," * * *");     //未测过
            }
            result2.close();
            stat2.close();

            //2
            //取班级须测评的课程及教师
            Statement stat3 = con.createStatement();
            ResultSet result3 = stat3.executeQuery("select sccourse.sono,soname,sccourse.stno,stname from sccourse,scourse,steacher where sccourse.sono = scourse.sono and sccourse.stno = steacher.stno and scno = '" + ssno.substring(0,4) + "'");
            i = 0;
            while(result3.next())
            {
                i++;
                sono = result3.getString("sono");
                soname = result3.getString("soname");
                stno = result3.getString("stno");
                stname = result3.getString("stname");

                rtn1.add(sono);
                rtn2.add(soname);
                rtn3.add(stno);
                rtn4.add(stname);
            }
            result3.close();
            stat3.close();

            //取已测评过的课程及教师
            Statement stat4 = con.createStatement();
            ResultSet result4 = stat4.executeQuery("select sono,stno from styn where ssno = '" + ssno + "'");
            j = 0;
            while(result4.next())
            {
                j++;
                sono = result4.getString("sono");
```

```
                stno = result4.getString("stno");
                rtn5.add(sono);
                rtn6.add(stno);
            }
            result4.close();
            stat4.close();

            //3
            if(flag && (i!=0 && i==j))          //已全部测评过
                rtn0.set(0,"yes");

        }
        catch(SQLException e)
        …//省略部分同 CheckAdminUser(username,userpwd)方法中该部分
        rtn.add(rtn0);
        rtn.add(rtn1);
        rtn.add(rtn2);
        rtn.add(rtn3);
        rtn.add(rtn4);
        rtn.add(rtn5);
        rtn.add(rtn6);

        return rtn;
    }
```

2. 修改动作类文件 StudentLoginAction.java

StudentLoginAction.java 类在前面的模块中已经定义过，现在只需对其修改，将调用 getTestInfoByStudentSsno()方法所返回的待测任课教师的相关信息取出并保存在会话级变量中。

将 if…else…结构中的 else 分支部分实现的代码修改如下：

```
…
Vector rtn = dbuser.getTestInfoByStudentSsno(username);
//1
    //取是否已全部测评过及待测评班级及班主任的相关信息
Vector rtn0 = (Vector)rtn.get(0);
//2
//取待测评课程的相关信息
Vector rtn1,rtn2,rtn3,rtn4,rtn5,rtn6;

rtn1 = (Vector)rtn.get(1);
rtn2 = (Vector)rtn.get(2);
rtn3 = (Vector)rtn.get(3);
rtn4 = (Vector)rtn.get(4);
rtn5 = (Vector)rtn.get(5);
rtn6 = (Vector)rtn.get(6);

if( (rtn0.get(0)).equals("yes"))
{
        errors.add(ActionMessages.GLOBAL_MESSAGE,new ActionMessage("errors.allTested"));
        if (!errors.isEmpty())
        {
                saveErrors(request,errors);
        }
        return mapping.findForward("AllTested");
}
else
{
    session.setAttribute(Constants.Test_CurSsno,rtn0.get(1));
    session.setAttribute(Constants.Test_CurScno,rtn0.get(2));
    session.setAttribute(Constants.Test_CurSnno,rtn0.get(3));
    session.setAttribute(Constants.Test_CurManagerNo,rtn0.get(4));
    session.setAttribute(Constants.Test_CurManagerName,rtn0.get(5));
```

项目六 教师测评系统（Struts模式）

```
            session.setAttribute(Constants.Test_ManagerNo,rtn0.get(6));
                //取班级须测评的课程号
                session.setAttribute(Constants.Test_ScCourseAll,rtn1);
                //取班级须测评的课程名
                session.setAttribute(Constants.Test_ScCourseNameAll,rtn2);
                //取班级须测评的教师号
            session.setAttribute(Constants.Test_TeacherAll,rtn3);
                //取班级须测评的教师名
                session.setAttribute(Constants.Test_TeacherNameAll,rtn4);
                //取已测评过的课程
                session.setAttribute(Constants.Test_ScCourseYes,rtn5);
                //取已测评过的教师
                session.setAttribute(Constants.Test_TeacherYes,rtn6);
                session.setAttribute(Constants.Test_ScourseCur,"***");
            return mapping.findForward("Success");//登录成功
        }
    }
```

3. 修改 main.jsp 文件

如图 6-55 所示，在页面的开始处，编制以下代码将所有的课程号代码及课程名称从会话级变量 Test_ScCourseAll、Test_ScCourseNameAll 中取至两个 Vector 类型的对象中：

```
…
Vector sono = (Vector)session.getAttribute(Constants.Test_ScCourseAll);
    Vector soname = (Vector)session.getAttribute(Constants.Test_ScCourseNameAll);
…
```

再将页面上"测评任课教师"菜单下的＜tr＞标签对改为如下代码，以便将各门待测的课程超链接以如图 6-62 所示的形式显示出来：

```
…
<%
                    for(int i = 0;i<sono.size();i++)
                    {
                    %>
                    <tr height = "25">
                        <td  align = "center">
                            <a href = "teacherpreaction.do?sono=<% = sono.get(i)%>" target = "inMain">
<% = soname.get(i)%></a>
                        </td>
                    </tr>
                    <%
                    }
                    %>
…
```

4. 发布并测试程序

启动 Tomcat 服务器，将 TeacherTest 应用发布至服务器，打开浏览器，在地址栏中输入 http://localhost:8080/TeacherTest，运行并测试学生测评子系统的测评任课教师部分的登录模块。

任务十五　编制测评任课教师模块

【任务描述】

本任务主要编制测评任课教师模块的相关文件并测试运行。单击主页面中各课程名称超链接后，如果该门课程已被测评过，出现如图 6-64 所示错误信息页面；如果未被测评过，出现如图 6-65 所示测评页面。选择各项（各项中必须选择一个）并且填入意见后，单击"提交"按钮，如保存成功，则出现如图 6-66 所示信息页面。

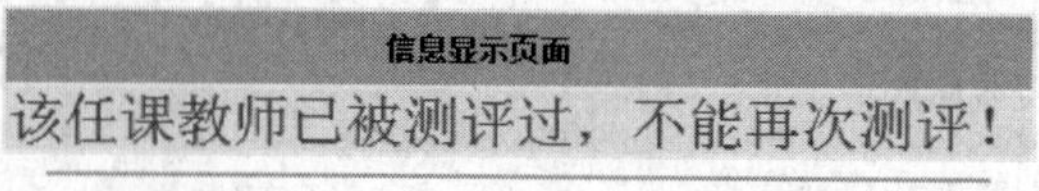

信息显示页面
该任课教师已被测评过，不能再次测评！

图 6-64　显示已测评信息的页面

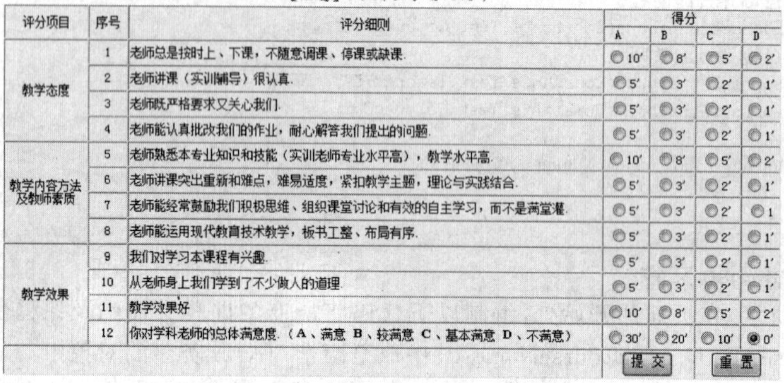

图 6-65　测评任课教师的页面

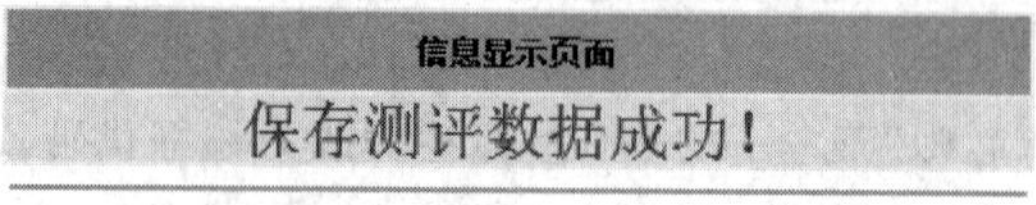

信息显示页面
保存测评数据成功！

图 6-66　显示测评成功信息的页面

【任务分析】

1. 程序运行基本流程

图 6-67 所示为测评任课教师模块运行流程图。

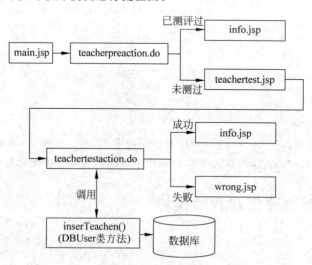

图 6-67　测评任课教师模块运行流程图

2. Action 映射表

测评任课教师模块中的 Action 映射表见表 6-28。

表 6-28 测评任课教师模块中的 Action 映射表

动作（Action）	入口	表单 bean（ActionForm）	出口
TeacherPreAction	main.jsp		teachertest.jsp info.jsp
TeacherTestAction	teachertest.jsp	TeacherTestForm	info.jsp wrong.jsp

【任务实施】

1. 新建并编制动作类文件 TeacherPreAction.java

TeacherPreAction.java 类的主要作用是在测评任课教师之前进行预判断处理。在该动作类中，先获取当前待测课程代码、会话级变量 Test_ScCourseYes 中所有已测评课程代码，查找并判断当前课程是否已被测过。如果已被测过，则转至 info.jsp 页面，显示如图 6-64 所示的信息；否则，取出待测课程的名称，进入图 6-65 所示测评页面。

实现代码如下：

```java
package com.teach.actions;

import java.util.Vector;
…//省略的包和类可参考 AdminLoginAction
public class TeacherPreAction extends Action
{
    public ActionForward execute(ActionMapping mapping, ActionForm form, HttpServletRequest request,
HttpServletResponse response) throws Exception
    {
        String      sono = request.getParameter("sono");

        HttpSession session = request.getSession();

        //
        Vector sonoyesall = (Vector)session.getAttribute(Constants.Test_ScCourseYes);

        boolean flag = false;
        if(sonoyesall!= null)
            for(int i = 0;i < sonoyesall.size();i++)
                if(sonoyesall.get(i).equals(sono))     //已测评过
                {
                    flag = true;
                    break;
                }

        if(flag)//已测评过
        {
            session.setAttribute(Constants.Show_Message,"该任课教师已被测评过,不能再次测评!");
            return mapping.findForward("Info");
        }
        else                    //未被测评过
        {
            String sonamecur = "";

            Vector sonoall = (Vector)session.getAttribute(Constants.Test_ScCourseAll);
            Vector sonameall = (Vector)session.getAttribute(Constants.Test_ScCourseNameAll);
            //取当前测评课程号对应的课程名
```

```
            if(sonoall!= null)
                for(int i = 0;i < sonoall.size();i++)
                {
                    if(sonoall.get(i).equals(sono))
                    {
                        sonamecur = (String)sonameall.get(i);
                    }
                }
            session.setAttribute(Constants.Test_ScourseCur,sono);
            session.setAttribute(Constants.Test_ScourseNameCur,sonamecur);
            return (mapping.findForward("TeacherTest"));
        }
    }
}
```

2．新建并编制 teachertest.jsp 文件

如图 6-65 所示，该页面为学生对任课教师进行测评的页面。编制方法类似于测评班主任页面，完全可参考 managertest.jsp 页面文件进行编制。

3．配置及动作类

配置 TeacherPreAction 动作类映射关系的代码应写在 struts-config.xml 文件的＜action-mappings＞元素对中。

配置代码如下：

```
<action path = "/test/teacherpreaction"    type = "com.teach.actions.TeacherPreAction"    scope = "request">
        <forward name = "TeacherTest"    path = "/test/teachertest.jsp"/>
</action>
```

4．发布并测试程序

启动 Tomcat 服务器，将 TeacherTest 应用发布至服务器，打开浏览器，在地址栏中输入 http://localhost:8080/TeacherTest，运行并测试是否能正确地打开测评任课教师的页面。

5．修改 DBUser 类

在其中定义一个方法，方法为 insertTeacher(String ssno,String scno,String snno,String sono,String soname,String stno,String stname,int r1,int r2,int r3,int r4,int r5,int r6,int r7,int r8,int r9,int r10,int r11,int r12)，该方法的作用是保存对任课教师的测评数据至数据库，成功返回 true，失败返回 false，其中用户 ID 号（stid）列的值（1,2,3…）是自动生成的。编制方法类似于测评班主任中的 insertManager()方法，完全可参考其进行编制。

6．新建并编制表单 bean 类文件 TeacherTestForm.java

该表单 bean 类中的属性应与 teachertest.jsp 页面表单中域的名称一致，有 12 个 int 型属性（rb1,rb2…），类中还有与属性配合的获取（getter）和设置（setter）属性的方法。具体编制可参考 ManagerTestForm.java。

7．新建并编制动作类文件 TeacherTestAction.java

TeacherTestAction.java 类的主要作用是处理测评数据的保存请求。在该类中，先将表单 bean 中的测评数据取出放入 int 型变量 rb1 至 rb12 中；从会话级变量中取出学号、班级、年级、当前课程代码以及所有课程代码、所有课程名称、所有任课教师代码、所有任课教师姓名、已测评课程代码相关数据；从已测评课程号中查找当前的测评课程号是否存在，如果存在，说明已被测评过，不能再次被测评，并显示相关提示信息；如果未被测评过，则根据当前课程代码查找出课程名称、任课教师代码、任课教师姓名；调用 insertTeacher()方法插入数据，保存成功则在 info.jsp 页面中显示"保存测评数据成功！"信息，插入数据失败时在 wrong.jsp 页面中显示错误信息。

项目六　教师测评系统(Struts模式)

实现代码如下：

```
package com.teach.actions;

import java.util.Vector;
…//省略的包和类可参考 AdminLoginAction
public class TeacherTestAction extends Action
{
    public ActionForward execute(ActionMapping mapping, ActionForm form, HttpServletRequest request,
HttpServletResponse response) throws Exception
    {
        TeacherTestForm teachertestform = (TeacherTestForm) form;
        int rb1 = teachertestform.getRb1();
        …
        int rb12 = teachertestform.getRb12();

        ServletContext context = servlet.getServletContext();
        DataSource dataSource = (DataSource)context.getAttribute("dbkey");

        HttpSession session = request.getSession();
        String ssno = (String)session.getAttribute(Constants.Test_CurSsno);          //学号
        String scno = (String)session.getAttribute(Constants.Test_CurScno);          //班级
        String snno = (String)session.getAttribute(Constants.Test_CurSnno);          //年级
        String sonocur = (String)session.getAttribute(Constants.Test_ScourseCur);    //当前课程号

        String sonamecur = "＊＊＊";
        String stnocur = "＊＊＊";
        String stnamecur = "＊＊＊";

        Vector sono = (Vector)session.getAttribute(Constants.Test_ScCourseAll);         //所有课程
        Vector soname = (Vector)session.getAttribute(Constants.Test_ScCourseNameAll);   //所有课程名
        Vector stno = (Vector)session.getAttribute(Constants.Test_TeacherAll);          //所有任课教师
        Vector stname = (Vector)session.getAttribute(Constants.Test_TeacherNameAll);    //所有任课教师名
        Vector scourseyesall = (Vector)session.getAttribute(Constants.Test_ScCourseYes); //已测评课程号

        boolean flag = false;                          //是否已测评过,以防后退页面时再次提交
        for(int i = 0;i < scourseyesall.size();i++)
        {
            if(scourseyesall.get(i).equals(sonocur))
            {
                flag = true;
                break;
            }
        }

        if(flag)                                       //已测评过
        {
            session.setAttribute(Constants.Show_Message,"该任课教师已被测评过,不能再次测评!");
            return mapping.findForward("info");
        }
        //取当前测评课程号对应的课程名,任课教师号及姓名
        for(int i = 0;i < sono.size();i++)
        {
        if(sono.get(i).equals(sonocur))
        {
            sonamecur = (String)soname.get(i);
            stnocur = (String)stno.get(i);
            stnamecur = (String)stname.get(i);
        }
        }

        DBUser dbuser = new DBUser(dataSource);
```

```
        flag = dbuser.insertTeacher(ssno,scno,snno,sonocur,sonamecur,stnocur,stnamecur,rb1,rb2,rb3,rb4,
    rb5,rb6,rb7,rb8,rb9,rb10,rb11,rb12);
        if(!flag)                          //插入记录失败
        {
            session.setAttribute(Constants.Error_Message,"保存测评数据失败!");
            return mapping.findForward("Wrong");
        }
        else                               //插入记录成功
        {
            scourseyesall.add(sonocur);
            //刷新已测评课程号
            session.setAttribute(Constants.Test_ScCourseYes,scourseyesall);

            session.setAttribute(Constants.Show_Message,"保存测评数据成功!");
            return (mapping.findForward("Info"));
        }
    }
}
```

8. 配置表单 bean 及动作 Action 类

配置 TeacherTestForm 表单 bean 的代码应写在 struts-config.xml 文件的＜form-beans＞元素对中。配置代码如下：

```
<form-bean name="teachertestform"     type="com.teach.forms.TeacherTestForm" />
```

配置 TeacherTestAction 动作类映射关系的代码应写在 struts-config.xml 文件的＜action-mappings＞元素对中。

配置代码如下：

```
<action path="/test/teachertestaction"          type="com.teach.actions.TeacherTestAction"
        name="teachertestform"    scope="request" validate="false"   />
```

9. 发布并测试程序

启动 Tomcat 服务器，将 TeacherTest 应用发布至服务器，打开浏览器，在地址栏中输入 http://localhost:8080/TeacherTest，运行并测试学生测评子系统的测评任课教师模块。

任务十六　编制数据查询子系统登录模块

【任务描述】

本任务主要编制数据查询子系统登录模块中的相关文件并测试运行。单击首页面中"数据查询子系统"超链接，出现如图 6-68 所示登录页面。输入用户名和密码后，如果不存在该用户，则显示如图 6-69 所示出错信息；否则，进入如图 6-71 所示数据查询子系统的主页面。

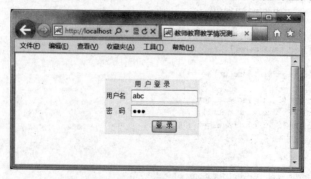

图 6-68　数据查询子系统登录页面

项目六 教师测评系统(Struts模式)

图 6-69 显示登录失败信息的页面

图 6-70 数据查询子系统主页面

【任务分析】

1. 程序运行基本流程

图 6-71 所示为数据查询子系统登录模块运行流程图。

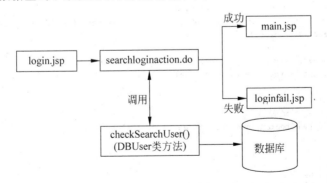

图 6-71 数据查询子系统登录模块运行流程图

2. Action 映射表

数据查询子系统登录模块中的 Action 映射表见表 6-29。

表 6-29 数据查询子系统登录模块中的 Action 映射表

动作(Action)	入口	表单 bean(ActionForm)	出口
SearchLoginAction	login.jsp	LoginForm	main.jsp loginfail.jsp

【任务实施】

1. 新建并编制 login.jsp 文件

如图 6-68 所示,该页面为教师查询时的登录页面。只需将系统管理子系统中的 login.jsp 页面复制到 search 文件夹中,修改两处,一是修改页面上的文字,二是将表单中 action 属性的值改为 "searchloginaction.do"。

2. 修改 DBUser 类

在其中定义一个方法,方法为 checkSearchUser(String username, String userpwd),参数为用户名及密码,该方法的作用是根据传入的用户名及密码,判断该用户是否存在,存在则返回一布尔值 true,否则返回 false。在本系统中,只要是系统用户表中的用户均具有查询权限,因此只要将前面已定义的方法 checkAdminUser(String username, String userpwd)复制后将方法名先改为 checkSearchUser(String username, String userpwd),再将其中 select 语句中的"isadmin=1"条件去掉即可。

3. 修改 DBUser 类

在其中定义一个方法,方法为 getSdnameAll(),该方法的作用是从班级表中取出所有不重复的部门名称,存放在一个 Vector 中返回。

实现代码如下:

```java
public Vector getSdnameAll()
{
    Connection con = null;
    Vector sdnameVector = new Vector();
    String sdname;

    try
    {
        con = dataSource.getConnection();
        Statement stat = con.createStatement();
        ResultSet result = stat.executeQuery("select distinct sdname from sclass");
        while(result.next())
        {
            sdname = result.getString("sdname");
            sdnameVector.addElement(sdname);
        }
    }
    catch(SQLException e)
    …//省略部分同 CheckAdminUser(username,userpwd)方法中该部分
    return sdnameVector;
}
```

4. 修改 DBUser 类

在其中定义一个方法,方法为 getSoAll(),该方法的作用是从教师教学情况测评表 styn 中取出所有不重复的课程代码及名称,存放在一个 Vector 中返回。

实现代码如下:

```java
public Vector getSoAll()
```

```
{
    Connection con = null;
    Vector rtn = new Vector();
    Vector sonoVector = new Vector();
    Vector sonameVector = new Vector();
    String sono,soname;

    try
    {
        con = dataSource.getConnection();
        Statement stat = con.createStatement();
        ResultSet result = stat.executeQuery("select distinct sono,soname from styn order by sono");
        while(result.next())
        {
            sono = result.getString("sono");
            soname = result.getString("soname");
            sonoVector.add(sono);
            sonameVector.add(soname);
        }
    }
    catch(SQLException e)
    …//省略部分同 CheckAdminUser(username,userpwd)方法中该部分
    rtn.add(sonoVector);
    rtn.add(sonameVector);
    return rtn;
}
```

5．新建并编制动作类文件 SearchLoginAction.java

SearchLoginAction.java 类的主要作用是处理查询用户的登录请求。在该动作类中，先取出表单 bean 中的用户名及密码，用 DBUser 类中的方法 checkSearchUser()验证该用户的合法性，不合法则转至 loginfail.jsp 页面，显示错误信息；合法则调用 getSdnameAll()、getScnoAll()、getSoAll() 方法取出所有部门信息、班级信息及已测课程信息，并转至主页面 main.jsp。

实现代码如下：

```
package com.teach.actions;

import java.util.Vector;
…//省略的包和类可参考 AdminLoginAction
public class SearchLoginAction extends Action
{
    public ActionForward execute (ActionMapping mapping, ActionForm form, HttpServletRequest request, HttpServletResponse response) throws Exception
    {
        LoginForm loginform = (LoginForm) form;
        String username = loginform.getUsername();
        String userpwd = loginform.getUserpwd();

        ServletContext context = servlet.getServletContext();
        DataSource dataSource = (DataSource)context.getAttribute("dbkey");
        DBUser dbuser = new DBUser(dataSource);

        ActionMessages errors = new ActionMessages();
        if (!dbuser.checkSearchUser(username,userpwd))
        {
            errors.add(ActionMessages.GLOBAL_MESSAGE,new ActionMessage("errors.loginFail"));
            if (!errors.isEmpty())
            {
                saveErrors(request,errors);
            }

            return mapping.findForward("LoginFail");
```

```
        }
    else
    {
            //取出全部系部
            Vector sdnameVector = dbuser.getSdnameAll();

            //取出全部班级
            Vector rtn1 = dbuser.getScnoAll();
            Vector scnoVector = (Vector)rtn1.get(0);;
            Vector stnameVector = (Vector)rtn1.get(1);;

            //取出全部课程信息
            Vector rtn2 = dbuser.getSoAll();
            Vector sonoVector = (Vector)rtn2.get(0);
            Vector sonameVector = (Vector)rtn2.get(1);

            HttpSession session = request.getSession();
            session.setAttribute(Constants.Login_Curuser,username);
            session.setAttribute(Constants.Search_SdnameAll,sdnameVector);
            session.setAttribute(Constants.Bas_ScnoAll,scnoVector);
            session.setAttribute(Constants.Bas_StnameAll,stnameVector);
            session.setAttribute(Constants.Search_SonoAll,sonoVector);
            session.setAttribute(Constants.Search_SonameAll,sonameVector);

            return mapping.findForward("Success");
        }
    }
}
```

6. 新建并编制 first.jsp 文件

如图 6-72 所示，该页面仅用来在登录成功后在框架右侧栏中显示欢迎信息。具体编制可参考系统管理子系统中 first.jsp 页面文件的实现代码。

欢迎使用数据查询子系统
请选择左边的栏目对相关数据进行查询！

图 6-72 登录成功后显示欢迎信息的页面

7. 新建并编制 main.jsp 文件

如图 6-70 所示，该页面为系统主框架页面，其编制方法与系统管理子系统中的 main.jsp 页面基本一致。在页面开始处，从会话级变量中取出所有部门、班级及课程数据，将其填充左侧栏的下拉列框中供选择。具体编制可参考系统管理子系统中 main.jsp 页面文件的实现代码。

8. 配置动作类

配置 SearchLoginAction 动作类映射关系的代码应写在 struts-config.xml 文件的＜action-mappings＞元素对中。

配置代码如下：

```
< action path = "/search/searchloginaction"    type = "com.teach.actions.SearchLoginAction"   name = "loginform"   scope = "request" validate = "false"   >
        < forward name = "Success"     path = "/search/main.jsp"/>
        < forward name = "LoginFail"     path = "/bas/loginfail.jsp"/>
</action>
```

9. 发布并测试程序

启动 Tomcat 服务器，将 TeacherTest 应用发布至服务器，打开浏览器，在地址栏中输入 http://localhost:8080/TeacherTest，运行并测试数据查询子系统的登录模块。

任务十七　编制查询班主任汇总数据模块

【任务描述】

本任务主要编制查询班主任测评情况汇总数据模块中的相关文件并测试运行。本模块主要是从班主任教育情况测评汇总表 scynsum 中查询出班主任测评情况汇总数据。单击主页面中“查询班主任测评情况”超链接,出现如图 6-73 所示结果页面。

班主任教育情况测评汇总表

序号	年级	班级	班主任	参测人数	总分	名次	分数1	分数2	分数3	分数4	分数5	分数6	分数7	分数8	分
1	10	1021	王健	46	100.00	1	10.00	10.00	15.00	5.00	10.00	10.00	10.00	10.00	5
2	10	1032	王登科	20	100.00	2	10.00	10.00	15.00	5.00	10.00	10.00	10.00	10.00	5
3	10	1042	浦晶晶	42	100.00	3	10.00	10.00	15.00	5.00	10.00	10.00	10.00	10.00	5
4	10	1051	杨国威	36	100.00	4	10.00	10.00	15.00	5.00	10.00	10.00	10.00	10.00	5
5	10	1053	杨晓俊	34	100.00	5	10.00	10.00	15.00	5.00	10.00	10.00	10.00	10.00	5
6	10	1055	许杭建	30	100.00	6	10.00	10.00	15.00	5.00	10.00	10.00	10.00	10.00	5
7	10	1063	严栋梁	35	100.00	7	10.00	10.00	15.00	5.00	10.00	10.00	10.00	10.00	5
8	11	1101	杨小琴	48	100.00	8	10.00	10.00	15.00	5.00	10.00	10.00	10.00	10.00	5
9	11	1113	毛艳青	39	100.00	9	10.00	10.00	15.00	5.00	10.00	10.00	10.00	10.00	5
10	11	1131	吴婷	30	100.00	10	10.00	10.00	15.00	5.00	10.00	10.00	10.00	10.00	5
11	11	1151	傅乃文	30	100.00	11	10.00	10.00	15.00	5.00	10.00	10.00	10.00	10.00	5
12	11	1153	沈钻科	23	100.00	12	10.00	10.00	15.00	5.00	10.00	10.00	10.00	10.00	5
13	11	1162	刘日帮	30	100.00	13	10.00	10.00	15.00	5.00	10.00	10.00	10.00	10.00	5
14	11	1164	王辉	26	100.00	14	10.00	10.00	15.00	5.00	10.00	10.00	10.00	10.00	5
15	12	1211	糜德萍	38	100.00	15	10.00	10.00	15.00	5.00	10.00	10.00	10.00	10.00	5
16	12	1233	殷存举	36	100.00	16	10.00	10.00	15.00	5.00	10.00	10.00	10.00	10.00	5
17	12	1263	赵菁	39	100.00	17	10.00	10.00	15.00	5.00	10.00	10.00	10.00	10.00	5
18	13	1304	陈腊梅	17	100.00	18	10.00	10.00	15.00	5.00	10.00	10.00	10.00	10.00	5
19	13	1344	沈乐	26	100.00	19	10.00	10.00	15.00	5.00	10.00	10.00	10.00	10.00	5
20	13	1352	徐建伟	35	100.00	20	10.00	10.00	15.00	5.00	10.00	10.00	10.00	10.00	5

共111条记录　第1页/共6页　GO　首页 上一页 下一页 末页

图 6-73　汇总的全部班主任测评数据

【任务分析】

1. 程序运行基本流程

图 6-74 所示为查询班主任测评情况汇总数据模块运行流程图。

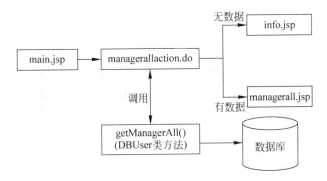

图 6-74　查询班主任测评情况汇总数据模块运行流程图

2. Action 映射表

查询班主任测评情况汇总数据模块中的 Action 映射表见表 6-30。

表 6-30　查询班主任测评情况汇总数据模块中的 Action 映射表

动作(Action)	入口	表单 bean(ActionForm)	出口
ManagerAllAction	main.jsp		managerAll.jsp info.jsp
PageAction	managerAll.jsp	DynaActionForm	managerAll.jsp

【任务实施】

1. 新建并编制 JavaBean 类文件 ManagerAll.java

这是一个用来表示班主任测评数据的实体 JavaBean 组件，其中的私有属性与班主任测评数据汇总表 scynsum 中的列一致，类中还要编写与属性配套的 get/set 方法。

实现代码如下：

```
package com.teach.comms;

import java.io.Serializable;
public class ManagerAll implements Serializable
{
    private String snno;
    private String scno;
    private String stname;
    private int num;

    private double itsum;
    private int    order;
    private double it1;
    …
    private double it10;

    public void setSnno(String snno)
    {
        this.snno = snno;
    }
    public String getSnno()
    {
        return snno;
    }
    …//省略与其余属性对应的 get 和 set 方法
}
```

2. 修改 DBUser 类

在其中定义一个方法，方法为 getManagerAll()，该方法的作用是从班主任测评数据汇总表 scynsum 中取出汇总数据，存放在一个 List 类型的对象中，并返回。

实现代码如下：

```
public List getManagerAll()
    {
        Connection con = null;
        List managerAll = new ArrayList();
        ManagerAll managerObj = null;
          //int i = 0;

        try
        {
            con = dataSource.getConnection();
            Statement stat = con.createStatement();
```

项目六　教师测评系统(Struts模式)

```java
            ResultSet result = stat.executeQuery("select * from scynsum order by itorder asc");
            while(result.next())
            {
                //i++;
                managerObj = new ManagerAll();
                managerObj.setSnno(result.getString("snno"));
                managerObj.setScno(result.getString("scno"));
                managerObj.setStname(result.getString("stname"));
                managerObj.setNum(result.getInt("num"));
                managerObj.setItsum(result.getDouble("itsum"));
                managerObj.setOrder(result.getInt("itorder"));
                managerObj.setIt1(result.getDouble("it1"));
                managerObj.setIt2(result.getDouble("it2"));
                managerObj.setIt3(result.getDouble("it3"));
                managerObj.setIt4(result.getDouble("it4"));
                managerObj.setIt5(result.getDouble("it5"));
                managerObj.setIt6(result.getDouble("it6"));
                managerObj.setIt7(result.getDouble("it7"));
                managerObj.setIt8(result.getDouble("it8"));
                managerObj.setIt9(result.getDouble("it9"));
                managerObj.setIt10(result.getDouble("it10"));

                managerAll.add(managerObj);
            }
        }
        catch(SQLException e)
        …//省略部分同 CheckAdminUser(username,userpwd)方法中该部分
        return managerAll;
}
```

3. 新建并编制动作类文件 ManagerAllAction.java

ManagerAllAction.java 类的主要作用是处理汇总班主任测评数据的请求。在该动作类中，调用 DBUser 类的 getManagerAll()方法查询数据，将数据保存至会话级变量 Search_AllList 中，如果取到数据，则转至 managerAll.jsp 页面，将数据分页显示出来；如果取不到数据，则显示"没有数据"信息。

实现代码如下：

```java
package com.teach.actions;

import java.util.List;
…//省略的包和类可参考 AdminLoginAction
public class ManagerAllAction extends Action
{
    public ActionForward execute(ActionMapping mapping, ActionForm form, HttpServletRequest request,
    HttpServletResponse response) throws Exception
    {
        ServletContext context = servlet.getServletContext();
        DataSource dataSource = (DataSource)context.getAttribute("dbkey");
        DBUser dbuser = new DBUser(dataSource);

        HttpSession session = request.getSession();

        //全部班主任测评情况
        List managerAll = dbuser.getManagerAll();
        session.setAttribute(Constants.Search_AllList,managerAll);
        if (managerAll.size() == 0)
        {
            session.setAttribute(Constants.Show_Message,"没有数据!");
            return mapping.findForward("Info");
        }
```

Java & JSP应用程序实例开发

```java
        else
        {
            PageControl pageControl = new PageControl();
            pageControl.setPageData(managerAll,session,0);

            return (mapping.findForward("ManagerAll"));
        }
    }
}
```

4．新建并编制 managerall.jsp 文件

如图 6-73 所示，用表格进行布局。在页面的开始处将数据从会话级变量 Search_PageList 中取出，用循环的方法将数据显示在各行上。

实现代码如下：

```jsp
<%@page contentType="text/html;charset=utf-8"%>
<%@page import="com.teach.comms.Constants,com.teach.comms.ManagerAll,java.util.List,java.text.DecimalFormat"%>
<%
    List managerAll = (List)session.getAttribute(Constants.Search_PageList);
    ManagerAll  managerObj = null;
    DecimalFormat precision2 = new DecimalFormat("0.00");
    Integer Records_All = (Integer)session.getAttribute(Constants.Records_All);
    Integer PageNo_Cur = (Integer)session.getAttribute(Constants.PageNo_Cur);
    Integer Pages_All = (Integer)session.getAttribute(Constants.Pages_All);

    int iRecords_All = Records_All.intValue();
    int iPageNo_Cur = PageNo_Cur.intValue();
    int iPages_All = Pages_All.intValue();
%>
<html>
    <head>
        <link href="../css/style_css.css" rel=stylesheet type="text/css">
        <script language="javascript">
          function Goto()
          {
              location.href = "pageaction.do?pageId=" + document.getElementById("pageId").value + "&forwardPage=ManagerAll";
          }
        </script>
    </head>

    <body>
        <table  width="100%" height="100%">
          <tr>
              <td  align="center" valign="top">
                <table width="850"  bgcolor="#dec3b5" cellspacing="1"  >
                    <caption>班主任教育情况测评汇总表</caption>
                    <tr height="20" align="center" bgcolor="#f3d5d5">
                        <td width="30">序号</td>
                        <td width="30">年级</td>
                        <td width="50">班级</td>
                        <td width="60">班主任</td>
                        <td width="60">参测人数</td>
                        <td width="50">总分</td>
                        <td width="50">名次</td>
                        <td width="45">分数 1</td>
                        <td width="45">分数 2</td>
                        <td width="45">分数 3</td>
                        <td width="45">分数 4</td>
                        <td width="45">分数 5</td>
                        <td width="45">分数 6</td>
```

```
                    <td width="45">分数 7</td>
                    <td width="45">分数 8</td>
                    <td width="45">分数 9</td>
                    <td width="45">分数 10</td>
                </tr>
              <%
              if(managerAll != null)
              {
                for(int i = 0;i < managerAll.size();i++)
                {
                    managerObj = (ManagerAll)managerAll.get(i);
                        %>
                <tr height="20" align="center" bgcolor="#f1fcf7">
                    <td><%=i+1+Constants.Rows_PerPapge*iPageNo_Cur%></td>
                    <td><%=managerObj.getSnno()%></td>
                    <td><%=managerObj.getScno()%></td>
                    <td><%=managerObj.getStname()%></td>
                    <td><%=managerObj.getNum()%></td>
                    <td><%=precision2.format(managerObj.getItsum())%></td>
                    <td><%=managerObj.getOrder()%></td>
                    <td><%=precision2.format(managerObj.getIt1())%></td>
                    <td><%=precision2.format(managerObj.getIt2())%></td>
                    <td><%=precision2.format(managerObj.getIt3())%></td>
                    <td><%=precision2.format(managerObj.getIt4())%></td>
                    <td><%=precision2.format(managerObj.getIt5())%></td>
                    <td><%=precision2.format(managerObj.getIt6())%></td>
                    <td><%=precision2.format(managerObj.getIt7())%></td>
                    <td><%=precision2.format(managerObj.getIt8())%></td>
                    <td><%=precision2.format(managerObj.getIt9())%></td>
                    <td><%=precision2.format(managerObj.getIt10())%></td>
                </tr>
              <%
                }
              }
                        %>
            </table>
            <table width="850" bgcolor="#f3d5d5">
                <tr height="20">
                    <td width="200">共<%=iRecords_All%>条记录   第<%=iPageNo_Cur+1%>页/共<%=iPages_All%>页</td>
                    <td align="left">
                        <input type="text" name="pageId" style="width:25px;height:20px;">
<a href="javascript:Goto()">GO</a>   
                        <a href="pageaction.do?pageId=1&forwardPage=ManagerAll">首页</a>
                        <a href="pageaction.do?pageId=<%=iPageNo_Cur%>&forwardPage=ManagerAll">上一页</a>
                        <a href="pageaction.do?pageId=<%=iPageNo_Cur+2%>&forwardPage=ManagerAll">下一页</a>
                        <a href="pageaction.do?pageId=<%=iPages_All%>&forwardPage=ManagerAll">末页</a>
                    </td>
                </tr>
            </table>
        </td>
    </tr>
</table>
</body>
</html>
```

5．配置动作类

配置 ManagerAllAction 动作类映射关系的代码应写在 struts-config.xml 文件的＜action-mappings＞元素对中。

配置代码如下：

```
<action path = "/search/managerallaction"    type = "com.teach.actions.ManagerAllAction"    scope = "request">
    <forward name = "ManagerAll"    path = "/search/managerall.jsp"/>
</action>
```

6. 发布并测试程序

启动 Tomcat 服务器，将 TeacherTest 应用发布至服务器，打开浏览器，在地址栏中输入 http://localhost:8080/TeacherTest，运行并测试数据查询子系统的班主任测评情况汇总数据模块。

任务十八　编制按系部查询班主任汇总数据模块

【任务描述】

本任务主要编制按系部查询班主任测评情况汇总数据模块中的相关文件并测试运行。本模块主要是从班主任教育情况测评汇总表 scynsum 中按系部查询出班主任测评情况汇总数据。从主页面"请选择系部"下拉列表框中选取一个系部，出现如图 6-75 所示结果页面。

图 6-75　按系部汇总的班主任测评数据

【任务分析】

1. 程序运行基本流程

图 6-76 所示为按系部查询班主任测评数据模块运行流程图。

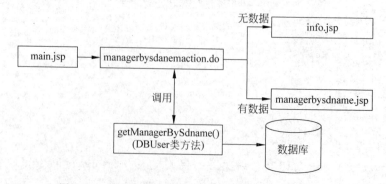

图 6-76　按系部查询班主任测评数据模块运行流程图

项目六 教师测评系统(Struts模式)

2. Action 映射表

按系部查询班主任测评数据模块中的 Action 映射表见表 6-31。

表 6-31 按系部查询班主任测评数据模块中的 Action 映射表

动作(Action)	入口	表单 bean(ActionForm)	出口
ManagerBySdnameAction	main.jsp		managerbysdname.jsp info.jsp
PageAction	managerbysdname.jsp	DynaActionForm	managerbysdname.jsp

【任务实施】

1. 修改 DBUser 类

在其中定义一个方法,方法为 getManagerBySdname(String sdname),该方法的作用是从班主任测评数据汇总表 scynsum 中取出当前系部的汇总数据,存放在一个 List 类型的对象中并返回。具体实现代码可参考本项目任务十七中编制的代码。

2. 新建并编制动作类文件 ManagerBySdnameAction.java

ManagerBySdnameAction.java 类的主要作用是处理汇总当前系部的班主任测评数据的请求。在该动作类中,先取出传递过来的部门索引号,根据部门索引号取到该部门名称,调用 DBUser 类的 getManagerBySdname()方法查询数据,将数据保存至会话级变量 Search_AllList 中。如果取到数据,则转至 managerbysdname.jsp 页面,将数据分页显示出来;如果取不到数据,则显示"没有数据"信息。具体实现代码可参考本项目任务十七编制的代码。

3. 新建并编制 managerbysdname.jsp 文件

如图 6-75 所示,编制方法与 managerall.jsp 文件的编制方法一致,只需将其另存后修改三处。

一是在页面开始处的 Java 代码中加入以下取出当前系部的语句:

```
<%
…
    String sdname = (String)session.getAttribute(Constants.Search_ManagerByCurSdname);
…
%>
```

二是将表格的<caption>标签改为如下代码,以便显示当前系部名称:

```
<caption><b><font face="华文新魏" color=#d71345>[<%=sdname%>]</font></b>班主任教育情况测评汇总表</caption>
```

三是将处理分页的动作类的 forwardPage 参数的值改为 ManagerBySdname。

4. 配置动作类

配置 ManagerBySdnameAction 动作类映射关系的代码应写在 struts-config.xml 文件的<action-mappings>元素对中。

配置代码如下:

```
<action path="/search/managerbysdnameaction"    type="com.teach.actions.ManagerBySdnameAction"   >
    <forward name="ManagerBySdname"    path="/search/managerbysdname.jsp"/>
</action>
```

5. 发布并测试程序

启动 Tomcat 服务器,将 TeacherTest 应用发布至服务器,打开浏览器,在地址栏中输入

http://localhost:8080/TeacherTest，运行并测试数据查询子系统的按系部查询班主任测评情况汇总数据模块。

任务十九 编制按班级查询任课教师汇总数据模块

【任务描述】

本任务主要编制按班级查询任课教师测评情况汇总数据模块中的相关文件并测试运行。本模块主要是从教师教学情况测评汇总表 stynsum 中按班级查询出任课教师测评情况汇总数据。从主页面"请选择班级"下拉列表框中选取一个班级，出现如图 6-77 所示结果页面。

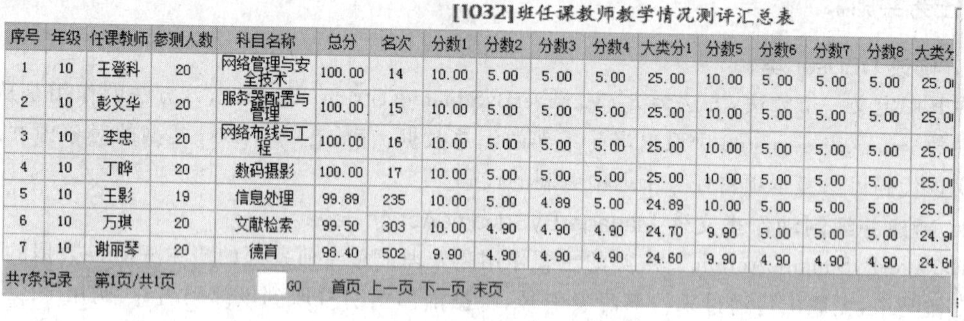

图 6-77 按班级汇总的任课教师测评数据

【任务分析】

1. 程序运行基本流程

图 6-78 所示为按班级查询的任课教师测评数据模块运行流程图。

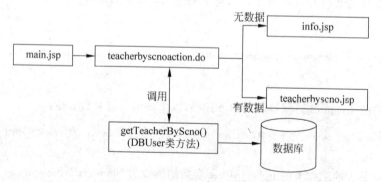

图 6-78 按班级查询的任课教师测评数据模块运行流程图

2. Action 映射表

按班级查询的任课教师测评数据模块中的 Action 映射表见表 6-32。

表 6-32 按班级查询的任课教师测评数据模块中的 Action 映射表

动作（Action）	入口	表单 bean（ActionForm）	出口
TeacherByScnoAction	main.jsp		teacherbyscno.jsp info.jsp
PageAction	teacherbyscno.jsp	DynaActionForm	teacherbyscno.jsp

· 310 ·

项目六 教师测评系统(Struts模式)

【任务实施】

1. 新建并编制 JavaBean 类文件 TeacherAll.java

这是一个用来表示任课教师测评数据的实体 JavaBean 组件,其中的私有属性与任课教师测评数据汇总表 stynsum 中的列一致,类中还要编写与属性配套的 get/set 方法。

实现代码如下:

```java
package com.teach.comms;

import java.io.Serializable;
public class TeacherAll implements Serializable
{
    private String snno;
    private String scno;
    private String stname;
    private String soname;
    private int num;

    private double itsum;
    private int order;
    private double it1;
    …
    private double it912;

    public void setSnno(String str)
    {
        snno = str;
    }
    public String getSnno()
    {
        return snno;
        …//省略与其余属性对应的 get 和 set 方法
}
```

2. 修改 DBUser 类

在其中定义一个方法,方法为 getTeacherByScno(String scno),该方法的作用是从任课教师测评数据汇总表 stynsum 中取出所选班级所有任课教师的汇总数据,存放在一个 List 类型的对象中并返回。

实现代码如下:

```java
public List getTeacherByScno(String scno)
{
        Connection con = null;
        List teacherByScno = new ArrayList();
        TeacherAll teacherObj = null;
        String strSql = "select * from stynsum "
             + "where scno = '" + scno + "'"
             + "order by itorder asc";
        try
        {
            con = dataSource.getConnection();
            Statement stat = con.createStatement();
            ResultSet result = stat.executeQuery(strSql);
            while(result.next())
            {
                teacherObj = new TeacherAll();
                teacherObj.setSnno(result.getString("snno"));
                teacherObj.setScno(result.getString("scno"));
```

```
            teacherObj.setStname(result.getString("stname"));
            teacherObj.setSoname(result.getString("soname"));
            teacherObj.setNum(result.getInt("num"));
            teacherObj.setItsum(result.getDouble("itsum"));
            teacherObj.setOrder(result.getInt("itorder"));
            teacherObj.setIt1(result.getDouble("it1"));
            teacherObj.setIt2(result.getDouble("it2"));
            teacherObj.setIt3(result.getDouble("it3"));
            teacherObj.setIt4(result.getDouble("it4"));
            teacherObj.setIt14(result.getDouble("it14"));
            teacherObj.setIt5(result.getDouble("it5"));
            teacherObj.setIt6(result.getDouble("it6"));
            teacherObj.setIt7(result.getDouble("it7"));
            teacherObj.setIt8(result.getDouble("it8"));
            teacherObj.setIt58(result.getDouble("it58"));
            teacherObj.setIt9(result.getDouble("it9"));
            teacherObj.setIt10(result.getDouble("it10"));
            teacherObj.setIt11(result.getDouble("it11"));
            teacherObj.setIt12(result.getDouble("it12"));
            teacherObj.setIt912(result.getDouble("it912"));

            teacherByScno.add(teacherObj);
        }

    }
    catch(SQLException e)
    …//省略部分同 CheckAdminUser(username,userpwd)方法中该部分
    return teacherByScno;
}
```

3. 新建并编制动作类文件 TeacherByScnoAction.java

TeacherByScnoAction.java 类的主要作用是处理汇总当前所选班级的所有任课教师测评数据的请求。在该动作类中，先取出传递过来的班级号 scno，它代表当前班级，调用 DBUser 类的 getTeacherByScno()方法查询数据，将数据保存至会话级变量 Search_AllList 中。如果取到数据，则转至 teacherbyscno.jsp 页面，将数据分页显示出来；如果取不到数据，则显示"没有数据"信息。

实现代码如下：

```
package com.teach.actions;

import java.util.List;
…//省略的包和类可参考 AdminLoginAction
public class TeacherByScnoAction extends Action
{
    public ActionForward execute(ActionMapping mapping, ActionForm form, HttpServletRequest request, HttpServletResponse response) throws Exception
    {
        String scno = request.getParameter("scno");

        ServletContext context = servlet.getServletContext();
        DataSource dataSource = (DataSource)context.getAttribute("Evaluation");
        DBUser dbuser = new DBUser(dataSource);

        HttpSession session = request.getSession();
        if(scno == null || scno.equals(""))
        {
            scno = (String)session.getAttribute(Constants.Search_TeacherByCurScno);
        }
        else
        {
            session.setAttribute(Constants.Search_TeacherByCurScno,scno);
```

项目六　教师测评系统(Struts模式)

```
    }
//根据班级查询任课教师测评情况
List teacherByScno = dbuser.getTeacherByScno(scno);
session.setAttribute(Constants.Search_AllList,teacherByScno);
if (teacherByScno.size() == 0)
{
    session.setAttribute(Constants.Show_Message,"没有数据!");
     return  mapping.findForward("Info");
}
else
{
    PageControl pageControl = new PageControl();
    pageControl.setPageData(teacherByScno,session,0);

    return (mapping.findForward("TeacherByScno"));
}
    }
}
```

4．新建并编制 teacherbyscno.jsp 文件

如图 6-77 所示，编制方法与 managerall.jsp 文件的编制方法一致。在页面的开始处将数据从会话级变量 Search_PageList 中取出，用循环的方法将数据显示在各行上。可参考前述模块的方法进行编制。

5．配置动作类

配置 TeacherByScnoAction 动作类映射关系的代码应写在 struts-config.xml 文件的＜action-mappings＞元素对中。

配置代码如下：

```
<action path = "/search/teacherbyscnoaction"    type = "com.teach.actions.TeacherByScnoAction"    scope = "request">
        <forward name = "TeacherByScno"    path = "/search/teacherbyscno.jsp"/>
</action>
```

6．发布并测试程序

启动 Tomcat 服务器，将 TeacherTest 应用发布至服务器，打开浏览器，在地址栏中输入 http://localhost:8080/TeacherTest，运行并测试数据查询子系统的按班级查询任课教师测评情况汇总数据模块。

任务二十　编制按学科查询任课教师汇总数据模块

【任务描述】

本任务主要编制按学科查询任课教师测评情况汇总数据模块中的相关文件并测试运行。本模块主要是从教师教学情况测评汇总表 stynsum 中按学科查询出任课教师测评情况汇总数据。从主页面"请选择学科"下拉列表框中选取一门学科，出现如图 6-79 所示页面。

【任务分析】

1．程序运行基本流程

图 6-80 所示为按学科查询任课教师测评数据模块运行流程图。

2．Action 映射表

按学科查询任课教师测评数据模块中的 Action 映射表见表 6-33。

· 313 ·

图 6-79 按学科汇总的任课教师测评数据

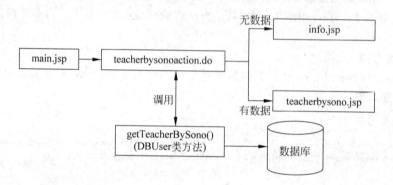

图 6-80 按学科查询任课教师测评数据模块运行流程图

表 6-33 按学科查询任课教师测评数据模块中的 Action 映射表

动作（Action）	入口	表单 bean（ActionForm）	出口
TeacherBySonoAction	main.jsp		teacherbysono.jsp info.jsp
PageAction	teacherbysono.jsp	DynaActionForm	teacherbysono.jsp

【任务实施】

1. 修改 DBUser 类

在其中定义一个方法，方法为 getTeacherBySono(String sono)，该方法的作用是从任课教师测评数据汇总表 stynsum 中取出所选课程所有班级的汇总数据，存放在一个 List 类型的对象中并返回。具体实现代码可参考前一模块。

2. 新建并编制动作类文件 TeacherBySonoAction.java

TeacherBySonoAction.java 类的主要作用是处理汇总当前所选课程的所有班级任课教师测评数据的请求。在该动作类中，先取出传递过来的课程代码 sono，它代表当前课程，调用 DBUser 类的 getTeacherBySono() 方法查询数据，将数据保存至会话级变量 Search_AllList 中。如果取到数据，则转至 teacherbysono.jsp 页面，将数据分页显示出来；如果取不到数据，则显示"没有数据"信

项目六 教师测评系统(Struts模式)

息。具体实现代码可参考前一模块。

3. 新建并编制 teacherbysono.jsp 文件

如图 6-79 所示,编制方法与 teacherbyscno.jsp 文件的编制方法一致,只需将其另存后修改三处。一是在页面开始处的 Java 代码中加入以下根据课程代码查找取出课程名称的语句:

```
<%
...
String sonocur = (String)session.getAttribute(Constants.Search_TeacherByCurSono);
    String sonamecur = "";
    Vector sonoVector = (Vector)session.getAttribute(Constants.Search_SonoAll);
    Vector sonameVector = (Vector)session.getAttribute(Constants.Search_SonameAll);
    for(int i = 0;i < sonoVector.size();i++)
    {
        String sono = (String)sonoVector.get(i);
        if(sono.equals(sonocur))
        {
            sonamecur = (String)sonameVector.get(i);
        }
    }
...
%>
```

二是将表格的<caption>标签改为如下代码,以便显示当前课程名称:

```
<caption><b><font face = "华文新魏" color = #d71345>[<% = sonamecur %>]</font></b>任课教师教学情况测评汇总</caption>
```

三是将处理分页的动作类的 forwardPage 参数的值改为 TeacherBySono。

4. 配置动作类

配置 TeacherBySonoAction 动作类映射关系的代码应写在 struts-config.xml 文件的<action-mappings>元素对中。

配置代码如下:

```
<action path = "/search/teacherbysonoaction"    type = "com.teach.actions.TeacherBySonoAction"    scope = "request">
        <forward name = "TeacherBySono"    path = "/search/teacherbysono.jsp"/>
</action>
```

5. 发布并测试程序

启动 Tomcat 服务器,将 TeacherTest 应用发布至服务器,打开浏览器,在地址栏中输入 http://localhost:8080/TeacherTest,运行并测试数据查询子系统的学科查询任课教师测评情况汇总数据模块。

任务二十一 编制按班级查询意见、建议和简要评价模块

【任务描述】

本任务主要编制按班级查询意见、建议和简要评价模块中的相关文件并测试运行。本模块主要是从教师教学情况测评情况表 styn 中按班级查询出意见、建议和简要评价数据。从主页面"请选择班级"下拉列表框中选取一个班级,出现如图 6-81 所示结果页面。

【任务分析】

1. 程序运行基本流程

图 6-82 所示为按班级查询意见、建议和简要评价模块运行流程图。

图 6-81 按班级查询的意见、建议和简要评价

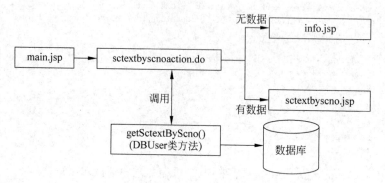

图 6-82 按班级查询意见、建议和简要评价模块运行流程图

2. Action 映射表

按班级查询意见、建议和简要评价模块中的 Action 映射表见表 6-34。

表 6-34 按班级查询意见、建议和简要评价模块中的 Action 映射表

动作(Action)	入口	表单 bean(ActionForm)	出口
SctextByScnoAction	main.jsp		sctextbyscno.jsp info.jsp
PageAction	sctextbyscno.jsp	DynaActionForm	sctextbyscno.jsp

【任务实施】

1. 修改 DBUser 类

在其中定义一个方法,方法为 getSctextByScno(String scno),该方法的作用是从班主任测评数据情况表 scyn 中取出所选班级的所有意见、建议和简要评价数据,存放在一个 List 类型的对象中并返回。

实现代码如下:

```
List getSctextByScno(String scno)
{
    Connection con = null;
    List sctextByScno = new ArrayList();
    try
    {
        con = dataSource.getConnection();
        Statement stat = con.createStatement();
        ResultSet result = stat.executeQuery("select sctext from scyn  where scno = '" + scno + "' and datalength(sctext)!= 0 order by scid ");
        while(result.next())
```

```
            {
                String sctext = result.getString("sctext");
                sctextByScno.add(sctext);
            }
        }
        catch(SQLException e)
        {
            e.printStackTrace();
        }
        catch(Exception e)
        {
            e.printStackTrace();
        }
        finally
        {
            if(con!= null)
            {
                try
                {
                    con.close();
                }
                catch(Exception e)
                {
                    e.printStackTrace();
                }
            }
        }
        return sctextByScno;
    }
}
```

2. 新建并编制动作类文件 SctextByScnoAction.java

SctextByScnoAction. 类的主要作用是处理根据当前班级号查询意见、建议、评价数据的请求。在该动作类中，先取出传递过来的班级号 scno，它代表当前班级，调用 DBUser 类的 getSctextByScno()方法查询数据，将数据保存至会话级变量 Search_AllList 中。如果取到数据，则转至 sctextbyscno.jsp 页面，将数据分页显示出来；如果取不到数据，则显示"没有数据"信息。

实现代码如下：

```
package com.teach.actions;

import java.util.List;
…//省略的包和类可参考 AdminLoginAction
public class SctextByScnoAction extends Action
{
    public ActionForward execute(ActionMapping mapping, ActionForm form, HttpServletRequest request, HttpServletResponse response) throws Exception
    {
        String    scno = request.getParameter("scno");

        ServletContext context = servlet.getServletContext();
        DataSource dataSource = (DataSource)context.getAttribute("dbkey");
        DBUser dbuser = new DBUser(dataSource);

        HttpSession session = request.getSession();
        //根据班级查询对任课教师、班主任、学校的意见、建议和简要评价
        if(scno == null || scno.equals(""))
        {
            scno = (String)session.getAttribute(Constants.Search_ScTextByCurScno);
        }
        else
        {
            session.setAttribute(Constants.Search_ScTextByCurScno,scno);
```

```
        }
        List sctextByScno = dbuser.getSctextByScno(scno);
        session.setAttribute(Constants.Search_AllList,sctextByScno);
        if (sctextByScno.size() == 0)
        {
            session.setAttribute(Constants.Show_Message,"没有数据!");
            return  mapping.findForward("Info");
        }
        else
        {
            PageControl pageControl = new PageControl();
            pageControl.setPageData(sctextByScno,session,0);
            return (mapping.findForward("SctextByScno"));
        }
    }
}
```

3．新建并编制 sctextbyscno.jsp 文件

如图 6-81 所示，编制方法与 managerall.jsp 文件的编制方法一致。在页面的开始处将数据从会话级变量 Search_PageList 中取出，用循环的方法将数据显示在各行上。可参考前述编制方法编制代码。

4．配置动作类

配置 SctextByScnoAction 动作类映射关系的代码应写在 struts-config.xml 文件的＜action-mappings＞元素对中。

配置代码如下：

```
<action path = "/search/sctextbyscnoaction"
type = "com.teach.actions.SctextByScnoAction"        scope = "request" >
         < forward name = "SctextByScno"
path = "/search/sctextbyscno.jsp"/>
</action>
```

5．发布并测试程序

启动 Tomcat 服务器，将 TeacherTest 应用发布至服务器，打开浏览器，在地址栏中输入 http://localhost:8080/ TeacherTest，运行并测试数据查询子系统的按班级查询意见、建议和简要评价模块。

项 目 小 结

教师测评系统是一个完整的基于 Struts 框架技术开发的项目，Struts 框架技术已经是 JSP 开发 Web 应用程序的一个标准，因此，通过本项目的学习与开发，可以比较全面、深入地掌握 Struts 框架技术的开发流程、方法及特点，也有利于读者进一步理解一个实用管理系统中的业务逻辑及处理方法，为以后开发更专业的系统打下基础。